T0257829

Encyclopedia of Earth Observations

Encyclopedia of Earth Observations

Edited by **Joe Carry**

New York

Published by Callisto Reference,
106 Park Avenue, Suite 200,
New York, NY 10016, USA
www.callistoreference.com

Encyclopedia of Earth Observations
Edited by Joe Carry

© 2015 Callisto Reference

International Standard Book Number: 978-1-63239-231-2 (Hardback)

This book contains information obtained from authentic and highly regarded sources. Copyright for all individual chapters remain with the respective authors as indicated. A wide variety of references are listed. Permission and sources are indicated; for detailed attributions, please refer to the permissions page. Reasonable efforts have been made to publish reliable data and information, but the authors, editors and publisher cannot assume any responsibility for the validity of all materials or the consequences of their use.

The publisher's policy is to use permanent paper from mills that operate a sustainable forestry policy. Furthermore, the publisher ensures that the text paper and cover boards used have met acceptable environmental accreditation standards.

Trademark Notice: Registered trademark of products or corporate names are used only for explanation and identification without intent to infringe.

Printed in the United States of America.

Contents

Preface

In recent times, space technology is being used as a significant tool for Earth observation applications. Satellites and other accessible remote sensing platforms are used for collecting the data. Wide ranges of electromagnetic energy, that are being reflected, transmitted or emitted from the surface of the Earth, can be detected through remote sensing data collection. Implementing further data processing requires suitable detection systems. Space technology has proved to be a successful application in the study of climatic change due to the dynamic comparison between current and past data. This book offers various aspects of climatic change and discusses applications of space technology.

This book unites the global concepts and researches in an organized manner for a comprehensive understanding of the subject. It is a ripe text for all researchers, students, scientists or anyone else who is interested in acquiring a better knowledge of this dynamic field.

I extend my sincere thanks to the contributors for such eloquent research chapters. Finally, I thank my family for being a source of support and help.

Editor

Part 1

Earth Observation – Capability and Advances

Earth Observation – Space Technology

Rustam B. Rustamov, Saida E. Salahova,
Maral H. Zeynalova and Sabina N. Hasanova

Institute of Physics, Azerbaijan National Academy of Sciences,
ENCOTEC –Engineering & Consulting Technologies,
Institute of Botany, Azerbaijan National Academy of Sciences,
Architecture and Construction University/ENCOTEC LLC,
Baku,
Azerbaijan

1. Introduction

For monitoring of the Earth thousands of satellites have been sent into space on missions to collect data related different spheres of the Earth investigations and studies. Today, the ability to forecast weather, climate, and natural hazards, environmental monitoring and ecological issues depend critically on these satellite-based observations. Based on this data it is possible to gather satellite images frequently enough to create the model of the changing planet, improving the understanding of Earth's dynamic processes and helping society to manage limited resources and environmental challenges. Earth observations from space open and makes requirement to address scientific and societal challenges of the future.

Space technologies play the significant role in the sustainable development in national, regional and global level. Modern and advances of the Earth observation techniques are taking a great importance amongst existing traditional technologies. Radar remote sensing is one of the new Earth observation technologies with promising results and future. Interferometric SAR (InSAR) is a sophisticated radar remote sensing technique for combining synthetic aperture radar (SAR) complex images to form interferogram and utilizing its phase contribution to land topography, surface movement and target velocity. Presently considerable applications of InSAR technique are developed. It is an established technique for precise assessment of land surface movements and generating high quality digital elevation models (DEM) from spaceborne and airborne data. InSAR is able to produce DEM with the precision of a couple of ten meters whereas its movement map results have sub-centimeter precision. The technique has many applications in the context of Earth sciences such as in topographic mapping, environmental modeling, rainfall-runoff studies, landslide hazard zonation, and seismic source modeling.

Making observations of the land, sea and air from space allow scientists to develop and improve their models of the Earth. Space instruments provide continuous global measurements of the Earth for many years at a time. Currently this includes to consider following issues:

- Expertise in obtaining information on the surface and atmosphere using remote sensing methods;
- Expertise in modeling environmental phenomena;
- Expertise in and provision of facilities for generating archiving and distributing environmental data;
- Expertise in and facilities for characterizing spectral properties of environmental components;
- Expertise in and facilities for atmospheric research using radars;
- Expertise in the technology and practice of remote sensing at mm and sub-mm wavelengths;
- Carrying out a program of research in aspects of environmental science;
- Developing e-science applications in environmental research.

2. Earth observation systems

It is necessary to emphasize that one of the most important and controversial uses of satellites today is that of monitoring the Earth's environment. Many satellites study features on the ground, the behavior of the oceans, or the characteristics of the atmosphere. Satellites that observe the Earth to collect scientific data are usually referred to as "Earth observation satellites." Sometimes the interpretation of their data has been controversial because the interpretation is difficult and people have used the data to call for substantial changes in human behavior.

One of the popular satellite for Earth observation the Envisat is an advanced polar-orbiting Earth-observation satellite that provides measurements of the atmosphere, ocean, land and ice. It was launched in March 2002 on an Ariane 5 rocket into an 800km polar orbit by the European Space Agency (ESA). Originally was planned for five years, the life of Envisat has been extended till 2013.

It is necessary to mention that the satellite also helps scientists access data for analyzing long-term climatic changes.

The recent advances and developments in information and communication technologies, education and health care, agriculture and agro-food processing, geo-strategic initiatives, infrastructure and energy and critical technologies and strategic industries have been realized in light of the space technologies. Earth observation techniques which apply optical and thermal spectra of the electromagnetic wavelengths have so far developed considerably. Although there is done a lot in this area beforehand, a long way is still ahead. The background of using microwaves for remote sensing goes far the decades ago while it was remaining in the experimental domain and exploratory status for years. It is only in the recent couple of decades that radar remote sensing techniques have been commercialized and used widely. Radar remote sensing is actually accounted for as a new earth observation technology with promising results and future. Its potentials and capacities by itself and being a strong complementary tool for optical and thermal remote sensing are undeniable currently.

i.   Radar and SAR techniques for remote sensing

Obviously, the use of radar systems opens a wide opportunity to reduce an obstacles existing in the traditionally used technologies. For the time being it became very interesting and important explorations and examining a new radar technologies, their unique possibilities to comply the needs and answering the questions that the classic optical and thermal remote sensing techniques have been unable or difficult to tackle has grown the expectation that radar technologies can take place due to a more flexibility in bridging the gaps for sustainable development for which the optical and thermal remote sensing is an important tool while the latter techniques show shortage in some cases and areas.

Currently, radar remote sensing that is mainly developed on the Synthetic Aperture Radar (SAR) technique represents its values and potentials increasingly. Radar is a useful tool for land and planetary surface mapping. It is a good mean for obtaining a general idea of the geological setting of the area before proceeding for field work. Time, incidence angle, resolutions and coverage area all play important role at the outcome.

ii.   InSAR techniques

SAR interferometry (InSAR), Differential InSAR (DInSAR), Persistent Scatterer (PSInSAR) is the a new achieved techniques in radar remote sensing systems. By using InSAR technique very precise digital elevation models (DEM) can be produced which privilege is high precision in comparison to the traditionally used methods. DEM refers to the process of demonstrating terrain elevation characteristics in 3-D space, but very often it specifically means the raster or regular grid of spot heights. DEM is the simplest form of digital representation of topography, while digital surface model (DSM) describes the visible surface of the Earth.

Considerable applications of InSAR have been developed leaving it an established technique for high-quality DEM generation from spaceborne and airborne data and that it has advantages over other methods for the large-area DEM generation. It is capable of producing DEMs with the precision of a couple of ten meters while its movement map results have sub-centimeter precision over time spans of days to years. Terrestrial use of InSAR for DEM generation was first reported in 1974. It is used for different means particularly in geo-hazards and disasters like earthquakes, volcanoes, landslides and land subsidence.

2.1 Earth observation satellites

The first satellite to be used for Earth observation purposes was Explorer VII, launched in October 1959. This satellite was equipped with an infrared sensor designed to measure the amount of heat reflected by the Earth. This measurement, referred to as the "radiation budget," is a key to understanding global environmental trends, for it represents the difference between the amount of incoming energy from the sun and the outgoing thermal and reflected energy from the Earth. But it was not until the launch of the Earth Radiation Budget Satellite (ERBS) in 1984 by the National Aeronautics and Space Administration (NASA) that more authoritative readings of this important figure were obtained. Many Earth observation satellites like ERBS use specialized sensors that operate in non-visible wavelengths like the infrared, allowing them to gather data on many different types of atmospheric and ground phenomena.

The most important early Earth observation satellites were members of the Nimbus series. NASA launched eight Nimbus satellites between 1964 and 1978, with only one failing to reach orbit. Although they started out as part of the weather satellite program, the Nimbus satellites were not weather satellites, but carried a number of instruments for measuring the temperature and humidity of the atmosphere. This was a major advance, for earlier weather satellites like Tiros (Television Infrared Observation Satellite) had only been capable of taking visible light photographs of clouds and could not provide the kinds of traditional weather measurements that meteorologists normally used. Eventually many of the instruments demonstrated on Nimbus, named "sounders," were incorporated into later weather satellites. Atmospheric sounders are now common on many meteorological satellites, as well as on scientific satellites and even planetary space probes (Belew & Stuhlinger, 1973) , (Covault, 1991).

In July 1972, NASA launched the Earth Resources Technology Satellite (ERTS-1) into orbit. ERTS-1 used advanced instruments to view the Earth's surface in several infrared wavelengths. These sensors enabled scientists to assess vegetation growth, monitor the spread of cities, and make many other measurements of how the Earth's surface was changing. ERTS was so successful that it was followed by two more satellites named Landsat. By the early 1980s, with the launch of Landsat 4, the satellites became an "operational" system rather than an experimental one and their data was heavily used around the world by farmers, urban planners, geologists and environmentalists. Landsat and similar satellites are often referred to as "remote sensing satellites," a term that is usually used to refer to satellites that focus on the ground rather than the oceans or atmosphere.

In the mid 1970s NASA also conducted numerous observation experiments aboard the Skylab space station. Skylab was equipped with handheld as well as fixed cameras using special film. It also had an array of other instruments. Data the crews obtained during their three visits to Skylab was used to refine the instruments on other satellites, such as Landsat. Skylab also demonstrated the value of other observations, such as tracking icebergs and the breakup of sea ice (Skylab, 1977) .

In 1978 NASA launched SeaSat, an ocean observation satellite with a synthetic aperture radar, or SAR. SAR works by taking several radar images from different positions and combining them to produce a more detailed single image. SeaSat's radar produced detailed images of the surface of the ocean, providing valuable data on waves and the interaction of the ocean's surface with the winds. Although SeaSat's mission ended prematurely due to a malfunction, it demonstrated the immense value of space-based SARs.

Approximately around the same time the United States was experimenting with SeaSat, the Soviet Union launched a similar series of satellites known as Okean. Later, during the late 1980s, the Soviet Union orbited several large radar satellites. These spacecraft, launched aboard Proton rockets, produced radar maps of the Earth's surface and were also used to measure waves on the oceans' surface. In 1991 the Soviet Union launched Almaz-1, which was another of this series of satellites but the first that the Soviet government openly acknowledged. Although they announced that this was a civilian Earth observation satellite and sought international customers, many experts speculated about the military uses of these satellites and their role in searching for objects such as submarines, which can create waves on the ocean surface when traveling at high speed at shallow depths. Because such data has military uses, SAR technology has always been sensitive. Although the Soviets attracted the attention of western military officials, they found no commercial customers for their satellite (SeaWifs projects).

During the 1980s, and 1990s NASA, along with German and Italian participants, conducted several Space Shuttle missions carrying a large SAR in the Shuttle's payload bay. This radar, called SIR (for Shuttle Imaging Radar) produced topographical maps of much of the Earth's surface. The radar equipment was modified several times to collect more accurate data during the latter missions. In February 2000, NASA flew another mission called SRTM, for Shuttle Radar Topography Mission (SRTM), with Italian and German participation. This time NASA used a modified version of the radar capable of obtaining much more precise altitude data. Three-dimensional electronic maps produced from the SRTM data are highly accurate and can be used in aviation to guide aircraft and missiles, even over rough terrain like mountain ranges. In 1991 and again in 1995, the European Space Agency launched the ERS-1 and ERS-2 (European Remote Sensing) satellites. Both were equipped with SARs and were highly successful [1].

In 1988 astronaut Dr. Sally Ride led a committee to evaluate America's future in space (Ride, 1987). One of her suggestions was that NASA focus more attention on environmental monitoring in response to increasing scientific discussion of global climate change, a program the agency called Mission to Planet Earth. As a result, NASA started the Earth Observing System (EOS). At the turn of the century, a number of EOS satellites were launched, most importantly Terra and Aqua, to be followed by Aqua's sister-satellite Aura. Terra, as its name implies, is focused upon monitoring the Earth's surface. It is equipped with instruments like MOPITT, the Measurements of Pollution in the Troposphere, and MISR, the Multi-Angle Imaging Spectroradiometer. Aqua has instruments such as microwave, infrared, and humidity sounders. These provide information on clouds, precipitation, snow, sea ice, and sea surface temperature.

In 1992, an Ariane 42P rocket launched a spacecraft named Topex/Poseidon. A joint French space agency (CNES-Centre National d'Etudes Spatiales) and NASA spacecraft, it was equipped with a radar altimeter to allow it to measure ocean topography, or surface features. Data gathered from Topex/Poseidon over years of operation have allowed scientists to accurately map ocean circulation, a key factor in understanding both global weather and climate change. In particular, Topex/Poseidon has been able to track the phenomenon known as El Nico, a warming of the ocean surface off the western coast of South America that occurs every four to twelve years. El Nico affects weather patterns in various parts of the world as well as fish and plankton populations. Another spacecraft, called SeaStar and carrying the Sea-viewing Wide Field-of-view Sensor, or SeaWiFs, was launched in 1997 to study biological organisms in the oceans such as algae and phytoplankton (microscopic marine plants).

In 2002, the European Space Agency launched a large environmental monitoring satellite named Envisat, aboard an Ariane 5 rocket. Envisat, the successor to ERS-1 and 2, is designed to take simultaneous readings of various atmospheric and terrestrial features and contribute to understanding of global change. The data from satellites like Envisat is used to develop complex computer models of how the Earth's environment works and how human activities, like burning down forests or operating automobiles, affects the environment.

For successful Earth observation issues sustainable development can be defined as maintaining a delicate balance between the human need to improve lifestyles and feeling of well-being on one hand, and preserving natural resources and ecosystems, on which we and future generations depend (Parviz, 2010).

In general the seven dimensions including spiritual, human, social, cultural, political, economic and ecological can be considered for the sustainable development where the main components are economy, society and environment. Approaching sustainable development requires establishing a continuous balance between three latter components. An effectiveness of the space technology applications on the environmental, economical and social issues are quite apparent. The recent developments in information and communication technologies, education and health care, agriculture and agro-food processing, geo-strategic initiatives, infrastructure and energy, and critical technologies and strategic industries, construction, engineering and engineering management have been realized in light of the space technologies. Earth observation techniques are considered of great importance amongst these technologies. Earth observation techniques which apply optical and thermal spectra of the electromagnetic wavelengths have so far developed considerably. Although there is done a lot in this area beforehand, a long way is still ahead. The background of using microwaves for remote sensing goes far the decades ago while it was remaining in the experimental domain and exploratory status for years. It is only in the recent couple of decades that radar remote sensing techniques have been commercialized and used widely. Radar remote sensing is actually accounted for as a new earth observation technology with promising results and future. Its potentials and capacities by itself and being a strong complementary tool for optical and thermal remote sensing are undeniable currently.

2.2 Application of radar remote sensing and SAR techniques

As it was previously indicated InSAR is a sophisticated processing of radar data for combining synthetic aperture radar (SAR) single look complex (SLC) images to form interferogram and utilizing its phase contribution to generate DEM, surface deformation and movement maps and target velocity. The interferogram contains phase difference of two images to which the imaging geometry, topography, surface displacement, atmospheric change and noise are the contributing factors.

Satellite-based InSAR began in the 1980s using Seasat data, although the technique's potential was expanded in the 1990s with launch of ERS-1 (1991), JERS-1 (1992), Radarsat-1 and ERS-2 (1995). They provided the stable well-defined orbits and short baselines necessary for InSAR. The 11-day NASA STS-99 mission in February 2000 used two SAR antennas with 60-m separation to collect data for the Shuttle Radar Topography Mission (SRTM). As a successor to ERS, in 2002 ESA launched the Advanced SAR (ASAR) aboard Envisat. Majority of InSAR systems has utilized the C-band sensors, but recent missions like ALOS PALSAR and TerraSAR-X are using L- and X-band. ERS and Radarsat use the frequency of 5.375GHz for instance. Numerous InSAR processing packages are also used commonly. IMAGINE-InSAR, EarthView-InSAR, ROI-PAC, DORIS, SAR-e2, Gamma, SARscape, Pulsar, IDIOT and DIAPASON are common for interferometry and DEM generation.

It is obvious that digital elevation model (DEM) is important for surveying and other applications in engineering. Its accuracy is paramount; for some applications high accuracy does not matter but for some others it does. Numerous DEM generation techniques with different accuracies for various means are used. DEMs can be generated through different methods which are classified in three groups that are DEM generation by:

i. geodesic measurements,
ii. photogrammetry and
iii. remote sensing.

In DEM generation by geodesic measurements, the planimetric coordinates and height values of each point of the feature are summed point-by-point and using the acquired data the topographic maps are generated with contour lines. The 1:25000-scale topographic maps are common example. The method uses contour-grid transfer to turn the vector data from the maps into digital data. For DEM generation by photogrammetry, the photographs are taken from an aircraft or spacecraft and evaluated as stereo-pairs and consequently 3-D height information is obtained.

DEM generation by remote sensing can be made in some ways, including stereo-pairs, laser scanning (LIDAR) and InSAR. There are three types of InSAR technique that is single-pass, double-pass and three-pass. In double-pass InSAR, a single SAR instrument passes over the same area two times while through the differences between these observations, height can be extracted. In three-pass interferometry (or DInSAR) the obtained interferogram of a double-pass InSAR for the commonly tandem image pairs is subtracted from the third image with wider temporal baseline respective to the two other images. In single-pass InSAR, space-craft has two SAR instrument aboard which acquire data for same area from different view angles at the same time. With single-pass, third dimension can be extracted and the phase difference between the first and second radar imaging instruments give the height value of the point of interest with some mathematical method. SRTM used the single-pass interferometry technique in C- and X-band. Earth's height model generated by InSAR-SRTM with 90-m horizontal resolution is available while the DEM with 4-to-4.5-m relative accuracy is also available for restricted areas around the world.

InSAR ability to generate topographic and displacement maps in wide applications like earthquakes, mining, landslide, volcanoes has been proven. Although other facilities like GPS, total stations, laser altimeters are also used, comparison between InSAR and these tools reveals its reliability. Laser altimeters can generate high resolution DEM and low resolution displacement maps in contrary to InSAR with the spatial resolution of 25m. However, most laser altimeters record narrow swaths. Therefore, for constructing a DEM by laser altimeter, more overlapping images are required. Displacement map precision obtained by terrestrial surveying using GPS and total stations is similar or better than InSAR. GPS generally provides better estimation of horizontal displacement and with permanent benchmarks slow deformations is monitored for years without being concerned about surface de-correlation. The most important advantage of InSAR over GPS and total stations are wide continuous coverage with no need for fieldwork. Therefore, wide and continuous coverage, high precision, cost effectiveness and feasibility of recording data in all weather conditions are its main privileges. However, it is important that the InSAR displacement result is in the line-of-the-sight direction and to decompose this vector to parallel and normal components the terrestrial data or extra interferograms with different imaging geometry are required. It is shown that DEM generated by photogrammetric method is more accurate than the others. It has approximately 5.5m accuracy for open and 6.5m for forest areas. SRTM X-band DSM is 4m less accurate for open and 4.5m less accurate for forest areas.

Data availability and atmospheric effects limit using InSAR, however processing of its data is challenging. For each selected image pair, several processing steps have to be performed. One of the current challenges is to bring the techniques to a level where DEM generation can be performed on an operational basis. This is important not only for commercial exploitation of InSAR data, but also for many government and scientific applications. Multi pass interferometry is affected by the atmospheric effects. Spatial and temporal changes due to the 20% of relative humidity produce an error of 10cm in deformation. Moreover, for the image pairs with inappropriate baseline the error introduced to the topographic maps is almost 100m. In topographic mapping this error can be reduced by choosing interferometric pairs with relatively long baselines, while in the displacement case the solution is to average independent interferograms.

InSAR DEM advantages: Distinction between SAR imaging and the optical systems are more profound than the ability of SAR to operate in conditions that would cause optical instruments to fail. There are basic differences in the physical principles dominating the two approaches. Optical sensors record the intensity of radiation beamed from the sun and reflected from the features. The intensity of the detected light characterizes each element of the resulting image or pixel. SAR antenna illuminates its target with coherent radiation. Since the crests and troughs of the emitted electromagnetic wave follow a regular sinusoidal pattern, both the intensity and the phase of returned waves can be measured.

InSAR has some similarities to stereo-optical imaging in that two images of the common area, viewed from different angles, are appropriately combined to extract the topographic information. The main difference between interferometry and stereo imaging is the way to obtain topography from stereo-optical images. Distance information is inherent in SAR data that enables the automatic generation of topography through interferometry. In other words DEMs can be generated by SAR interferometry with greater automation and less errors than optical techniques. Moreover, using DInSAR surface deformations can be measured accurately.

Different DEM generation methods of Advanced Spaceborne Thermal Emission and Reflection Radiometer (ASTER) stereoscopy, ERS tandem InSAR, and SRTM-InSAR are used. Both the ERS-InSAR and SRTM DEMs are free of weather conditions, but ASTER DEM quality may be affected by cloud coverage in some local areas. InSAR has the potential of providing DEMs with 1-10cm accuracy, which can be improved to millimeter level by DInSAR. Its developments are rapid however it is our requirements that say which one is better for use.

2.2.1 Earth observation for river flood issues

Rivers of Azerbaijan can be divided into the three main groups regarding their water flow specifications:

1. Perennial rivers;
2. Seasonal rivers that flow only during the melting of snow in spring;
3. Episodic rivers that flow in episodes after a downpour of rain of flash flood.

These three groups differ from each other for the volume of underwater supply to their streams. Perennial rivers are fed by a constantly flowing baseflow (groundwater). Seasonal rivers are fed by an elevated water table during the rainy period, while episodic rivers are not at all dependent on base flow.

Like in all other countries, rivers have different feeding sources in Azerbaijan. Most rivers are fed by snow, rainfalls and ground waters. Snow is the predominant feeding source for the rivers of the Major Caucasus, while ground waters contribute the most to water supply of rivers in the Minor Caucasus. The Kur and Araz rivers pass Azerbaijan in their lower and middle courses.

The Kura river is the largest river of Azerbaijan. It stretches for 1,515 kilometers and covers an area of 188 thousand sq. km. The Kura originates from the Hel River in Turkey, passes through Azerbaijan and flows into the Caspian Sea in south-eastern part of the country. The Araz River covers an area of 86 thousand sq. km until its junction with the Kura River. It originates from the Bingol mountains in Turkey at the altitude of 3300 meters. On the whole, the Araz River forms Azerbaijan's border with Turkey and Iran. It passes through Azerbaijan in its lower 80 kilometers and joins the Kura River near Sabirabad. These two rivers belong to the group of rivers, flowing at full under the influence of snow and rainfalls in spring and rainfalls in autumn.

Weather produces the greatest impact on the river flow in Azerbaijan. Intensive rise in temperature causes melting of snow at heights of over 1500. The melting of snow further intensifies after heavy rainfalls of April and May. Snow melts more intensively in the high altitudes (over 2500-3000 meters) from early April through May until June. The melting process influences river flow even in summer time. Thus, melted snow water, absorbed by soil, emerges on the surface and raises water level in rivers. Low river basins (except for those of the Talysh region) are less influenced by the precipitation in spring and summer periods. Winter and autumn rainfalls account for the most part of precipitations in the Talysh region. Rivers are less full of water in summer in Azerbaijan. Heavy rainfalls that may from time to time occur in July and August, lead to floods, causing agricultural damages. Severe floods have been registered in the rivers of southwestern slopes of Major Caucasus Zengezur part. Rivers of the Major and Minor Caucasus mainly flow in hot seasons, while rivers of the Talysh regions flow in colder seasons of year. Rivers, flowing in hot seasons account for most part of all rivers (60-80%).

Such seasonal flows are difficult for industrial use. On the whole, rivers of the Azerbaijan Republic are divided into two groups, according to their water regime:

1.   rivers of full-flowing regime;
2.   rivers of flood regime.

Flood rivers are the Lenkoran rivers and episodic rivers of Gobustan. Other rivers are included into the first group of rivers.

Complex topography and other natural factors cause a non-standard flow across the country. The flow increases with altitudes and reaches its top at a certain height (2800, on the north-eastern slope of the Major Caucasus, 2000-2200-on its southern slope and 2200-2400 on the Minor Caucasus). The flow starts to decline from above the indicated height. Due to the orographic specifications of the Talysh mountains, the flow is inconsistent with the average height. It decreases with the increase of altitude in the Talysh mountains, while in Peshteser and Burovar mountains it rises with the altitude.

The full-flowing rivers of the Azerbaijan Republic mainly flow on the southern slope of the Minor Caucasus. The average flow of such rivers exceeds 45 l-cm. The flow falls to 5 l-cm till the Alazan-Ayrichay lowland. The flow module of rivers of the north-eastern slope of the Major Caucasus 18 l-cm. The increase of flow with the increase of altitude is relatively uniform in this part of the Major Caucasus. The intensive increase in the module of flow is registered on the area between the Yah mountain chains and the Major Caucasus mountains. (upper Qusar, Qudyal and other rivers.). The Average annual module of flow is from swings hesitates from 10 to 20 l-cm.

The flow of rivers, originating in the slopes of the Yah mountains, differs from that of the rivers, flowing from the Major Caucasus. The flow increases intensively and reaches from 6 to 18 l-cm at a height of 1000-2000 meters, due to high level of precipitation. The flow gradually decreases till the Caspian Sea shore down to 0.5 l-cm. the flow decreases beginning from the north-west of till south east of the seaside lowland and reaches zero level on the Apsheron peninsula. Compared with the Major Caucasus, the flow in the Minor Caucasus is more complicated, due to its orographic complexity and differing location of mountain chains. The highest flow has been registered in the rivers flowing from the slopes of Gamish and Qapidjic mountains (over 28 l-cm).

In the Karabakh plateau precipitation is absorbed by soil rocks, thus turning the region into the arid area, while in some places it bursts onto the surface thus increasing the water level in the rivers. That is typical of the upper Terter, Hekeri and other regions as under water provides 70-80% of water to them. The flow fluctuates from 0.8 to 22 l-cm in south east of the Minor Caucasus (rivers, originating in the Caucasus mountains) and from 0.5 to 10 l-cm in the Nakhchivan Autonomous Republic. The flow gradually decreases to the level even lower than 0.5 l-cm on the plains on the side of Araz. In the Talish region the flow increases in the direction from the north to south and from the west to east. The flow reaches its peak (over 25 l-cm) in Tengerud and Astara river basins in the central part of the region, while it reaches its minimum north of the Vilesh river, as well as in the Lenkeran and Vilesh rivers. Gobustan, Nakhchevan and Kura-Araz plains account for the lesser part of water system in Azerbaijan.

Rivers of Azerbaijan carry large quantity of sediment, the result of erosion in the river basins. The rivers in Azerbaijan are the most polluted rivers in the world. Their average annual pollution rate changes from 0.07 to 9 kg-1 cubic mete per region. It reaches its top on the north slope of Major Caucasus and minimum-on the Karabakh plateau. The surface erosion is intensive in the north slope of the Major Caucasus(100-6800 t/sq/km) , and it becomes weaker on the Karabakh plateau (5-10 t/sq.km). The surface erosion in the rivers of the Major Caucasus (0.53 mm) is by 13 higher from that of the Minor Caucasus (0.03 mm per year) and Talish mountains (0.04 mm per year).

The hydrological system of the Azerbaijan Republic contains 10.3 billion cubic meters of water reserves. These water reserves together with those, entering Azerbaijan from neighbor countries (20.6 billion cubic meters) make up 30.9 billion cubic meters. Each square meters of the country receives 90 thousand cubic meters of reserves, while the annual per capita volume of water reserves total 1270 cubic meters. The basin of the river Kura accounts from most part of the water reserves. The nonunifomal distribution of water reserves across the region and around the year hammers the utilization of these reserves and as a result of that the reserves are not able to meet constantly growing demands for fresh water. The situation requires the regulation of water flow. 60 water reservoirs of the country with the capacity of over 1 million cubic meters account for 21 billion cubic meters of water reserves. Most part of these reserves are used in different spheres (irrigation, water supply, industry, fishery, etc). The establishment of water reservoirs of the Middle Kura plays the important role to meeting demands for water. Currently, serious measures are undertaken to preserve pure water reserves and to prevent their polluting with communal and industrial wastes.

The Canals of the Azerbaijan Republic are the main source of irrigation. The canals used for the said purpose extend to 47058.8 kilometers., with canals, used by several farms, accounting for 8580.3 kilometers and those, used only by one farm-for 38478.5 kilometers. The amount of 11 billion cubic meters of water is used in irrigation each year. Irrigated area of Azerbaijan totals 1.4 million hectares.

3. Space technology in disaster monitoring, mitigation and preparedness

3.1 Natural disaster in global change

One of the main impacts of the global changes is the natural disaster. Natural disaster can be playing a significant indicator for the foregoing issue. Natural disaster is increasingly of global concern and its impact and actions in one region can have an impact on disaster in another and vice versa. This compounded by increasing vulnerabilities related to climate change, climate variability as well as other contributions like changing demographic, technological and socio-economic conditions, environmental degradations etc.

There is a highly need for international acknowledgement that efforts to reduce disaster risks which must be systematically integrated into policies, plans and programmes for comprehensive approach of global change and endorsed through bilateral, regional and international cooperation, including partnership.

The importance of promoting of natural disaster impacts reduce efforts on the international and regional levels as well as the national and local levels has been recognized in the past few years in a number valuable and significant multilateral frameworks and declarations.

The following main areas can be covers the challenges of objectives of the natural disaster as a key element of the global changes:

i.   Governance – organizational, legal and policy framework;
ii.   Natural disaster identification, assessment, monitoring and early warning;
iii. Knowledge management and education;
iv.   Reducing underlying natural disaster factors;

Foregoing items can be discussed as a items for further developments. Given the close linkages between disaster risk factors and environmental and natural resource management issues, a huge potential exists for the exploitation of existing resources and established practices aiming at greater disaster reduction. The need for carefully drawn up forest, vegetation, soil, water and land management measures is increasingly recognized and such investigations are being effectively employed to learn the global change.

While countries valued the increased availability of advanced technologies, some were disappointed that their technical capabilities or data were insufficient to make more effective use of them. However, take advantage of space technology and its advance methodology applications for earth observation are being developed and will be executed through global and regional strategically partnerships. The United Nations Office for Outer Space Affairs and the action team of the Committee on the Peaceful Uses of Outer Space are proceeded to implement an integrated global system for the management of natural disaster. A global multilateral imitative, involving both developed and developing countries, including for the countries of the former Soviet Union and Southern European countries with the transit economy has developed a framework document for a 10-years plan to implement a Global Earth Observation Systems. One of the its objectives is the global observation of earth for the aim of global change, reduction of losses from natural disasters and improved understanding, assessment and prediction of weather and climate system variables.

The value of methodology and advanced technology for global change is widely recognized. Their use has increased as the tools have improved, costs have decreased and local access has increased. Methodology and techniques related to the remote sensing, geographical information systems, space-based observations, computer modeling and prediction and information and communication technologies have proved very useful, especially in earth observation systems, mapping, monitoring, territorial or local assessments and early warning activities in case of the natural disaster occurs.

The use of advance methodology and associated data sets in global observation suggests possibility for synergy and shared approaches with global change management. With decreasing costs, those tools have become much more readily available as routine capacities and more useful at local scales in many countries.

States and regional and international organizations should support and encourage the capacities of regional mechanisms and organizations to develop regional plans, policies and common practices, as appropriate, in support of networking coordination, exchange of information and experience, scientific monitoring of earth observation outcomes and institutional capacity development and to deal with natural disaster.

In view of the particular vulnerabilities and insufficient capacities of least developed countries to respond to and recover from natural disasters, support is needed by the least developed countries as a matter of priority, in executing substantive programmes and relevant institutional mechanism for the implementation of the framework of action, including through financial and technical assistance for and capacity building in natural disaster as an effective and sustainable means to prevent and respond to natural disaster.

There is a highly need of the establishing standards for the systematic collection and archiving of comprehensive national records pertaining to the many related aspects of earth observation. In the meantime evaluating country-wide assessments of earth observation and conducting natural disaster assessments, incorporating technical dimensions would be a significant contribution for this issue.

There is an important to assume that earth observation is a national and local priority with strong institutional bases for implementation. It has to be executed key activities within the national institutions and legislative framework as resources – assess existing human resource capacities, community participation – promote community participation through the adaptation of specific policies, the promotion of networking, strategic management of volunteer resources.

Global Earth observation is a voluntary partnership of governments and international organizations. It provides a framework within which these partners can develop new projects and coordinate their strategies, integrate research activities, share results for common interest and investments.

Remote sensing one of the key instrument of the Earth observation provides an important source of data for environmental monitoring and natural disaster mapping and in fact several satellites can service a map the terrain with one meter resolution.

Natural disaster monitoring with integration of space technology can be focused for following significant:

- indication of change throughout of Earth observation by means of natural disaster;
- reduce loss of life and property from natural disaster;
- satellite data evaluation with further understanding, assessing, predicting, mitigating and adapting to climate variability and change;
- effect of natural disaster factors on understanding of the human health

The use of remote sensing and development of GIS will increase the access of the developing world to global change data and harness global Earth observation efforts in support of global environmental challenges for natural disaster issues.

The ability to model potential flood inundation areas and map actual extent of inundation, timing, and intensity under different environmental conditions is central to understanding the dynamics between vegetation, soils, geomorphology, and land productivity in a floodplain. In many regions, the lack of hydrologic and spatial data, constrains the accurate delimitation of flood inundation zones. In spite of these factors, different techniques involving GIS and remote sensing could be used for rapid general zonations of areas susceptible to flooding to reduce costly monitoring infrastructure. This study showed the ability of a DEM-based surface and a wetness layer derived from a Landsat ETM image to identify potential areas to flood inundation in the Kura River Basin, Salyan districts of Azerbaijan. The analyses involved tests in relation to a map of flooded areas derived from soils and geomorphology maps. The statistical tests showed that there is a significant relationship between potential inundation areas derived from a DEM-based surface and satellite image-based dataset with potential inundation areas derived from existent cartographic information on soils and geomorphology. However, the relationships were weak. This analysis showed that the integration of ancillary geomorphologic and soils data, simple DEM-based surfaces, and satellite images maybe a useful first approach to characterize flood inundation areas.

3.2 Methods

The use and application of space technology in a huge case in particularly for the case of river flood reduction is a more suitable means due to the covering a large areas, high accuracy, availability of application in the unacceptability areas etc (Finkl,2000). Moreover, according to the created and developed database there is an advantage to be very sensitive to any available change occurred in the investigated sites.

The benefit analysis of disaster risk reduction involves a number of particular challenges, including:

- Little related information may be available on the frequency and intensity of the hazard event, particularly in a developing country context, implying uncertainty about the level of risk.

- Many of the benefits of any disaster risk reduction measures, whether undertaken in the context of a disaster risk reduction project or as part of another type of development project, are related to the direct and indirect losses that will not ensue should the related hazard event occur over the life of the project, rather than streams of positive benefits that will take place, as would be the case for other investments.

For carrying out of the goals undertaken within the framework of the project execution the following methods have been used:

- The use of ALOS space imagery to be created the land use / land cover basic map for the investigated area using urban, agriculture, garden, scrub, open area, river, stream, canal, road, railroad basic classes;

The use of Landsat ETM space imagery to be detected potential flood inundation areas within the Kura River watershed in the Salyan district of Azerbaijan using a tasseled cap transformation;

The derive 1 m Digital Elevation Model (DEM) from contour lines and elevation points of the investigated area to be generated a deterministic model of potential inundated areas for the region using the DEM and a convex-areas surface;

- The evaluate the sensitivity of each approach to be characterized the flood inundations through statistical tests involving comparison of flooding areas extracted from an inventory of soils and a geomorphology maps.

Investigated area description: The geographical area of interest is the Kura River basin in Saylan district of Azerbaijan (Figure 1). The area comprises approximately 24 km^2. The Kura watershed is one of Azerbaijan's most important agricultural production areas. During the last 10 years, it was affected by 5 excessive floods, causing a lot of damage to people and goods. The one of major source of Azerbaijan freshwater is the Kura River. The mean discharge of 1,144 m^3 sec-1 for the Kura River is the highest among the main rivers in the Azerbaijan, representing 39% of the total discharge from this lowland region. Mean precipitation in the Kura River drainage system is 885 mm year-1, which may range from less than 400 to more 1,800 mm during any one year.

3.3 Satellite data processing

ALOS imagery was acquired 10 June 2007 (Figure 2). The image was georeferenced to UTM zone 39 North, WGS84 using a first degree polynomial rectification algorithm with 30 ground control points (GCPs) extracted from a digitized topographic map at the scale of 1:100 000. The root mean square (RMS) error was equal to 0.5 pixel (5 m).

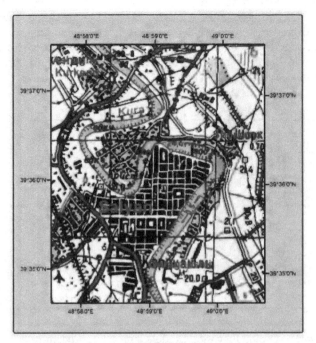

Fig. 1. 1:100 000 topographic map of the study area.

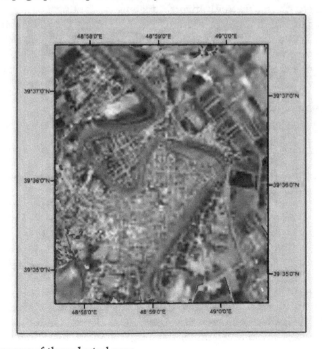

Fig. 2. ALOS imagery of the selected area.

Generation of a Digital Elevation Model: The digital elevation model (DEM) was generated from digitized contour lines and elevation points from topographic map (Figure 3). The digitized lines in shapefile format were converted to points in ArcGIS 9.2 using the "Feature to Point" transformation tools. The points were interpolated using the IDW – inverse distance weighting method.

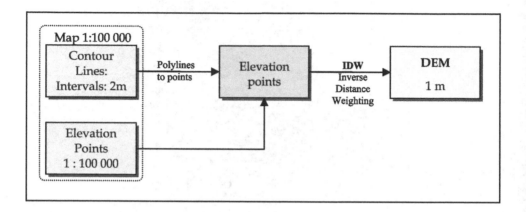

Fig. 3. The flowchart of Digital Elevation Model Generation procedure.

Inverse distance weighting method: Inverse distance weighting is a simple interpolation method, in which a neighborhood around the interpolated point is identified and a weighted average is taken of the observation values within this neighborhood. The weights are a decreasing function of distance. Generally, one can define the mathematical form of the weighting function and the size of the neighborhood expressed as a radius or a number of points.

The simplest weighting function (w) is the inverse power:

$$w(d) = \frac{1}{d^n}$$

with $n > 0$. The value of power can be specified depending upon data characteristics. The most common choice is $n = 2$.

The neighborhood size determines how many points are included in the inverse distance weighting. The neighborhood size can be specified in terms of its radius, the number of points, or a combination of the two. If a radius is specified, the user also can specify an override in terms of a minimum and/or maximum number of points. Invoking the override option will expand or contract the circle as needed. If the user specifies the number of points, an override of a minimum and/or maximum radius can be included. It also is possible to specify an average radius based upon a specified number of points. Again, there is an override to expand or contract the neighborhood to include a minimum and/or maximum number of points. For example, given the following distribution of points with a known value Z:

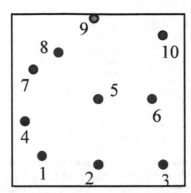

and we want to interpolate a grid surface based on the spatial distribution of the points and their values,

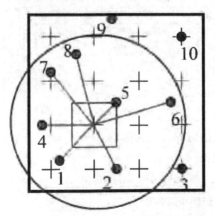

then, using IDW we would assign a value to particular cell based on a number of neighbors and their distance to this cell,

$$D = \frac{(1/d_1^n)V_1 + (1/d_2^n)V_2 + (1/d_4^n)V_4 + (1/d_5^n)V_5 + (1/d_6^n)V_6 + (1/d_7^n)V_7 + (1/d_8^n)V_8}{(1/d_1^n) + (1/d_2^n) + (1/d_4^n) + (1/d_5^n) + (1/d_6^n) + (1/d_7^n) + (1/d_8^n)}$$

Which can be generalized as

$$D = \frac{\sum_{i=1}^{n}\left(1/d_i^n\right) V_i}{\sum_{i=1}^{n}\left(1/d_i^n\right)}$$

where D is the interpolated value, di is the distance from the cell to a point with a known value, and Vi is the value of a particular point.

In this study, IDW with a second order power was used to interpolate the elevation values because of the coarse detail of the original data and the general objectives of the research. IDW is a fast and simple interpolation method, which can be used when the values of points are spatially auto correlated, like in the case of elevation points. Other interpolation methods such as Kriging, could be used when higher accuracy is required.

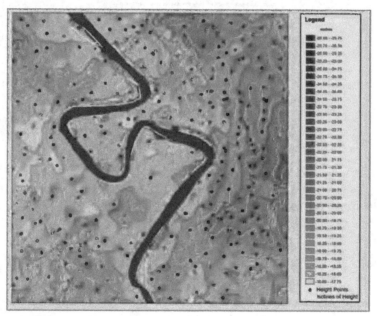

Fig. 4. Digital Elevation Model of the selected area with high points and isolines.

Identification of potential flood inundation areas: A convex surface was obtained with the formula:

Filled DEM – mean filled DEM

Where values < 0 where identified as convex zones (Figure 5). The mean DEM was calculated using standard GIS neighborhood operations. The areas selected as potential flooding areas where those that were convex and fall within an elevation range between -26 m and – 21 m, which is approximately the elevation range corresponding to the lower alluvial plain which is generally affected when severe flooding occurs.

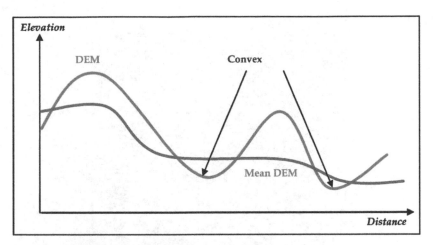

Fig. 5. Determination of convex areas based on the difference between the DEM and a mean DEM.

Potential flood inundation areas mapping: The study and identification of the potentially flood inundation areas in advance is a useful and important aspect of the natural disaster impact reduction.

For this reason the areas potentially flood inundation with a high probability of flooding has been developed and mapped. In this measurements and calculations the staring point has been undertaken as -26m.

The result reflects the potential flood inundation areas based on the height data supposed being as -22m. The result of data calculation and processing from DEM (Figure 4) has been demonstrated in a Figure 6. RF indicated zones reflect potentially flood inundation areas in case of the river level will be increased up to 4m.

This methodology can be successfully applied for potentially flood inundation areas after implementation of geodetic measurements related to the river level for acceptance of the high accuracy data.

Field trip measurements: The main aim of conducted field trips was identification of the inundation areas of the Kura river selected for investigation. One of the needs of this approach was defined due to the luck of the appropriate space data related to the seasonal date with a reach of flood impact of the area.

For the foregoing mentioned reason two field trips have been conducted for the selected area of investigation Salyan district of Azerbaijan. Those trips were implemented in summer season due to the heavy snow melting and autumn season due to the reach of raining when the river flood is more impacted among the all Kura river basin.

Field trips implementations have been scheduled and developed from the stage of the selection more sensitive areas of inundation in place. After those actions the counter of the river has been marked using the sticks installed among the river counter. Coordinates of the counters have been measured using GPS.

Fig. 6. Forecasting of the potentially flood inundation areas.

Based on those measurements all points of counters were installed on topographical map with further bounded of the space image.

The same actions have been applied for the seasons both summer and autumn. The results received from those measurements allow to compare the seasonal river level depends of the weather impacts. At the time it is the way to identify the expected inundation areas.

Based on those results as well as existed database for the river level change there is approach of study and identification of the dynamic change of the Kura river level. It is advantages of development of GIS technology which can be play a significant place on river flood problem solution especially valuable and extremely important instrument for local authority decision makers.

4. Conclusion

In this chapter have been reflected aspects of the use of space science and technology achievements in Earth observation systems. Furthermore it is described currently advances of space technology systems for Earth observation.

One of the main targets of this chapter is to develop of an advance tool for monitoring, data collection, data processing, review and report on progress and challenges in the implementation of disaster risk reduction and recovery actions undertaken at the national level. An advance tool has been undertaken of the use and application of modern achievements of space science and technology for the natural disaster events particularly the river flood.

Furthermore, the other target of project is to be undertaken to assist the local authorities to build up useful database in disaster risk reduction in particularly for the selected area with a more sensitively part of country in point of view the river flood in Azerbaijan. In the meantime the next issue was to demonstrate a contribution of the possibility and advantage of use of remote sensing methods and GIS technology based on space image data collection and data processing for application of similarity problem solving.

It was a highly desirable to create a favorable conditions and mechanisms to be able to develop the strengthened coordination and interaction for appropriate partners at the national level and facilitate explanation of the present status of the selected area and prioritization of strategic areas needed to be considered for purpose of natural disaster risk reduction.

Azerbaijan is the country of the Commonwealth of Independent States (CIS) with the transit economies. The Millennium Development with the eight Goals and Hyogo Framework Actions with three strategic goals and five priorities for actions have been related to the CIS countries.

The river flood is not a reason of damage impact of property and human life. The consequences are a huge as the eventually tracking with malaria, drinking water problem etc. The same problems with appropriate impact of scale occurs in case of Kura river when happens river flood. All this indicated accepts have to be undertaken for further successful management in order to be able to reduce the effect of natural disaster on river flood. An appropriate sufficient with high accuracy database has to be developed for local authorities for decision making.

The other very significant problem is the intended to be undertaken of diversion of the Kura river bed which plans to be started to construct in the upcoming period which will reduce of river flood impact for saving human life and properties.

5. References

Belew, Leland F. and Stuhlinger, Ernst. (1973). *Skylab, A Guidebook. NASA George C.* Marshall Space Flight Center

Covault, Craig. (1991). "Soviets Launch Largest Earth Resources Satellite on Modified Salyut Platform." *Aviation Week & Space Technology,* pp. 21-22

Finkl, C.W. 2000. Identification of Unseen Flood Hazard Impacts in Southeast Florida Through Integration of Remote Sensing and Geographic Information System Techniques. *Environmental Geosciences 7,* pp. 129-136

Parviz Tarikhi. (2010). The role of new Earth observation technologies on monitoring disasters and mitigating the effects; specific focus on radar techniques, *Proceedings of the International Conference on Geoinformation Technology for Natural Disaster Management (GIT4NDM),* Chiang Mai, Thailand, 19-20 October 2010.

Ride, Sally K. (1987). *Leadership and America's Future in Space,* Washington, D.C.: Government Printing Office.

"Sea-viewing Wide Field-of-view Sensor (SeaWiFS)". *NASA Facts on Line*. FS-97 (03)-004-
        GSFC.http://www.gsfc.nasa.gov/gsfc/service/gallery/fact_sheets/earthsci/seaw
        ifs.htm
Skylab Explores the Earth. (1977). Washington, DC: NASA SP-380.

2

Clarification of SAR Data Processing Systems and Data Availability to Support InSAR Applications in Thailand

Ussanai Nithirochananont and Anuphao Aobpaet

Geo-Informatics and Space Technology Development Agency, Thailand

1. Introduction

The Geo-Informatics and Space Technology Development Agency (GISTDA) was established since 2000, and it is the major organization in Thailand that responsible for geo-informatics and all space technology development activities under ministry of science and technology. Currently, GISTDA acquired data from Earth Observation Satellites such as THEOS, LANDSAT-5, RADARSAT-1 and -2, etc. by using remote sensing systems which extensively to be used in the past and tremendously useful from now on for acquiring the satellite data. Consequently, the recognition on the development of this technology and operation acceptant are very beneficial. Moreover, the users are necessary to understand the data processing procedures, as for their applications which depend on the satellite imageries and data processing quality. This article describes and discusses mainly about the data processing and production systems of SAR sensor, including the application example on Bangkok land subsidence using InSAR.

GISTDA has archives of many European, Canadian and Japanese SAR images of Thailand that are instantly available to InSAR applications in Thailand. With our capability to acquire the data direct down-link using 9- and 13-meter antennas, it provides the potential of times series SAR data available for the environmental change detection using InSAR techniques.

In Thailand, the land deformations are not a new phenomenon for major cities and some specific zone whose location lay on the tectonic plate. The applications such as land subsidence, flash flood induced land slide, coastal erosion and fault monitoring are subjected to the country apprehension. However, the irregular deformation patterns put severe demands to the traditional geodetic techniques such as levelling survey, GNSS etc. with respect to the number of stations and the time interval between consecutive measuring sessions. Therefore, to overcome the limitations, InSAR (Interferometric Synthetic Aperture Radar) techniques provide a high spatial resolution and accuracy at the sub-centimetre level. InSAR has all weather, day and night, capability, and the sampling rate of current space-borne systems is improving, 45 days (ALOS-PALSAR), 24 days (RADARSAT-1 and RADARSAT-2) to 11 days (TerraSAR-X), which is satisfactorily high to the monitoring of land deformations.

For SAR data, the production requests were submitted through a Product Generation System (PGS) interface for RADARSAT-1, RADARSAT-2 and APEX CMDR via Vexcel control processor system for ALOS-PALSAR at the ground receiving station facility. The necessarily data employed in most research for deformation is required to be in single look complex (SLC) products in CEOS format where generally each of them consists of five files containing various descriptive records. Each image pixel is represented by complex I and Q numbers to maintain the amplitude and phase information which makes it suitable for interferometric processing. Therefore, the clarification such as the processing algorithm, system configuration, data available to support applications will be provided to certify the potential of using SAR data in Thailand. Finally, a case study on using InSAR techniques for land subsidence monitoring in Bangkok and its vicinity area will show that the successful cooperation between data provider and the user will lead to conquer the best practice.

2. Brief background of satellite remote sensing in Thailand

Historically, Thailand Satellite Remote Sensing Program of the National Research Council of Thailand (NRCT) was established on September 14, 1971 (NRCT, 2000) with the main reason of participating in NASA Earth Resources Technology Satellite (ERTS) Program. The program was promoted to become the Remote Sensing Division under NRCT in 1979, and internationally known as the Thailand Remote Sensing Center (TRSC). Subsequently, in late 1981, the ground receiving station was set up to acquire Landsat-MSS data, and it was capable of receiving and processing data from major remote sensing satellites throughout consistent upgrading of the facilities. In 1982, Thailand Ground Receiving Station was set up as first of its kind in Southeast Asia with the available satellite data such as LANDSAT, SPOT, NOAA, ERS and MOS at that time.

On June 27, 2000, the Cabinet was approved the establishment of Geo-Informatics and Space Technology Development Agency (GISTDA) as the self-governing public organization for conducting technological research, development and applications of satellite remote sensing and GIS, related space technologies for providing relevant services to Thai and international community. Basically, GISTDA is the merging of the TRSC and the IGIS section of Information Center of MOST. Therefore, GISTDA is the national main organization implementation of remote sensing, GIS, and satellite development programs for Thailand. Due to the main mission, Thailand Earth Observation Center (TEOC) has become the common name of TRSC since then.

One of the big movement of space activity in Thailand has been recorded on October 1, 2008, that Thailand Earth Observation Satellite (THEOS) was successfully launched by Dnepr launcher from Yasny, Russian Federation. THEOS is the first operational earth observation satellite of Thailand. The THEOS program was developed by GISTDA, EADS Astrium, the prime contractor, initiated work on the satellite in 2004. Nowadays, GISTDA is developing a worldwide network of distributors to allow the users to use and access to all GISTDA products which is able primarily to access via web-site www.gistda.or.th.

On the other hand, Synthetic Aperture Radar (SAR) satellite systems formerly in function at TEOC include European Remote Sensing Satellite 1 (ERS-1) from the European Space Agency's (ESA) which was launched by July 1991, and the Japanese Earth Resources satellite (JERS-1), launched in February 1992. The ERS-1 sensor operated in the C-band frequency (approx. 5.6 cm wavelength) and JERS-1 operated in the L-band frequency (approx. 23 cm wavelength). Both sensors have a nominal spatial resolution of approximately 30 m. The ERS-1 satellite, with a projected lifespan of three years, was followed by an ERS-2 satellite to continue SAR data acquisition into the late 1990s.

The operations of SAR data at that time has been applied to several major applications such as land-use and land-cover information mapping, coastal monitoring, crop monitoring, etc. The mission record of SAR data had been started with ERS-1 in March 1993 after almost 2 year launched, and the contract had been expired in September 1995. In parallel, JERS-1 SAR ground system had been functioned from October 1993 until October 1998, respectively. Before the coming of RADARSAT-1 (Canadian Space Agency) in July 2000, TEOC had set up the new contract with ESA again for acquiring SAR data from ERS-2 mission which records from August 1996 to October 1999. Currently, the RADARSAT-1 (2000-present), RADARSAT-2 (2010-present) and ALOS-PALSAR (2007-2011) have been the main SAR satellite acquisition of TEOC. However, please note that, JAXA announced that ALOS satellite has been completed its operation due to power generation anomaly since May 12, 2011.

TEOC plays an important role in the area of remote sensing technology in the country and also in the Asian region. The center has some collaborative activities with several international agencies including NASA, JAXA, ESA, CSA, etc. TEOC is located at Ladkrabang district, Bangkok, which is about 4 kilometers from Suwanaphum International airport. It has radius coverage of 2,500 km, covering 17 countries such as Malaysia, Singapore, Philippines, Indonesia, Brunei, Myanmar, Laos, Vietnam, Cambodia, Thailand, Bangladesh, India, Nepal, Sri Lanka, Phutan, Taiwan, and South China and Hong Kong (see figure 1).

3. Fundamentals of synthetic aperture radar

Synthetic Aperture Radar (SAR) is a powerful active coherent imaging system that operates in the microwave frequency band. The system could be placed onboard an airbourne or a spacebourne plarform. It provides capabilities of working in daylight-independent and all-weather condition, and penetrating cloud cover. These capabilities allow SAR an attractive instrument for many applications i.e. change detection, disaster management and environmental monitoring. New applications increase as new technologies are developed.

SAR system imaging the Earth's surface by transmitting pulses and collecting echoes reflected from an illuminated area. To perform this, the transmitter generates pulses of electromagnetic energy at the regular time interval and sends to the antenna. Then the antenna radiates the energy from the transmitter in a directional beam. Each pulse travels at the speed of light to the target area. The returning echo energy are also picked up by the same antenna and passed to the receiver. By measuring the time delay between the transmitted pulses and the reflected return pulse or echo, SAR system is able to determine the distance of the target.

To construct an image, time delay of the received echo must be precisely measured in two orthogonal dimensions. One dimension is parallel to the antenna beam while another is orthogonal to the antenna beam. In the first dimension, parallel to the antenna beam, the SAR system places the received echo at the correct distance from the platform's sensor,

Fig. 1. TEOC Area Coverage for direct downlink.

along the x-axis of the image. The x-dimension is referred as range direction, or cross-track. For the second dimension, orthogonal to the antenna beam, the received echoes are placed in the y-axis of the image, according to the current position of the platform's sensor. The y-dimension is called azimuth direction, or along-track.

The basic geometry of imaging SAR is shown in figure 2. As illustrated, a platform, which could be an airplane or a satellite, travelling along the flight track with velocity V at altitude H. It carries a SAR antenna that illuminates the Earth's surface with pulse of electromagnetic energy. SAR antenna is typically rectangular with dimensions of length L and width W. The antenna is oriented parallel to the flight track and looking sideward to the area on the ground. The distance from the flight track to the target is denoted as range direction and direction along the flight track is referred as the azimuth direction. An area on the ground covered by the consecutive pulses is called swath. Antenna beam footprint is an area on the ground reflected by the pulse. θ is defined as the incident angle or look angle.

In fact the SAR system images an area on the ground but for simplicity, a single point on the ground is considered. This point is known as a point target. The data received from the SAR system are referred as raw data. The data are then demodulated to in-phase-quadrature-phase (I-Q) baseband data. The demodulated SAR signal, s, received from a point target can be modeled as (Cumming & Wong, 2005)

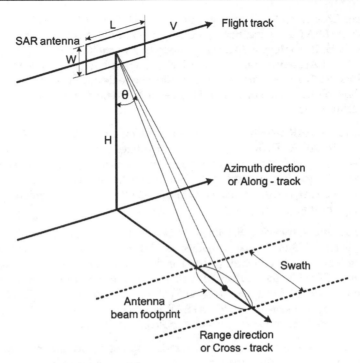

Fig. 2. Basic geometry of imaging SAR.

$$s(\tau,\eta) = A\omega_r[\tau - 2R(\eta)/c]\omega_a(\eta - \eta_c)$$
$$\times \exp(-j4\pi f_0 R(\eta)/c) \times \exp(j\pi K_r(\tau - 2R(\eta)/c)^2) \tag{1}$$

where
A = an arbitrary complex constant
τ = range time
η = azimuth time
η_c = beam center offset time
$\omega_r(\tau)$ = range envelope
$\omega_a(\eta)$ = azimuth envelope
f_0 = radar center frequency
K_r = range chirp FM rate
$R(\eta)$ = instantaneous slant range.

The raw data is not an image due to the point targets are spread out in range and azimuth direction. It will be compressed in two dimensions by SAR data processor, to produce the image. The purpose of SAR processing is to convert the raw data into an interpretable image. Several algorithms have been developed and each algorithm has its advantages in either imaging quality or computation efficient. In the following section two SAR image processing techniques will be briefly introduced: the range–Doppler and the sprectral analysis.

There are three SAR satellites acquiring data at the TEOC: RADARSAT-1, RADARSAT-2 and ALOS. RADARSAT-1 is Canadian first commercial Earth observation satellite launched on November 1995. It employs a SAR sensor operating in the C-band frequency (5.3 GHz). The RADARSAT-1 SAR sensor has two right-looking operational modes, Single Beam mode and ScanSAR mode. The modes of observation offer the real-time swath width ranging from a narrow high-resolution beam of 50-km, Fine beam in Single Beam mode, to a full 500-km swath in ScanSAR mode.

The Next-generation SAR satellite, RADARSAT-2, follow-on RADARSAT-1, was launched on December 2007. All RADARSAT-1 operational modes maintain in RADARSAT-2. The major extended capabilities are a new observation beam, ultra-fine with 3-meter resolution, a fully polarimetric imaging and ability to look either left or right side of satellite track. More details on the RADARSAT-1 and RADARSAT-2 satellites are provided by (Ahmed et al., 1990; Thompson et al., 2001).

The Advanced Land Observing Satellite (ALOS) is Japan's research earth observation satellite operated by JAXA. It was launched on January 2006. The ALOS carries three remote-sensing instruments onboard: (i) the Panchromatic Remote-sensing Instrument for Stereo Mapping (PRISM), the Advanced Visible and Near Infrared Radiometer type 2 (AVNIR-2) and the Phase Array type L-band Synthetic Aperture Radar (PALSAR). PRISM and AVNIR-2 are optical sensors while PALSAR is a microwave sensor. In this paper , we mainly focuses on the data processing system for PALSAR data only.

The PALSAR is L-band synthetic aperture radar operating in the microwave L-band frequency (1270 MHz). It was designed to achieve cloud-free, all-weather and day-and-night collecting high-resolution land observations data on a global scale. PALSAR has three imaging modes: single-polarimetric stripmap mode, ScanSAR mode, and multi-polarimetric mode. More information on the ALOS satellite can be found in (Japan Aerospace Exploration Agency [JAXA], 2008).

4. SAR processing algorithms

SAR processing algorithm is a processing tool used for transforming unfocused raw SAR signal data into a complex image data. Each processing algorithm is suitable for different SAR data types. For continuous SAR data such as data from the stripmap in SAR imaging mode, the most common algorithm is the Range-Doppler (RD) algorithm, but the burst data such as data from the scanning SAR imaging mode, the Spectral Analysis (SPECAN) algorithm, is best suitable.

The Range-Doppler algorithm is the most common algorithm used in most SAR processor. The algorithm was developed since SEASAT program. This algorithm has simplicity of one-dimensional operations and archive block processing efficiency by using frequency domain operations in both range and azimuth. These two directions processing can be independently performed by using range cell migration correction (RCMC) between the two one-dimensional operations.

Computation of the RD algorithm is divided into two processing steps: range compression and azimuth compression. The unfocused raw SAR data compression in each direction is first taking the fast Fourier transform (FFT), and then multiplied in frequency domain by the reference function and finally taking the inverse fast Fourier transform (IFFT). For azimuth compression the RCMC is applied after the azimuth FFT. The most important modification of this algorithm called secondary range compression (SRC) has been added to handle data with a moderate amount of squint.

The SPECAN algorithm was developed to produce a quick-look image for real-time SAR processing. It is the most efficient processing algorithm for ScanSAR data. The key property is computing efficiency which makes the algorithm require less memory than the RD algorithm does but may suffer from some image quality effects. The compression in range direction is the same as in the RD algorithm but different in the azimuth compression.

After range compression, the RCMC is applied before the azimuth compression. The RCMC is efficiency performed a linear correction only. The compression in azimuth direction performs deramping and FFT. Then there are two possible optional way, mutli-looking and phase compensation. The multilook processing is to be performed as the RD algorithm while the phase compensation replaces when single-look processing is to be performed. Reference [4] provides more details of these algorithms.

5. SAR data processing systems

SAR data processing system (SDPS) is used to transform unprocessed raw SAR data or signal data into georeferenced and geocoded image data. The TEOC has two SDPS: the RADARSAT SDPS for data from RADARSAT-1 and RADARSAT-2 SAR sensors and ALOS SDPS for data from ALOS PALSAR sensor. The RADARSAT SDPS is a sub-system of the Product Generation System (PGS) developed by MDA. ALOS SDPS is a sub-system of the ALOS Data Reception and Processing (ALOSRP) system developed by JAXA. The PGS and ALOSRP also have a capability to process data from optical sensor satellite such as LANDSAT TM for the PGS or ALOS AVNIR-2 for the ALOSRP.

5.1 RADARSAT SAR data processing system

The RADARSAT SAR data processing system is a sub-system of the Product Generation System used to transform RADARSAT-1 and RADARSAT-2 raw SAR data into the georeferenced and geocoded image data. This system is a hybrid computer system between UNIX and Windows platform. An advantage of this system is combining power, scalability and reliability of the UNIX with the ease of operation of the Windows. The physical architecture diagram of the RADARSAT SDPS is illustrated in figure 3.

In figure 3, the multi-CPU UNIX server is the SGI Origin 350 executes the core processing software of the RADARSAT SDPS. Its processors are based on Microprocessor without Interlocked Pipeline Stages (MIPS) architecture so that they can take advantage of the multiprocessor environment to parallelize the processing operations to provide efficient, scalable, and accurate data product generation. The Archive Management System (AMS) is a component to manage the archived data in Framed Raw Expanded Data (FRED) format. It tracks and retrieves a large volume of archived data in online, near-line and offline locations. The Windows terminals are Windows operating system HP Workstation used to control and monitor the processing.

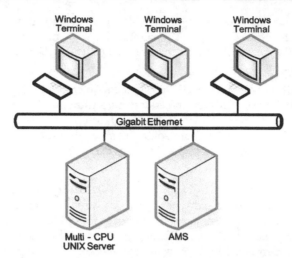

Fig. 3. RADARSAT SDPS Physical Architecture.

The RADARSAT SDPS is driven by a graphical user interface called Human Machine Interface (HMI) on the Windows terminal. The HMI provides the operator with full control over the product generation process via operator control panel. The product generation process is initiated by creating a work order. Work orders can be reviewed and edited via a work order editor panel. Multiple work orders executes in parallel, which each operator can customize to display only information of interest.

For image quality assessment, the HMI also provides the image viewer to perform visual quality assessment on an image. Image viewer tools includes map overlays, measuring distances, displaying average image intensity, displaying Doppler centroid plots and displaying product coverage. Map overlays turn on the overlays in the image to see various map features. Measuring distance allows operator to measure distance between any two points in the image. Average image intensity displays the image intensity in both range and azimuth direction. Doppler centroid plots display SAR Doppler centroid estimation results graphically. Product coverage used to check the product coverage against the expected product boundaries.

The RADARSAT SDPS software can be divided into four processing modules: Data Ingest module, SAR Processor module, Geocoded module and Product Formatting module. The logical architecture diagram of RADARSAT SDPS software is illustrated in figure 4.

The Data Ingest module is responsible for retrieving archived data in FRED format and transferring as signal data to the SAR processor module. The archived data sources could be (i) Magnetic Tape Device Storage (MTDS), (ii) Direct Archive System (DAS) or (iii) Robotic Tape Library (RTL). The MTDS is the offline storage, currently uses super digital linear tape (SDLT), the DAS is an online storage stores downlink data from RADARSAT satellites in the redundant array of independent disks (RAID), and RTL is the near-line storage using the automatic tape archive.

The SAR Processor module is used to focus the raw SAR data into single-look and multi-look image data. It consists of two major software-based SAR processors: the Single-Beam

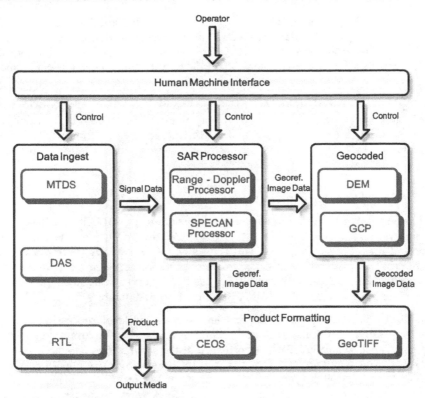

Fig. 4. RADARSAT SDPS Logical Architecture.

processor and the ScanSAR processor. The Single-Beam processor employs the Range-Doppler algorithm as a processing algorithm and suitable for processing Single Beam mode data while the ScanSAR processor employs the SPECAN algorithm as a processing algorithm and suitable for processing ScanSAR mode data. The processed data are georefernced image data stored on disk to be transferred to the Product Formatting module or the Geocoded module.

The Geocoded module is an optional module performs prior to the Product Formatting module. This module supports both systematic and precision geocoding. The digital elevation model (DEM) is employed to produces the systematic geocoded data. The ground truth sources in the form of ground control points (GCP) are used to refine a satellite acquisition model for the precision geocoded data. The output of the Geocoded module is geocoded image data stored on disk to be transferred to the Product Formatting module.

The Product Formatting module receives processed image data from the SAR processor module and the Geocoded module, formats the data, according to the MDA's data product specifications and then writes to output media. The data product format may be CEOS or GeoTIFF. Available output media of the data product can be in the form of disk, CD, DVD or electronics delivery i.e. FTP. The data product can be also archived back to the AMS.

Available data products generated from the RADARSAT SDPS are five georeferenced data products and two geocoded data products. There are single-look complex (SLC), SAR georeferenced fine resolution (SGF), SAR georeferenced extra-fine resolution (SGX), ScanSAR narrow (SCN), ScanSAR wide (SCW), SAR systematic geocoded (SSG) and SAR precision geocoded (SPG).

The throughput of the RADARSAT SDPS generates one standard georeferenced or systematic geocoded data product within twenty minutes. For the eight operation hours, minimum standard thirty SSG data products can be generated. The efficient resources sharing and parallel processing architecture of the system enabling up to twelve work orders can be processed at the same time.

5.2 ALOS SAR data processing system

The ALOS SAR data processing system is a sub-system of the ALOS Data Reception and Processing system used to transform ALOS raw PALSAR data into the standard data products and higher level data products. The ALOS SDPS consists of processing cluster servers, higher level processing servers, a product generation server, an archive server and a workstation terminal. All servers are Linux-based Dell server with Xeon processor. The physical architecture diagram of the ALOS SDPS is illustrated in figure 5.

In figure 5, the processing cluster servers and the higher level processing servers are multiple processors, multiple users and multiple work-order environments, so it can provide high capacity and excellent performance of the system. The data archive server is used to collect and maintain data received directly from ALOS satellite, and received as level 0 from JAXA, as well as higher level data products. All archived data are stored on the automatic tape archive in Sky Telemetry Format (STF). A workstation terminal is used for controlling and monitoring processing of data product via a product generation server. The throughput of the ALOS SDPS for each product and each sensor is at least ten scenes per eight working hours.

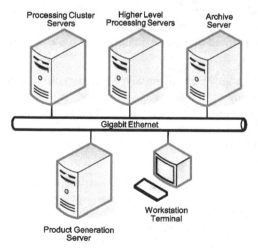

Fig. 5. ALOS SDPS Physical Architecture.

The ALOS SDPS software can be divided into four processing modules: Data Ingest module, PALSAR Processor module, Higher Level Processor module and Product Formatting module. The logical architecture diagram of the ALOS SDPS is illustrated in figure 6.

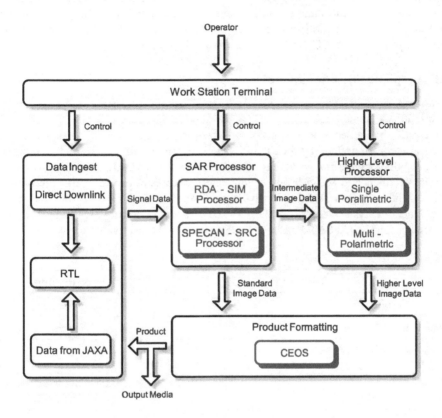

Fig. 6. ALOS SDPS Logical Architecture.

The Data Ingest module is used for retrieving archived data or level 0 data in STF format from the Robotic Tape Library (RTL). There are two possible archived data sources: (i) direct receiving ALOS PALSAR data received at the TEOC (Wide Area Observation Mode or WB1 only) and (ii) imported data from JAXA stored on DTF-2 and LTO-4. The archived data is then transferred to the PALSAR Processor module.

The PALSAR Processor module used to focus on the raw SAR data into standard image data and intermediate image data. It consists of two major software-based SAR processors: the Single Beam processor and the ScanSAR processor. The Single Beam processor employs the Range-Doppler algorithm with squint imaging mode (RDA-SIM) as a processing algorithm. It is suitable for processing Single Beam mode data. The ScanSAR processor employs SPECAN algorithm with chirp transform and secondary range compression (SPECAN-SRC) as a processing algorithm. It is suitable for processing Scanning SAR mode data.

When the STF archived data arrives at the PALSAR Processor module, the sky telemetry data and corresponding parameter are extracted and formatted into CEOS format. Then the formatted data are processed with Doppler parameter file by either of two SAR processors depending on the input data type. The processed data are stored on disk. The RDA-SIM processor can produce the standard SLC image data (L1.1), and level 1.5 (L1.5) image data referred to georeferenced and geocoded images. The data product is CEOS format with the available output media on CD, DVD or electronics delivery i.e. FTP.

6. SAR image quality characteristics

The image quality characteristic consists of a large variety of different parameters. The basic image quality parameters for general users are range resolution, azimuth resolution, peak side lobe ratio, integrated side lobe ratio and absolute location error. The specifications of these parameters are defined by the satellite operating agency and each satellite differently. The specifications of the image quality characteristics for RADARSAT-1 SLC Wide beam mode data products and ALOS level 1.1 Fine beam mode data products are summarized in table 1 and table 2. (MacDonald, Dettwiler and Associates [MDA], 2000; Earth Remote Sensing Data Analysis Center [ERSDAC], 2009) provide a full set of image quality characteristics for RADARSAT-1 and ALOS data products respectively.

Parameter	Specification
Range Resolution (RR)	15.7 m
Azimuth Resolution (AR)	8.9 m
Peak Side Lobe Ratio (PSLR)	< -20.0 dB
Integrated Side Lobe Ratio (ISLR)	-11.2 dB
Absolute Location Error (ALE)	< 750 m

Table 1. RADARSAT-1 SLC wide beam data products image quality characteristics.

Parameter	Specification
Range Resolution (RR)	16.0 m – 17.1 m
Azimuth Resolution (AR)	5.8 m
Peak Side Lobe Ratio (PSLR)	< -20.0 dB
Integrated Side Lobe Ratio (ISLR)	< -15.0 dB
Absolute Location Error (ALE)	< 750 m

Table 2. ALOS level 1.1 fine beam data products image quality characteristics.

Impulse response function is a two-dimensional signal appearing in a processed image as a result of the compression of returned energy from a point target. The width of the impulse response function at a power level 3 dB below the peak of the function is defined to be the impulse response width (IRW). The IRW is commonly referred to as the resolution, and its values are given separately for the two dimensions of the image. The IRW in the range direction is defined as the range resolution (RR), and the IRW in the azimuth direction is defined as azimuth resolution (AR). The azimuth resolution is constant within each beam.

A side lobe of the impulse response function is any local maximum other than those within the contour around the peak, which passes through points 3 dB below the main lobe peak. Side lobes are measured relatively to the main lobe peak. The peak side lobe ratio (PSLR) is defined to be the ratio of the maximum side lobe level and the main lobe level. The integrated side lobe ratio (ISLR) is defined to be the ratio of the integrated energy in the side lobe region of the two dimensional (range and azimuth) impulse response function relative to the integrated energy in the main lobe region. The absolute location error (ALE) is specified as the distance along the ground between the actual geographical location of a point within a processed image and the location as determined from the data product. It may be separated in two direction, range absolute location error (RALE) and azimuth absolute location error (AALE).

7. SAR interferometry

A more recent geodetic measurement technique is interferometric synthetic aperture radar (InSAR) which based on the combination of two radar images. It's earliest the measurement for allowing us to retrieve a Digital Elevation Model, and it has been developed to measure the large-scale surface deformation monitoring or so call Differential InSAR (D-InSAR). The principle of D-InSAR is to first obtain two interferograms of a study area, and then make a differential between for the detection of deformation information. Then, the topographic phase will be removed, and leave just only deformation phase. However, there are several limitations essentially due to temporal and geometric decorrelation. These limitations are well addressed in the time series InSAR techniques, which will be introduced in the following.

7.1 Permanent scatterer InSAR (PSI)

First algorithms of Permanent Scatterer technique were developed by (Ferretti et al., 2000, 2001). Similar processing strategies have been developed by (Crosetto et al., 2003; Lyons et al., 2003; Werner et al., 2003; Kampes, 2005). This method has been very successful for InSAR analysis of radar scenes containing large numbers of man-made structures. The numbers of differential interferograms are generated with respect to a single master (see figure 7). Pixels are selected based on its amplitude stability along the whole set of images, but the stable scatterers with low amplitude may not be detected.

In contrast, StaMPS (Hooper et al., 2007) algorithm uses spatial correlation of phase measurements, so it is applicable in areas undergoing non-steady deformation with no prior knowledge of the variation in deformation rate. PS pixels are defined by phase stability, so

PS candidates are selected on the basis of their phase characteristics. It takes advantage of pixels dominated by a single scatterer to reduce the influence of atmosphere and decorrelation. Then, the phase is corrected for non-spatially correlated errors and "unwrapped" using a statistical-cost approach (Hooper, 2010). After phase unwrapping, spatially-correlated DEM error is estimated from the correlation of phase with perpendicular baseline. The phase is the re-unwrapped with the DEM error subtracted, to improve unwrapping accuracy for larger baselines. Atmospheric artefacts are estimated by high-pass temporal filtering and low-pass spatial filtering. Finally, we can subtract this signal from the estimate value of phase and leave just deformation phase while spatial uncorrelated error terms can be modeled as noise.

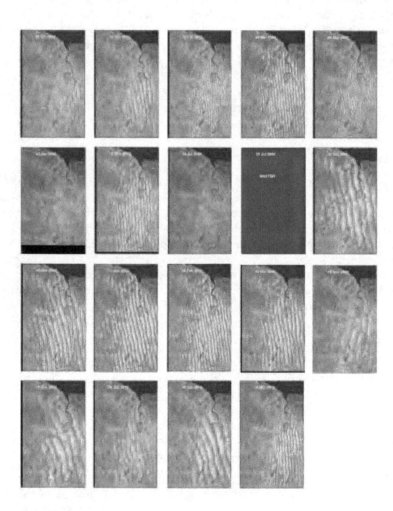

Fig. 7. Interferograms for PS using single master with no spectral filtering.

7.2 Small baseline subset (SBAS)

The Small Baseline Subset (SBAS) proposed by (Berardino et al., 2001, 2002) that the data pairs involved in the generation of the interferograms are carefully selected in order to minimize the spatial baseline. Thus, the mitigation of the decorrelation phenomenon and topography errors will be reduced. The SBAS method was initially exploited the investigation of large scale deformations by calculating the time sequence deformation and estimating DEM error and the atmospheric artifact in a similar way as PS. Noise is then further reduced by multilooking and applying range and azimuth filters (Just et al., 1994) with the aim of unwrapping them spatially. SB methods (Hooper, 2008) on the other hand seek to minimize the separation in time, in space and Doppler frequency of acquisition pairs to maximize the correlation of the interferograms formed. Slow-varying filtered phase (SFP) pixels are identified among the candidate pixels the same way as for PS pixels. For each pixel in the topographically corrected interferograms, its phase can be considered to the wrapped sum of five terms as (Hooper, 2008)

$$\phi_{int,x,i} = \phi_{def,x,i} + \phi_{top,x,i} + \phi_{atm,x,i} + \phi_{orb,x,i} + \phi_{n,x,i} \tag{2}$$

where $\phi_{def,x,i}$ is the deformation phase in the satellite line-of-sight (LOS) direction, $\phi_{top,x,i}$ is the topographic phase caused by uncertainty in the DEM, $\phi_{atm,x,i}$ is the atmospheric phase delay, $\phi_{orb,x,i}$ is orbital phase error, and $\phi_{n,x,i}$ is the noise term.

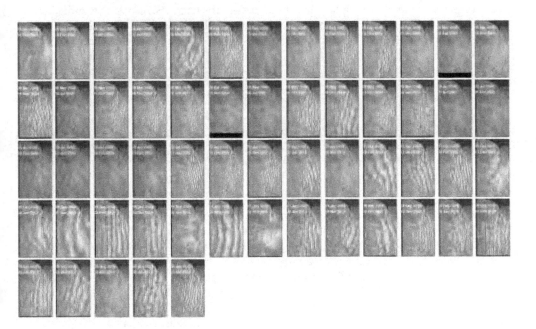

Fig. 8. Interferograms for SBAS using multiple masters with spectral filtering.

All phases error can be subtracted, and leave just deformation phase as the same algorithm used for PSI. Note that different sets of pixels are selected based on different sets of interferograms (single master with no spectral filtering vs. multiple masters with spectral filtering (see figure 8).

8. Application of SAR interferometry in Thailand

8.1 Land subsidence in Bangkok, Thailand

Land subsidence in Bangkok is caused primarily by groundwater over-pumping for the past decade. Monitoring has been carried out by levelling survey technique. The technique cannot provide many benchmarks due to the cost and the difficulty to maintain the overall benchmarks. The locations of the benchmarks are also limited by the urban development to access any area that should be considered. On the other hand, InSAR technology has become more interested since it is overcome the limitation of levelling survey technique, and it has been firstly applied by (Kuehn et al., 2004) during the time spanning February 1996 to October 1996. They reported the maximum subsidence rate -30 mm per year in the southeast and southwest alongside Chao Phraya River. However, with only 4 images and the short time span, it was difficult to estimate the deformation reliably due to decorrelation noise and variable atmospheric phase delay. Nevertheless, the maximum subsidence rates for this area agreed with the levelling survey.

Later on, (Worawattanamateekul, 2006) applied PSI technique using ERS1 and ERS2 data (16 and 10 interferograms) for the time period of 1992-2000. However, the limited number of interferograms made it difficult to achieve reliable results from PSI analysis, as indicated by the accuracy of -6 to -8 mm per year reported by the study. (Aobpaet et al., 2008) applied L-Band ALOS-PALSAR to evaluate the potential and possibility of land subsidence detection using the DInSAR technique. The subsidence map derived from ALOS PALSAR L-band between November 25, 2007 and April 11, 2008 for Bangkok revealed the spatial extent of the deformations and subsidence estimates. However, the subsidence might not reflect long-term subsidence rates because of the short temporal base line and the seasonal cycle of surface movement. (Aobpaet et al., 2009) showed the potential of time series analysis by detecting more than 200,000 pixels that could be used as monitoring points. The results showed a maximum subsidence rate of around -15 mm per year in eastern central Bangkok. However, the study area is preliminary study on sub-scene basis for the processing approach in order to reduce analyzing time and modifying parameters.

The latest study has been successes on apply InSAR time series algorithms, the Persistent Scatterer and Small Baseline, to remotely detect subsidence in Bangkok (Aobpaet et al., 2011). The data set is composed of 19 images acquired in fine beam mode by the RADARSAT-1 satellite (see figure 9a). More or less 300,000 pixels were successfully detected as monitoring points in the analysis, a two order of magnitude greater than the number of ground monitoring points (see figure 10). The average pixel density in the study area is 120 PS per km² with over 150 PS per km² in the urbanized areas. The subsidence velocities fall mostly between 0 to -24 mm per year (see figure 9b). Finally, the validation of the results against levelling surveys has been performed and found agreement at one standard deviation in 87% of cases. They concluded that InSAR time series analysis shows strong potential as an alternative tool for monitoring land subsidence in Bangkok.

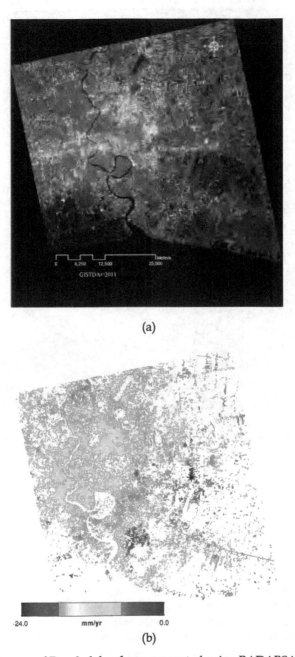

(a)

(b)

Fig. 9. (a) The study area of Bangkok has been presented using RADARSAT-1 data in Fine beam mode with the coverage area 2,500 km². (b) The subsidence rate from InSAR indicated that the maximum subsidence rate is -24 mm per year relative to all pixels in the whole scene with respect to the reference benchmark represented by black star.

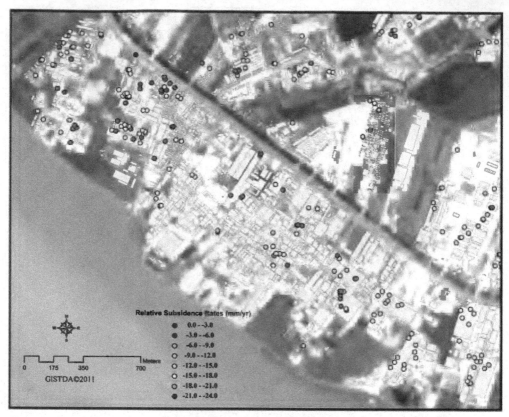

Fig. 10. The south-east Chao Phraya river estuary area which many permanent structures can serve as monitoring points which represent the subsidence rate in mm per year.

9. Conclusion

The establishment of the GISTDA and the long history of Thailand Earth Observation Center are the significants development and contribution to remote sensing activities in Thailand. From that time, Thailand has become one of the most successful countries for the space technology development program especially in remote sensing applications such as flood monitoring, fire monitoring, rice crop monitoring, disaster management, etc. Thus, the capability of direct acquisition in real-time data from SAR satellites such as RADARSAT-1 and RADARSAT-2 make the user who interested in InSAR can set up the plan to acquiring the data from current SAR satellite in time series analysis. Moreover, TEOC was the ALOS sub node, so the large ALOS data archive is still very attractive for the users' intent to study the past disaster or relate applications that may helpful for the prediction model creation.

Finally, the fully operational of TEOC can provide the customers and the users with rapid real-time satellite data for various applications to meet the country's requirement. The application of land subsidence in Bangkok reveals the potential of InSAR time series analysis, but the knowledge how to get the data is much challenged since the large amount of data is required. With the clarification of TEOC systems for especially SAR user, we believe that TEOC will be able to serve as a complimentary component to the development of remote sensing technology and space activities in Thailand and international.

10. References

Cumming, I.G., & Wong, F.H. (2005). *Digital Processing of Synthetic Aperture Radar Data,* Artech House, ISBN 978-1580530583, Norwood, MA, USA

Ahmed, S., Warren, H.R., Symonds, M.D., & Cox, R.P. (1990). The Radarsat system. *IEEE Trans. Geoscience. Remote Sensing,* Vol.28, No.4, (July 1990), pp. 598–602, ISSN 0196-2892

Thompson, A.A., Luscombe, A.P., James, K., & Fox, P. (2001). New Modes and Techniques of the RADARSAT-2 SAR, *Proceedings of IEEE 2001 International Geoscience and Remote Sensing Symposium,* pp. 485-487, ISBN 0-7803-7031-7, Sydney, Australia, July 9-13, 2001

Japan Aerospace Exploration Agency. (March 2008). ALOS Data Users Handbook, In: *World Wide Web,* 01.05.2011, Available from:
www.eorc.jaxa.jp/ALOS/en/doc/fdata/ALOS_HB_RevC_EN.pdf

MacDonald, Dettwiler and Associates. (May 2000). RADARSAT Data Products Specifications, In: *World Wide Web,* 01.05.2011, Available from:
http://www.hatfieldgroup.com/UserFiles/File/GISRemoteSensing/ResellerInfo/RSat/RADARSAT-1 Products.pdf

Earth Remote Sensing Data Analysis Center. (May 2009). ERSDAC PALSAR CEOS Format Specification, In: *World Wide Web,* 01.05.2011, Available from:
www.palsar.ersdac.or.jp/e/guide/pdf/ERSDAC-VX-CEOS-004_en.pdf

Aobpaet, A., Cuenca, M.C., Hooper, A. and Trisirisatayawong, I., 2009, Land Subsidence Evaluation using InSAR Time Series Analysis in Bangkok Metropolitan Area. In the FRINGE Workshop 09, 30 November-4 December 2009, ESA-ESRIN, Frascati (Rome), Italy.

Aobpaet, A., Cuenca, M.C. and Trisirisatayawong, I., 2009, PS-InSAR Measurement of Land Subsidence in Bangkok Metropolitan Area. In the 30th Asian Conference on Remote Sensing, 18-23 October 2009, Beijing, China.

Aobpaet, A., Keeratikasikorn C. and Trisirisatayawong, I., 2008, Evaluation on the Potential of L-Band ALOS Palsar imageries for DEM Generation and Land Subsidence Detection using InSAR/DInSARTechniques. In the 29th Asian Conference on Remote Sensing, 10-14 November 2008, Colombo, Sri Lanka.

Berardino, P., Fornaro, G., Lanari, R. and Sansosti, E., 2002, A new algorithm for surface deformationmonitoring based on small baseline differential SAR interferograms. IEEE Transactions on Geoscience and Remote Sensing, 40(11), pp. 2375-2383.

Crosetto, M., Arnaud, A., Duro, J., Biescas, E. and Agudo, M., 2003, Deformation monitoring using remotely sensed radar interferometric data. In the 11th FIG Symposium on Deformation Measurements, Santorini, Italy.

Ferretti, A., Prati, C. and Rocca, F., 2001, Permanent scatterers in SAR interferometry. IEEE Transactions on Geosciences and Remote Sensing, 39(1), pp. 8-20.

Ferretti, A., Rocca F. and Prati, C., 2000, Nonlinear subsidence rate estimation using permanent scatterers in differential SAR Interferometry. IEEE Transactions on Geoscience and Remote Sensing, 38(5), pp. 2202-2212.

Hooper, A., 2010, A Statistical-Cost Approach to Unwrapping the Phase of InSAR Time Series. In the FRINGE Workshop 09, 30 November-4 December 2009, ESA-ESRIN, Frascati (Rome), Italy.

Hooper, A., 2008, A multi-temporal InSAR method incorporating both persistent scatterer and small baseline approaches. Geophysical Research Letters, Vol. 35, L16302, pp. 5 10.1029/2008 GL034654.

Hooper, A., Segall, P. and Zebker, H., 2007, Persistent scatterer interferometric synthetic aperture radar for crustal deformation analysis, with application to Volcáno Alcedo. Journal of Geophysical Research, Vol. 112, B07407, doi:10.1029/2006JB004763.

Just, D. and Bamler, R., 1994, Phase statistics of interferograms with applications to synthetic aperture radar, Applied Optics, 33(20), pp. 4361-4368.

Kampes, B.M., 2005, Displacement Parameter Estimation Using Permanent Scatterer Interferometry. PhD thesis, Delft University of Technology, The Netherlands.

Kuehn, F., Margane, A., Tatong, T. and Wever, T., 2004, SAR-Based Land Subsidence Map for Bangkok, Thailand. Zeitschrift für Angewandte Geologie, Germany.

Lyons, S., and Sandwell, D., 2003, Fault creep along the southern San Andreas from interferometric synthetic aperture radar, permanent scatterers, and stacking. Journal of Geophysical Research, 108(B1), pp. 2047-2070.

National Research Council of Thailand., 2000. Remote Sensing and GIS Activities in Thailand. Paper present at the 1st Earth Observation Satellites Workshop for Earth Resources Monitoring between 11-12 September 2000, AIT, Bangkok.

Werner, C., Wegmuller, U., Strozzi, T. and Wiesmann, A., 2003, Interferometric point target analysis for deformation mapping. In International Geoscience and Remote Sensing Symposium (IGARSS), 21-25 July 2003, Toulouse, France.

Worawattanamateekul, J., 2006, The Application of Advanced Interferometric Radar Analysis for Monitoring Ground Subsidence: A Case Study in Bangkok. Ph.D Thesis, Technical University of Munich, Germany, pp. 169.

3

Nanosatellites: The Tool for Earth Observation and Near Earth Environment Monitoring

Marius Trusculescu[1], Mugurel Balan[1], Claudiu Dragasanu[2],
Alexandru Pandele[1] and Marius-Ioan Piso[2]
[1]Institute for Space Sciences
[2]Romanian Space Agency
Romania

1. Introduction

Large satellites continue to be affordable only to big national projects or extremely wealthy organizations. As such, emerging countries and small organizations are adopting smaller spacecrafts as means to their space exploration endeavours by forcing the miniaturization age to the space industry. In this chapter we evaluate the possibilities of using nanosatellites with the aim of achieving the best return of scientific output.

Adopted almost exclusively by small organization with limited budgets (universities, private firms or research institutes), nanosatellites have as their main requirement the maintaining off the overall costs at minimum. Unlike the traditional space missions, the nanosatellites use commercial off the shelf components - COTS - in order to decrease costs and fast track the design. This was identified as a liability since the space industry generally requires extensive qualification campaigns for flight hardware. But this is also a strong point since satellites can be designed, built, launched and operated in a fraction of the time required for conventional spacecrafts and at costs orders of magnitude lower.

The small scale counter parts of the traditional space missions represent the tool for Earth observation and near Earth space monitoring in the new age of space explorations. Almost 10 years ago the beginning of this new age became clear with the introduction of the CubeSat standard.

Generally, the nanosatellite term designates satellites in the 1 – 10 kg mass range. However, the most representative for this class is the CubeSat which restricts developers to a volume of approximately $10 \times 10 \times 10$ cm³ (Cal Poly SLO, 2009). Recently there have been developments of sub-nano (pico class) spacecrafts weighing several hundred grams, or even smaller to femtosats – the so called satellites on a chip. However, their characteristics are yet unknown as they are only in the early design phase at present.

Although there are many representative of the nano class, the standardization of the launcher interface and the deployer (P-POD) has helped the CubeSat to receive general acceptance as the de facto standard. Previous experience with small satellites existed before the CubeSats, but their introduction marks the moment when a critical mass of developers begun working on similar designs using similar components. The simultaneous introduction of the P-POD also brought a standardized interface to various rockets. As such it became easier for the developers to address launching organizations for a group of small satellites.

As nanosatellite developers, we propose the adoption of these types of spacecrafts to support Earth observation, space environment monitoring and space qualification efforts at minimal costs.

1.1 Typical characteristics of nanosatellites

The definition of the satellite classes is not very rigid. Contrary to general perception, the exterior dimensions are not defining the nanosatellite. Typically, when speaking of a nano class spacecraft we refer to a sub 10 kg satellite. Consequently, the mass restriction is also a size restriction limiting the exterior dimensions to tens of centimetres. The only standard that imposes restrictions on dimensions is the CubeSat – a cube with a 100 mm edge length permitting small protuberances up to 6.5 mm on each side. The standard also limits the mass of the spacecraft at 1.33 kg – recently upgraded from 1 kg. A deviation from the initial standard allows the use of the space equivalent for two or three CubeSats (or even halves) for a single satellite extending the maximum length at more than 200/300 mm but maintaining the other two dimensions unchanged. These variations from the standard are named double or triple CubeSats to differentiate them from the single cube models. It is worth mentioning that even if the standard permits it, there have been no double CubeSats launched but only single or triple units.

The main characteristic of the nanosatellite are given by their size, which is in the order of tens of centimetres. All the other subsystems need to be scaled down to accommodate the design requirements. There are two approaches in designing a spacecraft of the nano class: either start from the payload and scale the satellite to that payload (traditional method very unusual for small satellites) or scale the payload to the overall dimensions and try to accommodate the other subsystems. The later is the new method that involves setting a design for the payload and revisiting it if after adding the rest of the subsystems the overall restrictions are not met. This might require going into many iterations for the design of the payload and the subsystems.

1.1.1 Electrical power

The accessible power on board a satellite depends on the total surface area available for solar cells. Using the formula in equation (1) we can compute the maximum power one square side can generate. The first term is the solar constant (the power from the Sun light available on Earth's orbit on a dm^2), the second term is the surface area exposed to the Sun light, while the third term is the conversion coefficient between light and electricity.

$$P_{side}^{max}[W] = 13.68 \left[\frac{W}{dm^2}\right] \cdot S_{side}[dm^2] \cdot \frac{C[\%]}{100} \tag{1}$$

For the 10 kg satellite a gross estimation of the size is a cube with the edge length of 200 mm. If we presume that no deployable solar panels are used, the total surface available for photovoltaic cells is 4 dm^2 for each of the 6 sides. Considering the solar constant at 13.68 W/dm^2, and the average conversion coefficient 25%, the total power available when not in eclipse must be lower than 18 W. This value does not take into account Earth's albedo.

The single unit CubeSat is situated at the lower limit of the nano scale according to the definition, so the available surface and electric power is even lower. Repeating the previous calculations for a 10 cm cube gives a value of 4.5 W for the maximum instantaneous power available without deployable solar panels. Just like with the previous estimate we assume no variation of the conversion coefficient associated with the increase of the temperature on the photovoltaic cells and we presumed the satellite in an orientation corresponding to the maximum surface area directly exposed to the Sun. Orbit averages for the power will be significantly lower than the computed values if we take into account the time the satellite spends on eclipse – typically 30% of the orbit period. Deployable solar panels have been included in launched CubeSats, especially for triple units, but for single unit as well (Nakaya et al., 2003; Genbrugge et al., 2009).

1.1.2 Orbit

Due the average power being on the order of watts or tens of watts the nanosatellites are constrained on accessible orbits as well. The limited power available for the transceivers restricts the range between the ground station and the spacecraft. Consequently, nanosatellites are launched on low Earth orbits (LEO). The typical orbit is circular at almost 90º inclination and its altitude is near 700 km. The second, less encountered orbit class is also circular but at 300 – 350 km and its inclination much lower – Genesat-1 and satellites launched from the ISS or the Shuttle. These orbits are at the lower limit of the trapped radiation belts and although the particle fluxes are higher than at sea level they are inferior to those on higher altitude orbits. This is the main reason that COTS components are feasible to be used on board nanosatellites.

9 CubeSat class satellites will be launched on a non characteristic orbit on board the VEGA maiden flight. The orbit has changed several times but the current values for the perigee and the apogee are 300 km and 1450 km with the inclination at 69.5º. The higher altitude of the apogee takes the satellites inside the proton belt. The satellites launched on this mission would further evaluate the possibility of using COTS at high radiation fluxes.

The orbit of the nanosatellite also impacts the communication between the ground station and the spacecraft. For the orbits we previously mentioned a full period is approximately 90 minutes and each day there are between 3 and 5 windows of communications when the satellite is in range of the spacecraft and 3-10 minutes on each interval. These values are averages for a location at 45º latitude. There is daily re-visitation for satellites on LEO and this fits well into the objective of using nanosatellites for Earth observation applications.

1.2 Currently available technologies

Being in development for over a decade, different technologies have been adopted by the nanosatellite designers and advances have been made for increasing the capability of these spacecrafts. We are now at a time when the efforts are starting to show results and in-mission demonstrations of these technologies are beginning.

1.2.1 Processing power

A second important restriction imposed by the energy available on board is the processing power that can be feasibly accommodated on small satellites. Hence the on board computers typically found on nanosatellites launched in the past decade are microcontrollers functioning at frequencies of several MHz. The reason is not the lack of advanced processors that could be integrated, but the need to limit the functioning periods for them as they drain the batteries rapidly. The proposed solution is to use a mixed approach: low power microcontrollers for general functions and high power processor for demanding tasks like attitude determination and control systems (AOCS) or data processing in payload units. This method has already been applied by the integration of units functioning at hundreds of MHz on board nanosatellites already launched or being scheduled for launch.

Launched in 2008, the Japanese nanosatellite Cute-1.7 + APD II used the main boards of two commercial off the shelf (COTS) personal device assistant (PDA) running at 400 MHz as the main components of the on board computer and data handling system (OBDH) (Ashida et al., 2008). Scheduled for launch on the VEGA maiden flight, the Goliat CubeSat integrates a dual core 600 MHz digital signal processor (DSP) for on board image compression (Balan et al., 2008).

Fig. 1. Hitachi NPD-20JWL PDA on board Cute 1.7 + APDII (left) and the DSP board on board Goliat (right).

The trend of adapting commercial portable devices like PDAs and smartphones for use on board nanosatellites fits the general guidelines of low cost design through the use of COTS subsystems. Additionally, mass produced mobile devices are benefiting from extensive research in miniaturization and reduction of power consumption, levels that can't be achieved with the limited budgets of a small satellite research project. Therefore the orientation of nanosatellites developers toward using smartphone processor boards as part of theirs satellite's OBDH system is natural.

The most popular mobile platforms of the moment, iPhone and Android, have proven flight experience at the edge of the atmosphere, on board weather balloons at altitudes higher than 30 km. Taking the idea a step further, a team of researchers in UK plans on building and launching a triple unit CubeSat that will fly a complete smartphone (Surrey Satellite Technology Ltd, 2011). The smartphone will be the payload and the demonstration of its orbit functioning is intended. Part of the test also implies switching off the main microcontroller of the satellite and passing all the OBDH functions to the smartphone.

Besides costs and power optimization there are other benefits of adapting the processors of mobile devices to satellites: better development tools for software with better version control, usability of the same code among several devices facilitating the upgrade of the hardware with minimal software changes, a single low voltage power supply (typically 3.3 V) and a single data interface, numerous integrated peripherals (magnetometers, accelerometers, gyroscopes, temperature sensors). These benefits also come with the loss of some of the customization as there is little possibility to intervene on the hardware (sensor calibration, removing unnecessary modules) and some parts of the software. The number of additional interfaces is also limited and typically a single serial connection exists: Bluetooth. Additionally, USB host mode connection is being proposed as standard for smartphones running the next release of Android OS (version 3.1).

As part of our research, we propose the use of the on board data connections – mainly Wi-Fi, but GPRS or 3G also – as communication platforms for nanosatellites flying in close or dispersed orbital formations. If Wi-Fi devices allow ad-hoc networking, the use of the mobile phone data connections will necessitate the existence of a cell node managing the network.

1.2.2 Attitude and orbit control systems

Most advanced applications require precise determination of the orbit and the attitude of the satellite. Others also need capabilities to change the orientation and some even the position of the satellite. This is the technology field where most nanosatellite research is focused. Miniaturized attitude determination sensors existed at the time nanosatellites started being launched and various sensors were rapidly integrated: Sun sensors, magnetometers, Earth horizon sensors, star trackers.

Beside early attempts at using permanent magnets or magneto-torquers to stabilize the satellite or change its orientation, recent developments have been made at integrating reaction/momentum/inertial wheels on board even the CubeSats – see Fig. 2 (Balan et al, 2008; Bozovic et al, 2008). The CanX-2 was developed and launch for testing some of the critical components of the AOCS system required in the formation flying demonstration mission of CanX-4 and CanX-5. As such, the triple unit CubeSat included a complex attitude determination system based on multiple sun sensors and a magnetometer. It also integrated a single reaction wheel for evaluation purposes together with a propulsion system evaluation unit. The team reported successful operation for all the AOCS subsystems evaluated (Sarda et al., 2010).

Fig. 2. Motors and reaction wheels on the mechanical structure of Goliat (left), motor and the inertial wheel assembly for the SwissCube (right).

For nanosatellites bigger than single unit CubeSats, different commercial solutions have emerged recently. One such example is the MAI-x00 series which offer complete attitude determination and control for small satellites in packages from half a CubeSat to 1 CubeSat (Maryland Aerospace Inc., 2011). Position actuator products are not as advanced for small satellite, and either cold gas or micro thrusters are considered. A different approach is the use of aerodynamic breaking in close orbital formation scenarios. For two or even more CubeSats launched from the same deployer, the initial velocities are the same. Any change in the orientation results in a change of the surface area normal to the trajectory and in a change of the aerodynamic drag. Such a solution will work only in preventing the spacecraft separation and it actuates only in the direction of the orbit. Any difference in the velocities of the two spacecrafts for the other two axes would render the method unusable (Balan et al., 2009).

After a decade of nanosatellites missions the technologies have evolved enabling the exploitation of the new class of spacecrafts for more complex applications. As the subsystems available have evolved, sufficient flight data has been gathered for essential components and their reliability is guaranteed.

2. Earth observation and near Earth environment monitoring

The objectives of small spacecrafts were initially only educational while science and Earth observation were just viewed as secondary goals. However the nanosatellites' missions have quickly begun to evolve to more complex science with increased demand for reliability. From the industry perspective, nanosatellites now represent an easy access to space for simple instruments or for test bed applications. Among the instruments best suited are the sensors for monitoring the radiation environment on LEO, the magnetic field and some of the upper atmosphere phenomena. The inclusion of digital cameras on board nanosatellites did not have Earth observation objectives at first. Initially the imaging experiments were included for their public outreach potential.

The Earth observation potential of nanosatellites is still disregarded since optic instruments are considered too large for integration on nanosatellites. However as the exploitation potential of the new class of spacecrafts was revealed, the idea of Earth observation even on CubeSats starts to gain more general acceptance with every new launch. A camera having one of the highest focal lengths mounted on a CubeSat is part of the Goliat mission. Its integration proved very difficult as the optical lens and sensor assembly occupy almost half of the interior of the spacecraft.

One of the advantages of LEO is the proximity to the surface and to the upper atmosphere. Earth Observation doesn't target only the monitoring of the land or water masses, but also the monitoring of phenomena in the atmosphere. Small focal distance cameras are ideal at imaging the movement of large cloud formations (like with tropical storms or large scale meteorological manifestations). Also, we mentioned earlier the re-visitation interval of approximately 12 hours which is important for events with high dynamicity. These time intervals can be further decreased if several nanosatellites (a constellation) are deployed on the same orbit in successive launches. The satellites cover the same area at time intervals several hours apart with the actual timing depending on the number of spacecrafts launched.

A special application for low resolution image acquisition that could be implemented on nanosatellites involves multi-spectral imaging on board satellites flying in a close orbital formation. An identically built satellite is to be repeated and the optical systems will be the same among all the members of the orbital formation. Unlike the large spacecrafts, the imaging sensors on each satellite can be single-spectral, and the wavelength for the maximum sensitivity is the one that differs. For redundancy multiple spacecrafts will monitor each spectral band and the image acquisition will be commanded to all satellites. Multi-band images can be reconstructed either on ground or on the network on orbit. However, for each band a single image will be sent to the ground station, resulted from the fusion of all the images taken by satellites with the same spectral band sensitivity – see Fig. 3 (Balan et al., 2009).

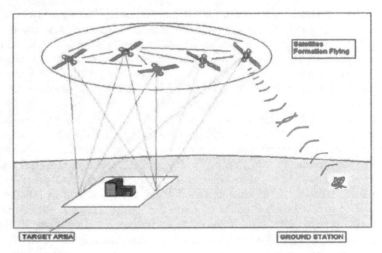

Fig. 3. Formation flying scenario with distributed sensors, in flight data processing and single data stream communications.

One of the key application of nanosatellites is as support in disaster management efforts. In these situations low re-visitation periods are required to monitor major floods, fires or other large scale natural disasters. For these types of conditions, rapid information delivery is more important than resolution as there is an immediate need to roughly identify the areas already affected and the ones most exposed to danger. Nanosatellites can therefore be used in conjunction with large spacecrafts to identify precisely the locations where higher resolution images are required and request the specific areas to be monitored.

Several approaches have been proposed to address the problem of the size of the optical systems. Among them, worth mentioning are the use of complex deployable lens mounts and the use of multiple sensors. A nanosatellite that successfully demonstrated deployable optics is the 8 kg, 19 cm x 19 cm x 30 cm PRISM nanosatellite developed by the Intelligent Space Systems Laboratory (ISSL) of University of Tokyo (Komatsu & Nakasuka, 2009).

The advantage of nanosatellites is their reduced costs. If multiple identical such spacecraft are to be built, the costs are decreased even more. As such, it is natural to consider multiple satellites scenarios in which the imaging of the same area, or adjacent sectors would result in a representation of higher resolution. The solution is not complete if the image processing is set to be conducted on ground as all the raw data from the sensors must be forwarded to the ground station. This situation is not feasible for nanosatellites as there is a limited data rate caused by the limited power. Therefore the use of on board processing for all the data acquired by the distributed sensors is a necessity. As resources are limited on nanosatellites, the ideal method for implementing complex data processing is by using the hardware on each of the spacecrafts and dividing tasks among processors based on their availability, like in grid computing. This complex image processing method was not yet implemented on launched satellites. The main issue is with scaling down the data fusion algorithms so they can be implemented on the limited hardware resource on board nanosatellites.

Precise Earth observation requires the use of key technologies identified in the previous section. The most obvious among the requirements is the need to determine the position and orientation of the satellite with the accuracy needed by the application – approximately 10% of the ground target size. The same resolution is required when controlling the orientation actuators. Once the image has been stored on board, the data must be sent to the ground station. The reported data rate in nanosatellite to ground station communications has increased in the last couple of years with the use of S-band transceivers and the utilization of the experience acquired during the operations of the first spacecrafts. Given the limited emission power, the data throughput can be increased if directive antennas shall be developed for use on the nanosatellites. Furthermore, even if the data rate is not increased, the amount of data transferred to the ground station can be increased by optimizing the radio communications windows. At present, with mid-latitude ground stations, the communications windows are less than 10% of the orbital period. A second ground station could increase the percentage, but either the separation among ground stations must be of hundreds to thousands kilometres, or each ground station must target a different satellite and different data streams are to be transferred. Single ground stations that can have greater communication windows must be situated in the Polar Regions if the polar orbits remain the custom for nanosatellites. An alternative is represented by the ground station networks currently being proposed – GENSO – but these are tailored for educational purposes and need to be adapted to the different needs of the commercial applications.

It is expected that the time from design to delivery for a nanosatellite missions to further decrease, and the mission costs to continue to go down together with it, due to the rapidly increase in the nanosatellites subsystems and components market.

3. Multiple satellites mission for Space Situational Awareness (SSA)

The multi satellite missions are best suited for small spacecrafts due to the small costs and rapid production associated with them. We present distributed measurements as a new way to better and faster understand complex phenomena by using simultaneous data gathering in the target environment. A group of nanosatellites (constellations or formations) is the most cost-effective way to implement this approach in space. Furthermore, distributed data collection can be correlated with distributed processing to enable single data stream transmissions between the spacecrafts in orbit and the ground station as opposed to the multiple streams associated with independent multiple satellites. This solution better addresses the issues of limited data rate in small satellites communications caused by the low available power and not using directive antennas. Unlike with imaging applications, the amount of data from multiple instruments in a close orbital formation can easily be transmitted from a single satellite even if measurements from each sensor are included. Raw signals from every event will however have the same impact as images on the size of data to be transmitted, but in the case of unusual results, the actual values recorded can be sent in multiple transmissions without impacting the stream of on board processed data.

The potential of small satellites, nanosatellites and CubeSats especially, to contribute valuable data necessary to the modelling and the prediction of the space environment in the context of the SSA has recently begun being recognized and the need to aggregate all the data from recent small satellites launches is identified (Holm et al, 2009). Extrapolating on this trend we consider there is a further need for a unified data collection structure with multiple points of acquisition and multiple similar – identical or complementary – sets of sensors. Nanosatellites are the perfect propositions for demonstrating the benefits of this type of missions due to the reduced mission costs and their rapid development.

One of the main directions in the field of near Earth space monitoring is the research and development of spacecrafts built for multi-satellite missions. Space weather's influence on our daily life increases constantly with the miniaturization as devices become more sensible to outside interferences. Within the context of a new maximum in the solar activity, the perturbations of space supported services are becoming more frequent so we base our mission proposition on the need to investigate this domain. Multiple spacecrafts missions, in either constellation or formation configurations, will serve as points of observations for the evolution of the complex environment of nuclear particles in conjunction with the dynamic magnetic field of the planet.

Based on the experience in developing the radiation detection experiment on board Goliat, we proposed the further investigation of the nuclear particles in LEO and the magnetic field, in order to identify correlations between local variations of the two. Observations on the dynamics of the phenomena are possible by using the distributed sensors and the short re-visitation intervals. All spacecrafts are to be identical from the hardware point of view. The minimal requirements for the radiation sensors are the need for differentiation based on particle type and the capability of measuring the energy of each event so as to obtain the representation of the radiation spectrum at each satellite. Precise magnetic field measurements require caution in separating the interferences generated by the spacecraft's own subsystems. This is why magnetometers need to be mounted as far from the satellite as possible, usually at the end of a deployable boom. Each spacecraft needs also to integrate precise attitude determination for both position and orientation of the magnetometer's axes with respect to the Earth.

Space weather monitor nanosatellites can be launched in solitary missions as demonstrators, but greater value can be added by launching several in a close orbital formation. In the first months of their mission, they will synchronize data collection between them and the data transmissions to the ground station are centralized through a single point of contact – one member of the formation. As the atmospheric drag starts affecting each satellite differently, their relative velocities change and the distances among satellites will increase. The formation transforms into a constellation and the phenomena recorded are no longer local, but become global.

The same approach can be applied to multiple applications in the context of SSA. The mixed configuration mission can theoretically fulfil both roles: being launched as a close orbital formation and, once the fuel has run out, gradually migrating to a dispersed formation and then becoming a constellation. The simplest demonstration would require launching three identical single unit CubeSats from the same PPOD and then test the formation flying capabilities on board these three spacecrafts. Such a mission can serve as a test bed for larger nanosatellites. During the demonstration various hardware and, equally important, software can be tested to facilitate future missions.

4. Case study: Goliat, building a CubeSat for Earth observation & near Earth environment monitoring

The authors of this chapter worked at developing Romania's first CubeSat class satellite - Goliat. Among its goals an important part is the demonstration of Earth observation and near Earth environment monitoring capabilities on board nanosatellites.

4.1 Goliat platform subsystems

Goliat is a single unit CubeSat developed by a Romanian consortium led by the Romanian Space Agency. The project was directed toward students at two universities in Bucharest that were tasked at designing and building the satellite in order to have them educated in the work practices of the space industry. The project involved not only building the satellite, but also setting up a ground station infrastructure at two locations near two major cities in Romania: Bucharest and Cluj-Napoca.

Fig. 4. Goliat Flight Model.

The satellite was selected to be launched on Vega's inaugural flight on an elliptical orbit having the perigee at 300 km and the apogee at 1450 km. The satellite's life on this orbit is between 1 and 3 years due to rapid altitude decay caused by atmospheric drag.

4.1.1 Mechanical structure

Goliat was built in accordance with the CubeSat specification as a single unit satellite. The skeletonized version of Pumpkin's mechanical structure is the basis of Goliat's design. The +Z side of the satellite was full metal and not skeletonized as optics mounting and several other components required a harder fixture. The structure is made out of aluminium alloys with the rails hard anodized.

4.1.2 OBDH

Two MSP430F1612 microcontrollers are the backbone of the satellite. One of the onboard computer (OBC) units was acquired from Pumpkin, while the other one was a custom solution built on an internal design. The two processors are running at 7.2 MHz and communicate with each other via a serial peripheral interface (SPI). The OBC board also includes a SD card interfaced on SPI as well. Other subsystems are also communicating using the SPI link: the camera processor board and the control unit of the UHF radio. Additionally each microcontroller connects on a serial interface to various components: camera processor board, 2.4 GHz transceiver, magnetometers, GPS. Data from two experiments (radiation measurement and micro-meteoroid impact instrument) and from the housekeeping sensors is collected at the microcontrollers on the built-in ADC channels. An independent microcontroller unit was implemented on the electronic power supply (EPS) board to manage this subsystem.

4.1.3 Radio communications

Goliat has two data links for radio communications. The primary data link unit uses a 1 W commercial transceiver operating in the 2.4 GHz band. This unit is controlled and it is directly interfaced to one of the MSP430 microprocessors. It is scheduled to operate only when in range of the ground station and its main purpose is to transmit data from the experiments and to receive commands from the operators in the control room.

The secondary transceiver is a beacon operating in the 70 cm radio amateur UHF band. It is built from a portable radio-amateur transceiver and a custom built AFSK modem controlled by a third MSP430F1612 microcontroller. This radio module is meant only at transmitting but receiving capabilities have been added to act as back-up for the main radio unit. The data transmission on this link will be continuous on the entire orbit and both Morse code and AFSK packets with housekeeping data will be transmitted. This unit is controlled by a different OBC than the 2.4 GHz transceiver so full redundancy is available on the spacecraft.

4.1.4 Electronic power supply

The EPS subsystem features the power generation, energy storage and voltage conditioning functions of the satellite. The first component of the subsystem is made up by the solar panels. 18 photovoltaic cells measuring 41 mm x 42.2 mm and having an efficiency of approximately 25% are distributed on the 6 sides of the satellite. Three sides contain 4 cells each while three sides contain only two cells each. The estimated average power from the solar panels is a little over 2 W. The cells are grouped so the voltage reaching the main EPS board is 4 V.

Due to the noise sensitive nature of one of the on board experiments, the main requirement of the EPS design was that no switching power supply should be present on the satellite's supply lines. This imposes the use of LDO regulators which are highly ineffective. More so, the need of a 5 V supply line, coupled with the less than 5 V output voltage of the solar panels, requires the use of a battery pack with the nominal voltage above 5 V. Li-Ion batteries were selected due to having the highest energy density per mass. The ping-pong architecture of the EPS uses two Li-Ion battery packs with their nominal voltage at 7.2 V. A battery pack always supplies the satellite, while the other is charging from a step-up converter that has the voltage from the solar panels as its input.

4.1.5 ADCS

For the determination of Goliat's position there are two independent methods. First uses a commercial GPS receiver while the second one involves sending the orbital parameters as a *.tle file (two line elements) and then calculate the position using an orbit propagator implemented on one of the microcontrollers. For orientation the satellite uses a triple axis magnetometer and an IGRF implementation on the same microcontroller to compare the data for the actual position and determine the orientation of the satellite with respect to the Earth.

Goliat is meant to demonstrate a simple reaction wheel system for changing the orientation of CubeSats. Due to the mission constraints only two wheels were able to be included in the satellite design. The attitude control system is made of two high precision reaction wheels mounted on top of two micro-motors and the assemblies are attached to the aluminium structure in the centre of two perpendicular sides of the satellite.

4.2 Payload

The payload of Goliat consists of three independent experiments for near Earth environment monitoring and Earth observation.

The first of them is named SAMIS and it is a micro-meteoroid detection instrument that uses a thin film piezo-element to measure the energy of the impact between the satellite and the micrometer sized particles on LEO. The measurement of the flux of particles encountered by the satellite will take place continuously after the commissioning of the spacecraft.

Dose-N is the second on board experiment and it targets the measurement of the total ionizing dose on Goliat's orbit. The experiment's added value increases with the new Vega orbit since the satellite's trajectory is no longer circular and a range of altitudes in the radiation environment is to be mapped. If the 700 km altitude orbit was at the lower limit at the trapped proton belt, the elliptical orbit enters the region and exposes the satellite's components to higher radiation fluxes. The radiation detection instrument uses a scintillating material that generates visible radiation when interacting with nuclear particles. The light is detected by a photodiode that has its maximum sensitivity at the same wave length as the photons emitted by the scintillators (430 nm). The signal from the photodiode is integrated and the amplitude of the output signal measured by the microcontrollers as the total energy deposited in the integration time frame. Measurements will be taken at equally distanced positions along the trajectory of the spacecraft and dose measurements will be correlated with resets and other errors in the functioning of the satellite.

Fig. 5. Micro-meteoroid impact sensor (left) and radiation detector (right). Integrated on Goliat.

The third and the last of the experiments on board Goliat is a narrow angle camera (NAC). The sensor of the camera consists of a 2048 x 1536 matrix of pixels, the highest resolution fitted on a single unit CubeSat. The pixel size is 3.2 μm x 3.2 μm. For the electronics of the experiment a commercial solution with the sensor board stacked on top the processor board was used. The processor board features a Blackfin ADSP-BF561 dual core DSP running at 600 MHz. A μClinux operating system is installed on the microcontroller and software written in C/C++ can be compiled on the device. A dual interface, serial and SPI, is used to communicate with the other microcontrollers on the satellite and with the SD card. The power consumption for the two stacked boards is typically at 1 W and does not exceed 2.25 W according to the manufacturer.

For a typical nanosatellite orbit – circular at 700 km altitude – the expected equivalent area in a 3 mega pixel image is a 50 x 70 km region. The expected pixel resolution is tens of meters, enabling the identification of geographical features and even of large constructions at the ground. The elliptical orbit for the Vega launch will make possible testing the camera at various altitudes in the 300 to 1450 km range. For the project a special lens mount was designed and built at PRO Optica in Bucharest. The optics had to be accommodated inside the satellite and compliance with the CubeSat standard was desired. The optics had to meet the restrictions of accommodating the other subsystems while maximizing the focal length. A 6° field of view was achieved at a 57 mm focal length.

The main objective of the Goliat satellite is to demonstrate the potential of nanosatellites to execute complex experiments at low costs. An auxiliary objective was the development of a flight proven satellite platform that could be adapted for future application oriented space missions.

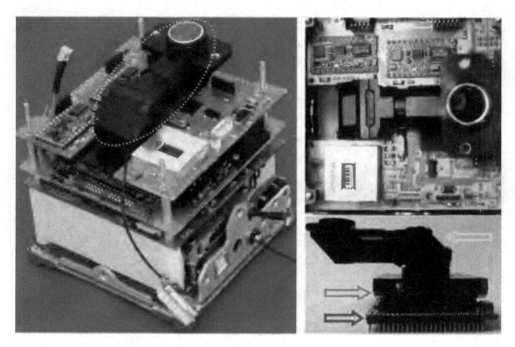

Fig. 6. The narrow angle camera on board Goliat: processor board (red), sensor board (blue), optics (yellow).

5. Conclusions

Nanosatellites are definitely the most rapid changing sector of the space industry in the last decade. Their development has taken many by surprise and their momentum is just starting to grow now that technologies essential for better exploiting their potential are becoming available. We are expecting their growth to continue due to the further reduction in costs and the decrease of the development cycle associated with the trend of standardizing the bus of the spacecraft.

At first missing, technologies like small scale AOCS systems, OBDH modules, and even low power, high data rate transceivers have rapidly evolved driven by their requirement in building complex subsystems. It is now cheap to build more than one satellite and satellites are becoming smarter when connecting them in a network. Furthermore, the applications proposed for the new types of spacecrafts and missions promise to revolutionize space operations with the outside of the box thinking associated with doing things at a smaller scale.

Space is finally becoming accessible to projects with limited budgets, through nanosatellites, the new tools for near Earth explorations.

6. References

Ashida, H.; Fujihashi, K.; Inagawa, S.; Miura, Y.; Omagari, K.; Konda, Y.; Miyashita, N; Matunaga, S. (2008). Design of Tokyo Tech Nano-satellite Cute-1.7 + APD II and Its Operation, *Proceedings of the 59th International Astronautical Congress*, 1995-6258, Glasgow, Scotland, October 2008.

Balan, M.; Piso, M.I.; Stoica, A.M.; Dragasanu, C.G.; Trusculescu, M.F.; Dumitru, C.M. (2008). Goliat Space Mission: Earth Obervation and Near Earth Environment Monitoring Using Nanosatellites, *Proceedings of the 59th International Astronautical Congress*, 1995-6258, Glasgow, Scotland, October 2008.

Balan, M.; Piso, M.I.; Trusculescu, M.F.; Dragasanu, C.G.; Pandele, A.C. (2009). Pluribus - Nanosatellites Formation Flying In A Networked Environment, *Proceedings of the 60th International Astronautical Congress*, 1995-6258, Daejeon, South Korea, October 2009.

Bozovic, G.; Scaglione, O.; Koechli, C.; Noca, M.; Perriard, Y. (2008). SwissCube: Development of an Ultra-Light Efficient Inertia Wheel for the Attitude Control and Stabilization of CubeSat Class Satellites, *Proceedings of the 59th International Astronautical Congress*, 1995-6258, Glasgow, Scotland, October 2008.

Cal Poly SLO (Aug. 2009). CubeSat Design Specification Rev.12, 15.05.2011 Available from: http://cubesat.org/index.php/documents/developers

Genbrugge, M.E.A.; Teuling, R., Kuiper, J.M.; Brouwer, G.W.; Bouwmeester, J. (2009). Configuration Management In Nanosatellites Projects; Evaluation of Delfi-C3 and Consequent Adaptation for Delfi-N3xt, *Proceedings of the 60th International Astronautical Congress*, 1995-6258, Daejeon, South Korea, October 2009.

Holm. J; Paxton, L.; Rogers, A.; Morrison, D.; Weiss, M.; Schaefer, R.; Darrin, A. (2009). Small Space Weather Satellites as a Component of a Space Situational Awareness Virtual Organization, *Proceedings of the 59th International Astronautical Congress*, 1995-6258, Glasgow, Scotland, October 2008.

Komatsu, M.; Nakasuka, S. (2009). Univesity of Tokyo Nano Satellite Project "PRISM", *Trans. JSASS Space Tech. Japan*, Vol. 7, ists26 (2009), pp Tf_19 - Tf_24, 1347-3840

Maryland Aerospace Inc. miniADACS Home, Accessed on 28.08.2011, Available from: http://www.miniadacs.com/index.html

Nakaya, K.; Konoue, K.; Sawada, H.; Ui, K.; Okada, H.; Miyashita, N.; Iai, M.; Urabe, T.; Yamaguchi, N.; Kashiwa, M.; Omagaru, K.; Morita, K., Matunaga, S.(2003). Tokyo Tech CubeSat: CUTE-I – Design & Development of Flight Model and Future Plan -,

Proceedings of the 21st International Communications Satellite Systems Conference, AIAA 2003-2388, Yokohama, Japan, 2003

Sarda, K; Grant, C; Eagleson, S.; Kekez, D.D; Zee, R.E. (2010). Canadian Advance Nanospace Experiment 2 Orbit Operations: Two Years of Pushing the Nanosatellite Performance Envelope, *Proceedings of the Symposium on Small Satellite Systems and Services (4S)*, , Funchal, Madeira, Portugal, May 31 – June 4 2010

Surrey Satellite Technology Ltd. SSTL STRaND smartphone nanosatellite, Accessed on 28.08.2011, Available from: http://www.sstl.co.uk/divisions/earth-observation-science/science-missions/strand-nanosatellite

Part 2

Approaches of Earth Observation Monitoring

4

Vision Goes Symbolic Without Loss of Information Within the Preattentive Vision Phase: The Need to Shift the Learning Paradigm from Machine-Learning (from Examples) to Machine-Teaching (by Rules) at the First Stage of a Two-Stage Hybrid Remote Sensing Image Understanding System, Part I: Introduction

Andrea Baraldi

Department of Geography, University of Maryland,
College Park, Maryland,
USA

1. Introduction

One traditional, although visionary goal of the remote sensing (RS) community is the development of operational satellite-based measurement systems suitable for automating the quantitative analysis of large-scale spaceborne multi-source multi-resolution image databases (Gutman et al., 2004). In past years this goal was almost exclusively dealt with by research programs focused on land cover (LC) and land cover change (LCC) detection at global scale (Gutman et al., 2004) (pp. 451, 452). In recent years the objective of developing operational satellite-based measurement systems has become increasingly urgent due to multiple drivers. While cost-free access to large-scale low spatial resolution (SR) (above 40 m) and medium SR (from 40 to 20 m) spaceborne image databases has become a reality (GEO, 2005; GEO, 2008a; GEO, 2008b; Gutman et al., 2004; Sart et al., 2001; Sjahputera et al., 2008), in parallel, the demand for high SR (between 20 and 5 m) and very high SR (VHR, below 5 m) commercial satellite imagery has continued to increase in terms of data quantity and quality, which has boosted the rapid growth of the commercial VHR satellite industry (Sjahputera et al., 2008). In this scientific and commercial context an increasing number of on-going international research projects aim at the development of operational services requiring harmonization and interoperability of Earth observation (EO) data and derived information products generated from a variety of spaceborne imaging sensors at all scales - global, regional and local. Among these on-going programs it is worth mentioning the Global EO System of Systems (GEOSS) conceived by the Group on Earth Observations (GEO) (GEO, 2005; GEO, 2008b), the Global Monitoring for the Environment and Security (GMES), which is an initiative led by the European Union (EU) in partnership with the

European Space Agency (ESA) (ESA, 2008; GMES, 2011), the National Aeronautics and Space Administration (NASA) Land Cover and Land Use Change (LCLUC) program (Gutman et al., 2004) (p. 3) and the U.S. Geological Survey (USGS)-NASA Web-Enabled Landsat Data (WELD) project (USGS & NASA, 2011).

Unfortunately, to date, the increasing rate of collection of EO imagery of enhanced spatial, spectral and temporal quality outpaces the automatic or semi-automatic capability of generating information from huge amounts of multi-source multi-resolution RS data sets (Gutman et al., 2004). This may explain why the percentage of data downloaded by stakeholders from the ESA EO image archives is estimated at about 10% or less (D'Elia, 2009).

If productivity in terms of quality, quantity and value of high-level output products generated from input EO imagery is low, this is tantamount to saying that existing scientific and commercial RS image understanding (classification) systems (RS-IUSs), such as (Definiens Imaging GmbH, 2004; Esch et al., 2008; Richter, 2006), score poorly in operational contexts (Tapsall et al., 2010). For example, RS-IUSs capable of proving their competitiveness at local/regional scale, such as the inductive supervised (labeled) data learning Support Vector Machines (SVMs) (Bruzzone & Carlin, 2006; Bruzzone & Persello, 2009), typically lack robustness and scalability for seamless application to LC and LCC problems at national, continental and global scale. As an example of these difficulties the interested reader may refer to (Chengquan Huang et al., 2008), where an SVM training algorithm and model selection strategies are applied to every image of a multi-temporal image mosaic at global scale. If the conjecture that existing RS-IUSs are affected by low productivity holds in general, it applies in particular to two-stage segment-based RS-IUSs which have recently gained widespread popularity and are currently considered the state-of-the-art in both scientific and commercial RS image mapping applications (Castilla et al., 2009; Mather, 1994). In literature the conceptual foundation of two-stage segment-based RS-IUSs is well known as geographic (2-D) object-based image analysis (GEOBIA), including a so-called iterative geographic OO image analysis (GEOOIA) approach (Baatz et al., 2008) (Hay & Castilla, 2006), also called object-oriented (image) analysis (OOA) (Castilla et al., 2008).

To summarize, in operational contexts (other than toy problems at small spatial scale and coarse semantic granularity) a RS-IUS can be considered as a low performer when at least one among several operational quality indicators (QIs) scores low. In (Baraldi et al., 2010a), a set of QIs eligible for use with an operational RS-IUS comprises the following: degree of automation (equivalent to ease of use; it is monotonically decreasing with the number of system-free parameters to be user-defined), classification and spatial accuracies (Baraldi et al., 2005), efficiency (e.g., computational time, memory occupation), robustness to changes in input parameters, robustness to changes in the input data set, scalability, timeliness (defined as the time span between data acquisition and high-level product delivery to the end user; it increases monotonically with manpower and computing time) and economy. In RS common practice, one or many of the aforementioned QIs of existing RS-IUSs tend to score low at local to global scale. This observation appears in line with a well-known opinion by Zamperoni according to which computer vision (CV) remains, to date, far more problematic than might be reasonably expected (Zamperoni, 1996). In addition to CV, other scientific disciplines such as Artificial Intelligence (AI)/Machine Intelligence (MAI) and Cybernetics/Machine Learning (MAL), whose origins date back to the late 1950s, still remain unable to provide their ambitious cognitive objectives with operational solutions (Diamant, 2005; Diamant, 2008; Diamant, 2010a; Diamant, 2010b).[1]

To outperform existing scientific and commercial image understanding approaches, a new trend of research and development is found in both CV (Cootes and Taylor, 2004) and RS literature (Mather, 1994; Matsuyama & Shang-Shouq Hwang, 1990; Pekkarinen et al., 2009). This new trend aims at developing novel hybrid models for retrieving sub-symbolic (sensory, non-semantic, objective) continuous variables (e.g., leaf area index, LAI) and symbolic (categorical, semantic, subjective) discrete variables (e.g., land cover types) from optical multi-spectral (MS) imagery. By definition, hybrid models combine both statistical (inductive, bottom-up, fine-to-coarse, driven-without-knowledge, learning-from-examples) and physical (deductive, top-down, coarse-to-fine, prior knowledge-based, learning-by-rules) models to take advantage of the unique features of each and overcome their shortcomings (Matsuyama & Shang-Shouq Hwang, 1990; Shunlin Liang, 2004).

The original contribution of this work is to revise, integrate and enrich previous analyses found in related papers about recent developments in the design and implementation of an operational automatic multi-sensor multi-resolution near real-time two-stage hybrid stratified hierarchical RS-IUS (Baraldi et al., 2006a; Baraldi et al., 2010a; Baraldi et al., 2010b; Baraldi et al., 2010c; Baraldi, 2011a; Baraldi, 2011b). These novel developments encompass the four levels of analysis of an information processing system (Baraldi, 2011a; Marr, 1982), namely: (i) computational theory (system architecture), (ii) knowledge/information representation, (iii) algorithm design and (iv) implementation.

Starting from these recent achievements the present work provides an in-depth analysis of Emanuel Diamant's works including original speculations on the conceptual framework of MAI together with image segmentation and edge detection algorithms provided as proofs of his concepts (Diamant, 2005; Diamant, 2008; Diamant, 2010a; Diamant, 2010b). To overcome the conceptual and algorithmic drawbacks highlighted in Diamant's works, this manuscript proposes revised/new definitions of the following concepts: objective continuous sub-symbolic sensory data, continuous physical information, subjective discrete semi-symbolic data structure, discrete semantic-square (semantic[2]) information and prior knowledge base. Continuous physical information is defined as a hierarchical description (multi-scale encoding/decoding or intra-scale transcoding) of an objective continuous sensory data set based on a given mathematical vocabulary/language, e.g., a fast Fourier transform (FFT) of a time signal. Discrete semantic[2] information is naturally (automatically, instantaneously) generated from the simultaneous combination of three components: (I) an objective continuous sensory data set, (II) an external subjective supervisor (observer) and (III) his/her own subjective prior ontology (model of the (3-D) world existing before looking at the objective sensory data at hand) whose hierarchical form is equivalent to that of a story in a natural language, comprising a title, an abstract, sections, paragraphs, sentences and words. In practical contexts these definitions imply the following.

---

[1] In Italian, acronym AI reminds of the English expression: 'ouch'. Acronym MAI means 'never'. Acronym MAL means 'pain'. Acronym MAT means 'fool'. These choices are arbitrary, but not by chance. Ancient Latins used to say: Nomen est omen... (meaning: 'true to its name').

a. It is impossible to *extract* semantic[2] information from objective continuous sensory data because the latter, *per se*, are provided with no semantics at all.

b. It is possible to *correlate* discrete semantic[2] information to objective continuous sensory data. Unfortunately, correlation between continuous sensory data and a finite and discrete set of categorical variables, corresponding to independent random variables generating separable data structures (data aggregations, data clusters, data objects), is low in real-world RS image mapping problems at large data scale or fine semantic granularity, other than toy problems at small data scale and coarse semantic granularity. This low correlation effect is due to the combination of two factors.

  i. According to the *central limit theorem* the distribution of the sample average of *n* independent and identically distributed (iid) random variables (corresponding to, say, categorical variables) approaches the normal distribution, featuring no "distinguishable" data sub-structure, as the sample size *n* increases. In other words, the separability of "distinguishable" data structures in a given measurement space of a given objective sensory data set is monotonically non-increasing (i.e., it decreases or remains equal) with the finite number of discrete semantic concepts (e.g., land cover classes) involved with the cognitive (classification) problem at hand.

  ii. In a given measurement space, within-class variability (vice versa, inter-class separability) is monotonically non-decreasing (i.e., it increases or remains equal) (vice versa, non-increasing) with the magnitude of the sample set per categorical variable when this variable-specific sample set size is "large" according to large-sample statistics (although large sample is a synonym for 'asymptotic' rather than a reference to an actual sample magnitude, a sample set cardinality of 30÷50 samples per random variable is typically considered sufficiently large that, according to a special case of the central limit theorem, the distribution of many sample statistics becomes approximately normal). For example, in (Chengquan Huang et al., 2008), where a time-consuming SVM training and classification model selection strategies are applied to every image of a world-wide RS image mosaic to separate forest from non-forest pixels, a so-called training data automation (TDA) procedure identifies a forest peak in a one-band first-order statistic (histogram) of a local image window. The size of this local image window must be fine-tuned based on heuristics because the inter-class spectral separability between classes forest and non-forest (vice versa, within-class variability) decreases (vice versa, increases) monotonically with the local window size above a certain (empirical) threshold (minimum window size, below which the collected sample is not statistically significant).

Some practical conclusions of potential interest to the RS, CV, AI and MAL communities stem from these speculations. Firstly, in operational contexts (e.g., RS image classification problems at national, continental and global scale), other than toy problems (e.g., RS image mapping at coarse spatial resolution and local/regional scale), inductive classifiers capable of learning from a finite labeled data set should be considered structurally inadequate to correlate (rather than extract, see this text above) discrete semantic[2] information with objective sensory data provided, *per se*, with no semantics at all.

Secondly, to increase the operational QIs of existing two-stage hybrid RS-IUSs, any first-stage inductive MAL-from-examples approach should be replaced by a deductive Machine Teaching (MAT)-by-rules sub-system capable of generating a preliminary classification first stage in the Marr sense (Baraldi et al., 2006a; Baraldi et al., 2010a; Baraldi et al., 2010b; Baraldi et al., 2010c; Baraldi, 2011a; Baraldi, 2011b; Marr, 1982). As a proof of this concept the operational automatic prior knowledge-based multi-sensor multi-resolution near real-time Satellite Image Automatic Mapper™ (SIAM™) is selected from existing literature (Baraldi et al., 2006a; Baraldi et al., 2010a; Baraldi et al., 2010b; 1 Baraldi et al., 2010c; Baraldi, 2011a; Baraldi, 2011b).

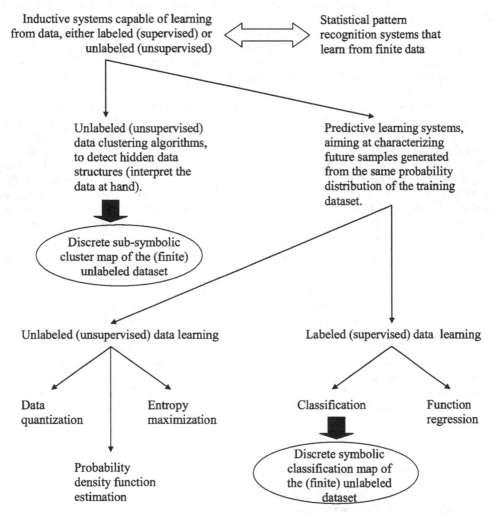

Fig. 1. The taxonomy of statistical pattern recognition systems proposed in (Baraldi et al., 2006b). Clustering algorithms and classification systems map an unlabeled input data sample into a discrete and finite set of sub-symbolic and symbolic labels, respectively. These discrete output maps are called (sub-symbolic) cluster maps (consisting of, say, cluster 1, cluster 2, etc.) and (symbolic) classification maps (consisting of, say, symbolic labels such as land cover classes broad-leaf forest, needle-leaf forest, etc.), respectively.

Thirdly, in RS-IUSs, MAL-from-data algorithms, either labeled (supervised) or unlabeled (unsupervised), either context-insensitive (e.g., pixel-based) or context-sensitive (e.g., 2-D object-based), should be adapted to work on a driven-by-knowledge stratified (semantic masked/layered) basis and moved to the second stage of a novel two-stage stratified hierarchical hybrid RS-IUS architecture recently proposed in RS literature (Baraldi et al., 2006a; Baraldi et al., 2010a; Baraldi et al., 2010b; Baraldi et al., 2010c; Baraldi, 2011a; Baraldi, 2011b).

The rest of this work is organized as follows. For publication reasons it consists of Part I and Part II. In Part I Section 2 related works, concepts and definitions are revised to provide this multi-disciplinary study with a significant survey value and make it self-contained. Part I Section 2 includes the following sub-sections: definitions and synonyms involved with inductive and deductive inference mechanisms (see Part I Section 2.1), a critical review of the history of AI/MAI and Cybernetics/MAL including a summary of Diamant's definitions of objective data, physical information, semantic information, knowledge and intelligence (refer to Part I Section 2.2), a definition of the cognitive process of vision (see Part I Section 2.3), a critical analysis of the inherent ill-posedness of inductive data learning algorithms (see Part I Section 2.4), a review of Diamant's image segmentation and contour detections algorithms presented as proofs of his concepts summarized in Part I Section 2.2 (refer to Part I Section 2.5), a discussion of the four levels of understanding of a RS-IUS (see Part I Section 2.6), a presentation (see Part I Section 2.7) of the Quality Assurance Framework for EO (QA4EO) guidelines (GEO/CEOSS, 2008) delivered by the Working Group on Calibration and Validation (WGCV) of the Committee of Earth Observations (CEOS), the space arm of the Group on Earth Observations (GEO) (GEO, 2005; GEO, 2008b), and a list of operational QIs of an RS-IUS (refer to Part I Section 2.8).

Part II includes a review session (see Part II Section 2) and an original contribution (from Part II Section 3 to Part II Section 7). In Part II Section 2 different families of existing RS-IUSs, namely, multi-agent hybrid RS-IUSs, two-stage segment-based RS-IUSs and two-stage stratified hierarchical hybrid RS-IUSs, are compared at the architectural level of analysis (refer to Part I Section 2.6). Part II Section 3 discusses theoretical inconsistencies and algorithmic drawbacks found in Diamant's works (discussed in Part I Section 2.2 and Part I Section 2.5, respectively). Revised/novel definitions of objective continuous sensory data, continuous physical information, discrete semantic[2] information and prior knowledge are provided in Part II Section 4. In Part II Section 5 practical consequences of the novel definitions provided in Part II Section 4 are considered for CV, AI and MAL applications. Part II Section 6 presents the operational automatic multi-sensor multi-resolution near real-time SIAM™ as a proof of the original concepts proposed in this work. Conclusions are reported in Part II Section 7.

2. Related works, concepts, definitions and synonyms

To provide this multi-disciplinary paper with a significant survey value and make it self-contained, a variety of related works, concepts and definitions collected from AI, MAL, CV and RS literature are revised in this section.

2.1 Inference mechanisms: Deductive top-down coarse-to-fine physical models and inductive bottom-up fine-to-coarse statistical models

Starting from classical philosophy to end up with MAL it is well known that the general notion of inference (learning) comprises two types of learning mechanisms.

1. "Induction, i.e., progressing from particular cases (e.g., training data) to general (e.g., estimated dependency or model)" (Cherkassky & Mulier, 2006). *Inductive inference systems* are also called *inference systems capable of learning-from-examples, bottom-up, fine-to-coarse, data-driven, driven-without-knowledge, statistical models, statistical pattern recognition systems* (Matsuyama & Shang-Shouq Hwang, 1990; Shunlin Liang, 2004). Statistical models are capable of learning from either *labeled* (*supervised*) or *unlabeled* (*unsupervised*) data, refer to Fig. 1. An inductive learning-from-data approach is not influenced by prior knowledge concerning the cognitive problem (e.g., our prior knowledge about an image), subjective desires (e.g., what information we are aiming to extract from an image) or subjective expectations (e.g., what we expect to see in an image). In the words of Cherkassky and Mulier, "induction amounts to forming generalizations from particular true facts. This is an inherently difficult (ill-posed) problem and its solution requires *a priori* knowledge in addition to data" (Cherkassky & Mulier, 2006) (p. 39). To summarize, inductive data learning problems are inherently ill-posed and require *a priori* knowledge in addition to (either labeled or unlabeled) sensory data to become better posed.

2. "Deduction, i.e., progressing from general (e.g., model) to particular cases (e.g., output values)" (Cherkassky & Mulier, 2006). *Deductive inference systems* are also called *inference systems capable of learning-by-rules, top-down, coarse-to-fine, model-driven, prior knowledge-based, driven-by-knowledge, physical models, physical pattern recognition systems* (Matsuyama & Shang-Shouq Hwang, 1990; Shunlin Liang, 2004), see Fig. 1. Physical models are abstracts of reality. They consist of prior knowledge of the physical laws of the (3-D) world which is available before (prior to) looking at the objective sensory data at hand.

As output, statistical and physical quantitative models of the (3-D) world (e.g., quantitative models of land surfaces observed from space) generate either *continuous sub-symbolic variables* (e.g., LAI) or *discrete symbolic (categorical) variables* (e.g., land cover types).

In addition to the synonyms presented above, the following terms are considered synonyms in the rest of this paper (Matsuyama & Shang-Shouq Hwang, 1990; Shunlin Liang, 2004).

*   *Sub-symbolic, non-semantic, sensory, instantaneous, continuous, numerical, quantitative, objective, absolute, varying variables or sensations.*
*   *Symbolic, discrete and semantic, categorical, linguistic, qualitative, subjective, abstract, vague, persistent, stable variables or percepts, concepts, classes of (3-D) objects in the (3-D) world, (3-D) object-models.*

In RS data applications, quantitative models are traditionally sorted into three major categories: *statistical, physical* and *hybrid*, whose main advantages and limitations are so well known in existing literature as to be summarized by Shunlin Liang in the following few words (Shunlin Liang, 2004).

a. Statistical models are inductive data learning systems (refer to this text above). Therefore, they are inherently difficult to solve (ill-posed) and their solution requires *a priori* knowledge in addition to data (Cherkassky & Mulier, 2006). Statistical pattern recognition systems are based on *correlation relationships* between objective sensory data (e.g., RS imagery) and either continuous (e.g., LAI) or categorical (e.g., land surface) variables. Statistical models are easy to develop, e.g., a human expert is not required to search for an explicit deterministic function, if any, between, say, a target physical variable (e.g., LAI) and sensory data. However, they are effective for summarizing local data exclusively, i.e., they are usually (always?) site-specific (Shunlin Liang, 2004). For example, in RS common practice no machine capable of learning from either unlabeled or labeled data scores high in operational contexts such as satellite image mapping at national/ continental/ global scale. As a proof of this concept, in (Chengquan Huang et al., 2008), a time-consuming SVM (Bruzzone & Carlin, 2006) training and classification model selection strategies are enforced for every RS image in a world-wide image mosaic. In addition, supervised data learning algorithms, either context-insensitive (e.g., pixel-based) or context-sensitive (e.g., (2-D) object-based (Definiens Imaging GmbH, 2004; Esch et al., 2008)), require the collection of reference training samples which are typically scene-specific, expensive, tedious, difficult or impossible to collect (Gutman et al., 2004). This means that in practical RS data applications where supervised data learning algorithms are employed, the cost, timeliness, quality and availability of adequate reference (training/testing) datasets derived from field sites, existing maps and tabular data have turned out to be the most limiting factors on RS data product generation and validation (Gutman et al., 2004). Finally, since statistical models are inherently ill-posed, they are difficult to maintain, adapt, modify and scale according to changing input data sets, sensor specifications and/or user requirements. For example, the free parameter selection phase of any image segmentation algorithm tends to be difficult because: (i) it is based on heuristic (empirical) criteria (correlation relationships) and (ii) due to its inherent ill-posedness (artificial insufficiency (Matsuyama & Shang-Shouq Hwang, 1990)), any image segmentation algorithm is site-specific and simultaneously affected by both omission and commission segmentation errors within each image at hand (Burr & Morrone, 1992; Corcoran & Winstanley, 2007; Corcoran et al., 2010; Delves et al., 1992; Hay & Castilla, 2006; Matsuyama & Shang-Shouq Hwang, 1990; Petrou & Sevilla, 2006; Vecera & Farah, 1997).

b. Physical models consist of prior knowledge concerning the physical laws of the (3-D) world which is available before looking at the objective sensory data at hand. They follow the physical laws of the real (3-D) world to establish *cause-effect relationships*. They have to be learnt by a human expert based on intuition, expertise and evidence from data observation. Thus, unfortunately, it takes a long time for human experts to learn physical laws of the real (3-D) world and tune physical models (Mather, 1994; Shunlin Liang, 2004). On the other hand, physical models are more intuitive to debug, maintain and modify than statistical models. In other words, if the initial physical model does not perform well, then the system developer knows exactly where to improve it by incorporating the latest knowledge and information. For example, with a non-adaptive decision-tree classifier it is easy to find the node of the decision process in which a misclassification error occurs. In practice, a non-adaptive decision-tree classifier is well-posed (i.e., every data sample is assigned a semantic label according to a specific rule set), but subjective (i.e., different system developers may generate different non-adaptive decision-tree classifiers in the same application domain), refer to this text above.

Vision Goes Symbolic Without Loss of Information Within the Preattentive Vision Phase: The Need to Shift the Learning
Paradigm from Machine-Learning (from Examples) to Machine-Teaching (by Rules) at the First Stage of a Two-Stage
Hybrid Remote Sensing Image Understanding System, Part I: Introduction   71

c. Hybrid models combine both statistical and physical models to take advantage of the unique features of each and overcome their shortcomings (refer to the two previous paragraphs) (Matsuyama & Shang-Shouq Hwang, 1990; Shunlin Liang, 2004).

2.2 Brief history of AI/MAI and Cybernetics/MAL

In every ML textbook and in the world wide web it is easy to find historical information on the multiple rises and falls of expectations and achievements in scientific disciplines such as Cybernetics/MAL and AI/MAI related to the inductive and deductive inference paradigms respectively (refer to Part I Section 2.1).

2.2.1 1940s, 1950s and 1980s: Bottom-up inductive Cybernetics/MAL

In the 1940s and 1950s, a number of researchers, mostly located at Princeton University and the Ratio Club in England, started exploring the connection between neurology and information theory to develop electronic networks capable of exhibiting rudimentary intelligence conceived as self-organizing network properties. This new scientific discipline, called Cybernetics, investigates the capability of complex distributed processing systems, consisting of multiple processing elements (agents) dynamically interacting in multiple ways based on simple local rules, to display emergent macro behaviors and persistent network structures from an input data flow, i.e., local rules lead to global network properties. For example, data regularities detected by a self-organizing network of processing elements are equivalent to a compression of input information with which the distributed system can provide an abstract representation of the external environment.

The key features of complex network systems adaptive to data are that: (i) to understand how it works, a self-organizing network must be run (learning by doing), which is to say that learning, intended as self-organizing network capability, emerges without anyone needing to define what learning and intelligence are all about, (ii) the global behavior outlasts any of the network processing elements (persistence of the whole over time), (iii) it is the competition among processing elements and their (lateral) connections which leads to the emergence of specialized network (sub-)structures; without competition all processing units would behave alike and no specializations of the units would evolve (Fritzke, 1997; Lawley, 2003; Martinetz & Schulten, 1994).

By the late 1950s, in spite of the low technological development of electronic devices, electronic networks such as W. Grey Walter's turtles and the Johns Hopkins Beast were considered eligible for proving the cybernetic concepts. However, during the 1960s, symbolic AI approaches had achieved great success at simulating high-level thinking in small demonstration programs. So, by 1960 approaches based on cybernetics were abandoned or pushed into the background.

Next, by the 1980s progress in symbolic AI seemed to stall. Many researchers started believing that symbolic systems would never be able to imitate all the processes of human cognition, such as perception, learning and pattern recognition. Again, a number of researchers looked for a "sub-symbolic" distributed approach capable of solving specific AI sub-problems. The basic idea was: "Why trouble oneself trying to grasp the principles of intelligence? Let us give the machine the chance to find (in a bottom-up approach) the best way to mimic intelligence" (Diamant, 2010b). In the middle 1980s interest in "connectionism" in general and so-called artificial neural networks in particular was revived by the works of David Rumelhart and others who focused on Multi-Layer Perceptrons (MLPs) and their Back-Propagation (BP) parameter adaptation algorithm. These and other distributed processing approaches, such as fuzzy learning systems and evolutionary computation, are now studied collectively by the emerging discipline of MAL (also called computational intelligence).

Finally, from the 1990s to date, MAL has achieved its greatest successes due to a combination of factors: the increasing computational power and memory capacity of computers, a greater emphasis on solving specific "tractable" MAL sub-problems and a new commitment by researchers to solid mathematical/statistical methods (Alpaydin, 2010; Bishop, 1995; Cherkassky & Mulier, 2006; Duda et al., 2001; Mitchell, 1997). In practice, once its first idealistic objective failed, MAL has been "broken into pieces, disintegrated and fragmented into many partial tasks and goals" to make its problem domain more "tractable" (Diamant, 2010b).

2.2.2 1956-1974, 1980s to date: Top-down deductive AI/MAI

Starting from the seminal work of Turing in 1950, the origin of AI dates back to the summer of 1956 when a conference on the campus of Dartmouth College was attended by John McCarthy, Marvin Minsky, Allen Newell and Herbert Simon who became the leaders of AI research for many decades. John McCarthy, who coined the term in 1956, defines AI as "the science and engineering of making intelligent machines" (Diamant, 2010b).

Intelligent agents must be able to set goals and achieve them by making choices that maximize the utility (or "value") of the available choices. To be termed intelligent these agents must be able to make predictions about how their actions will affect the present status of the world. This means they need a way to represent the current status of the world, to make predictions about the world's future status as a consequence of their actions, to have a periodical check to see if the world status matches their predictions and to change their plan as this becomes necessary, thus requiring the agent to reason under uncertainty.

Back in 1956 the excitement and hopes to reach AI goals in a short time were quite high. Herbert Simon predicted that "machines will be capable, within twenty years, of doing any work a man can do" (Diamant, 2010b). Marvin Minsky agreed by writing that "within a generation ... the problem of creating 'artificial intelligence' will substantially be solved". Reported by Diamant (Diamant, 2010b), Steve Grand sayed that "Rodney Brooks has a copy of a memo from Marvin Minsky in which he suggested charging an undergraduate for a summer project with the task of solving vision. I don't know where that undergraduate is now, but I guess he hasn't finished yet".

Many of the cognitive problems AI was expected to solve require extensive prior knowledge of the (3-D) world. A representation of "what exists in the (3-D) world" pertaining to the cognitive problem at hand is called *world model* (Matsuyama & Shang-Shouq Hwang, 1990) or *ontology* (borrowing a word from traditional philosophy). The graphical representation and implementation of an ontology is twofold.

- An *inverted tree* whose leaves are at the bottom level (layer 0), where semantic primitives (hereafter called *semi-concepts*) are found (Diamant, 2005; Diamant, 2010a; Diamant, 2010b; Diamant, 2008).

Vision Goes Symbolic Without Loss of Information Within the Preattentive Vision Phase: The Need to Shift the Learning
Paradigm from Machine-Learning (from Examples) to Machine-Teaching (by Rules) at the First Stage of a Two-Stage
Hybrid Remote Sensing Image Understanding System, Part I: Introduction   73

- A *semantic net* (*concept net*) is defined as a graph, either directed or non-oriented, either cyclic or acyclic, consisting of nodes linked by edges. Nodes represent concepts, i.e., classes of (3-D) objects in the world (see Part I Section 2.1), while edges represent relations, e.g., PART-OF, A-KIND-OF, spatial relations either topological (e.g., adjacency, inclusion) or non-topological (e.g., distance, angle), temporal transitions between nodes, physical model-based relationships between causes and effects, etc. (Hudelot et al., 2008; Matsuyama & Shang-Shouq Hwang, 1990; Pakzad et al., 1999).

Unfortunately, the number of atomic facts about the world that an average person knows is astronomical. It means that AI projects whose goal is to build a complete knowledge base of commonsense knowledge would require enormous amounts of laborious ontological engineering where one abstract concept must be built, by hand, at a time. In practice, it takes a long time for human experts to define ontologies, learn physical laws of the real (3-D) world and tune physical models based on human intuition, domain expertise and evidence from data observation. Within a decade or so it became clear that AI problems were immense, maybe even intractable. In 1974, in response to ongoing criticism and pressure to fund more productive projects, the U.S. and British governments cut off all exploratory research related to AI.

However, in the 1970s, computers with large memories became available. This drove AI researchers to began building prior knowledge into AI problem-specific "tractable" applications. In the early 1980s this led to the first commercial success of expert systems, a form of AI programs that simulated the knowledge base and analytical skills of human experts. By 1985 the market for AI reached over a billion dollars. At the same time, Japan's fifth generation computer project inspired the U.S and British governments to restore funding for academic research in the AI field. However, beginning with the collapse of the Lisp Machine market in 1987, AI once again fell into disrepute and a second, longer lasting, AI winter began.

Finally, from the 1990s to date, AI achieved its greatest successes, albeit somewhat behind the scenes. This success was due to a combination of factors, which are not surprisingly the same as those working in favor of the recent achievements of MAL (also refer to Part I Section 2.2.1), namely: the increasing computational power and memory capacity of computers, a greater emphasis on solving specific "tractable" AI sub-problems, a new commitment by researchers to solid mathematical/statistical methods and more rigorous scientific standards (Alpaydin, 2010; Bishop, 1995; Cherkassky & Mulier, 2006; Duda et al., 2001; Mitchell, 1997), and the creation of new ties between AI and other fields working on similar problems, such as MAL, knowledge representation (e.g., fuzzy logic) and uncertainty engineering (e.g., sensitivity analysis, error propagation). For example, a major goal of contemporary AI is to have the computer understand enough concepts to be able to learn by reading from sources like the internet, and thus be able to add to its own ontology. This is called Natural Language Processing, which gives machines the ability to read and understand the languages that humans speak.

Among the longest-standing AI questions that have remained unanswered, consider the following.

- Should AI simulate natural intelligence by studying psychology or neurology? Or is human biology as irrelevant to AI research as bird biology is to aeronautical engineering?

- In the attempt to develop hybrid inference systems where both statistical and physical models are combined to overcome their shortcomings (see Section 2.1), how, when and where do continuous sensory objective sub-symbolic data become discrete symbolic subjective information? This is the well-known *information gap* existing between (sub-symbolic, sensory, instantaneous, numerical, quantitative, absolute, non-semantic) sensations and (symbolic, linguistic, qualitative, vague, discrete and semantic, persistent, stable) percepts (refer to Part I Section 2.1), which has been thoroughly investigated in both philosophy and psychophysical studies of perception (Matsuyama & Shang-Shouq Hwang, 1990). In practice, "we are always seeing objects we have never seen before at the sensation level, while we perceive familiar objects everywhere at the perception level" (Matsuyama & Shang-Shouq Hwang, 1990).

2.2.3 Fundamental flaws responsible for AI and MAL derailment: The Diamant perspective

When did AI and MAL derail from their original and ambitious goals? Diamant's answer is: They did it right at their origin dating back to the late 1950s (refer to Part I Section 2.2.1 and Part I Section 2.2.2, respectively) due to the following fundamental flows (Diamant, 2010b).

a. The lack of proper definitions to distinguish between objective data, physical information, semantic information, knowledge and intelligence. These definitions deal with the well-known *information gap* between physical and semantic information thoroughly investigated in both philosophy and psychophysical studies of perception (see Part I Section 2.2.2). In Diamant's words: "In my view, philosophy is not a swear-word. Philosophy is a keen attempt to approach the problem from a more general standpoint, to see the problem from a wider perspective, and to yield, in such a way, a better comprehension of the problem's specificity and its interaction with other world realities. Otherwise we are ... prone to dead-ends and local traps" (Diamant, 2010b).

b. Misunderstanding of the very nature of semantic information. Unlike physical information, semantics is not a property of the raw data, but the property of an external observer who observes and scrutinizes the data. Since semantics is assigned to physical data structures by an external observer, it cannot be learned from the sensory data.

The Diamant explanations of these concepts are quoted below (Diamant, 2005; Diamant, 2008; Diamant, 2010a; Diamant, 2010b).

2.2.3.1 Kolmogorov's complexity theory

Among definitions of "data", "information", and "knowledge", the definition of information is the most controversial. To provide it, Diamant relies on Kolmogorov's complexity theory (actually developed independently by Kolmogorov, Chaitin, and Solomonoff), whose concern is: What is the best way to represent a single data object? What are the laws of minimizing the length of a description of a single data object? Such a short-length compressed description is the information that we are seeking about a particular data object.

Theoretically two extreme cases can be distinguished: (1) the elements of a data set are absolutely random and (2) the elements of a data set form "observable" data structures. In the first case the data set can be represented only by the original sequence of its data elements. In the second case the presence of observable data structures consisting of data elements can be taken into account, which leads to a more compact and concise (compressed) description. In terms of Kolmogorov's theory, this compressed description (encoding) must be a trustworthy (which does not mean lossless) abstract (summary) of the original data set such that: (i) the abstract description length is definitely shorter than the original uncompressed data set description, (ii) the abstract description is sufficient to reconstruct (reproduce, re-establish, decode) the salient properties or regularities or distinguishable data structures or data objects in the original data set.

Kolmogorov's theory prescribes the way in which a data set description has to be created: Firstly, the most simplified and generalized data structures must be described. (Recall the Occam's Razor principle: Among all hypotheses consistent with the observation, choose the simplest one that is coherent with the data (Mitchell, 1997)). Then, as the level of generalization (vice versa, granularity) is gradually decreased (vice versa, becomes finer), more and more fine-grained data details (structures) can be revealed and described.

2.2.3.2 Diamant's definitions of objective data, physical information, semantic information, knowledge and intelligence

Diamant reviews two survey papers (Legg & Hutter, 2007; Zins, 2007), published in the year 2007, where definitions of data, information, knowledge and MAI are collected from existing literature for comparison purposes. In (Zins, 2007), 130 definitions of data, information and knowledge are provided by 45 scholars. In (Legg & Hutter, 2007), more than 70 definitions of MAI are collected. According to Diamant, "what these two collections undoubtedly exhibit... is that definitions offered by the leading scholars in each field have nothing in common among them, and therefore are of little use when it comes to our practical problem-solving" (Diamant, 2010b). As a result, Diamant is forced to search for his own definitions.

Starting from the Kolmogorov complexity theory (see Section 2.2.3.1), Diamant provides the following definitions about data, information and knowledge.

1. (Objective) "data is an agglomeration of elementary facts" (Diamant, 2010a).
2. (Physical and semantic) "information is a description" (based on a) "language and/or alphabet" (Diamant, 2010a).
3. (Physical and semantic) "information is a hierarchy of decreasing level descriptions" (Diamant, 2010a).
4. (Physical?) "information elicitation (extraction) does not require incorporation of any high-level knowledge" (Diamant, 2010b; Diamant, 2008).
5. "Two kinds of information must be distinguished: objective (physical) information and subjective (semantic) information.
   a. By physical information we mean the description of data structures that are discernable in a data set" (Diamant, 2010b). (Noteworthy,) "successful recovery and description of image structures (e.g., successful image segmentation) does not lead to image understanding. The (data) structures that are observed in an image reflect aggregations of nearby data elements on the basis of similarity among their physical attributes (e.g., color or brightness in visual signals, frequency and intensity in audio signals). These (are called) 'primary (data) structures' or 'physical (data) structures'" (Diamant, 2010a). "Physical information, being a natural property of the data, can be extracted instantly from the data and no special rule is needed for such a task accomplishment" (Diamant, 2010b). (It is) "physical information... the only information present in an image, and therefore the only

information that can be extracted from an image " (Diamant, 2008). (In other words,) "defining (primary data structures) is certainly a well-grounded procedure that does not raise any objections, because objective (physical) nature laws underpin such a procedure" (Diamant, 2010a) (refer to point 4. above).

To summarize, according to Diamant, *physical information*, non-semantic *primary data structures* and discernable non-semantic *image segments* are synonyms.

b.   "By semantic information we mean the description of the relationships that may exist between the physical (data) structures of a given data set" (Diamant, 2010b). (In other words,) "'primary (data) structures'... undergo a further grouping and aggregation, which leads to formation of 'secondary (data) structures' (consisting of primary data structures) that can be called... 'semantic (data) structures'" (Diamant, 2010a)."Unlike physical information, semantics is not a property of the raw data. Semantics is assigned to physical data structures by an external observer who watches and scrutinizes the data... Semantics is a shared convention, a mutual agreement between the members of a particular group of viewers or users. Its assignment (to the primary data structures) has to be made on the basis of a consensus knowledge that is shared among the group members, and which an artificial semantic-processing system has to possess at its disposal... Therefore semantics cannot be learned straightforwardly from the raw data" (Diamant, 2010b). (In other words,) "the knowledge about the rules that underpin secondary (data) structures formation is a property of human observers and not an inherent property of the data" (Diamant, 2010a). (Since) "semantic information is a convention, an agreement, a property shared between a company of particular observers, it cannot be learned (from physical data) by any means. It can be exchanged, transferred, relocated between the group members, or between humans and intelligent machines (robots) collaborating with them in a working group, but it cannot be learned (from data)" (Diamant, 2010b). (This implies that) "MAL techniques are ... not applicable for the purposes of semantic information extraction (from the raw data set)... (Acquisition) of this knowledge presumes availability of a different and usually overlooked special learning technique, which would be best defined as Machine Teaching (MAT) – a technique that would facilitate externally-prepared-knowledge transfer to the system's disposal" (Diamant, 2010b).

To summarize, according to Diamant *semantic information* and semantic *secondary data structures*, generated from subjective aggregation (semantic labeling) of non-semantic primary data structures, e.g., *image segments*, by an external observer, are synonyms. In addition, what Diamant calls MAT is known in traditional AI as knowledge engineering, which is a process of codifying human knowledge into an expert system (Laurini and Thompson, 1992).

6.   "Both physical and semantic information descriptions are similar in that: (1) they are character strings, (2) they are top-down coarse-to-fine hierarchies, and (3) they are implemented according to a certain vocabulary/language. There is only a small difference – physical information can be described in a variety of languages while semantic information can be represented only in a human natural language... Therefore the most suitable form of semantic information representation should be a narrative, a

story, a tale. The usual top-down hierarchical structure of such a story (a narrative, a tale) is well known from other linguistic studies. Moving top-down, a story comprises a story title, abstract, chapter or section partition, paragraph subdivision, separate phrases and sentences which end up with single words (congregations of letters) actually composing a phrase. Further structural descent leads in linguistics to syntaxes. But in our case – the lowest level of a semantic structure is stuffed with physical information which represents the physical structure of a meaningful object designated by the word in a phrase... At the lowest level of a semantic description (hierarchy) a physical information sub-hierarchy is always present" (Diamant, 2010a).

To summarize, according to Diamant *semantic information* comprises *physical information* at the lowest level of a semantic description (hierarchy) equivalent to an inverted tree (see Part I Section 2.2.2).

7.   (Prior) "knowledge is memorized (semantic) information (stored in the system's memory, which incorporates physical information)" (Diamant, 2010b).
8.   "Data is not information, but knowledge is information (semantic information memorized in system's memory)" (Diamant, 2010b).
9.   "Intelligence (cognition) is the system's ability to process (semantic) information" (Diamant, 2010b).

Together with the aforementioned theoretical considerations, Diamant presents an unlabeled (unsupervised) multi-scale image segmentation algorithm and a single-scale unlabeled (unsupervised) image contour detector as proofs of his concepts (Diamant, 2005). A critical analysis of these theoretical and algorithmic contributions by Diamant can be found in Part II Section 3.

2.3 Vision as an ill-posed image understanding problem

The main role of a biological or artificial visual system is to backproject the information in the (2-D) image domain to that in the (3-D) scene domain (Matsuyama & Shang-Shouq Hwang, 1990). In greater detail, the goal of a visual system is to provide plausible (multiple) symbolic description(s) of the scene depicted in an image by finding associations between sub-symbolic (non-semantic, sensory, instantaneous, numerical, absolute, quantitative, varying, objective, see Part I Section 2.1) (2-D) image features or sensations with symbolic (semantic, subjective, linguistic, qualitative, vague, abstract, persistent, stable, see Part I Section 2.1) (3-D) objects (concepts or percepts) in the scene (e.g., a building, a road, etc.). Sub-symbolic (2-D) image features are either points or regions or, vice versa, region boundaries, i.e., edges, provided with no semantic meaning. In literature, (2-D) image regions are also called *segments, (2-D) objects, patches, parcels,* or *blobs* (Carson et al., 1997; Lindeberg, 1993; Yang & Wang, 2007).

There is a well-known *information gap* between symbolic information in the (3-D) scene and sub-symbolic information in the (2-D) image, e.g., due to dimensionality reduction and occlusion phenomena, see Fig. 2 (also refer to Part I Section 2.2.2 and Part I Section 2.2.3). This is called the *intrinsic insufficiency* of image features (Matsuyama & Shang-Shouq Hwang, 1990). This information gap is also related to the inherent ill-posedness of inductive inference (see Part I Section 2.1). It means that the problem of image understanding is inherently ill-posed and, consequently, very difficult to solve (Matsuyama & Shang-Shouq Hwang, 1990; Cherkassky & Mulier, 2006).

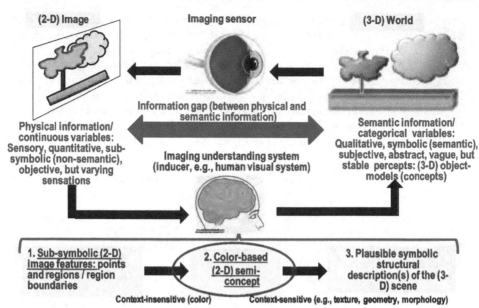

Fig. 2. Inherently ill-posed image understanding problem (vision). There is a well-known information gap between physical information and semantic information. This is the same information gap existing between (sub-symbolic, sensory, instantaneous, numerical, quantitative, absolute, non-semantic) sensations and (symbolic, linguistic, qualitative, vague, discrete and semantic, persistent, stable) percepts (concepts) which has been thoroughly investigated in both philosophy and psychophysical studies of perception. In practice, "we are always seeing objects we have never seen before at the sensation level, while we perceive familiar objects everywhere at the perception level" (Matsuyama & Shang-Shouq Hwang, 1990). The original automatic SIAM™ software button (executable), adopted as preliminary classification first stage of a novel two-stage stratified hierarchical hybrid RS-IUS architecture (see Part II, Section 2), generates as output a mutually exclusive and totally exhaustive set of symbolic spectral-based semi-concepts, also called spectral categories or land cover class sets, e.g., 'vegetation' (Baraldi et al., 2006a; Baraldi et al., 2010a; Baraldi et al., 2010b; Baraldi et al., 2010c; Baraldi, 2011a; Baraldi, 2011b). The semantic meaning of a spectral-based semi-concept is: (a) superior to zero, which is the semantic value of traditional sub-symbolic image features, namely, pixels, (2-D) image segments or edges, and (b) equal or inferior to the semantic meaning of target (3-D) land cover classes (e.g., needle-leaf forest), also called concepts or (3-D) object-models in the (3-D) world.

The aforementioned information gap coincides with the well-known *information gap* existing between (sub-symbolic, sensory, quantitative, objective, varying) sensations and (symbolic, semantic, qualitative, subjective, stable) percepts, traditionally investigated in both philosophy and psychophysical studies of perception (Matsuyama & Shang-Shouq Hwang, 1990) (see Part I Section 2.2.2).

In functional terms, biological vision combines preattentive (low-level) visual perception with an attentive (high-level) vision mechanism (Gouras, 1991; Kandel, 1991; Mason & Kandel, 1991).

Vision Goes Symbolic Without Loss of Information Within the Preattentive Vision Phase: The Need to Shift the Learning
Paradigm from Machine-Learning (from Examples) to Machine-Teaching (by Rules) at the First Stage of a Two-Stage
Hybrid Remote Sensing Image Understanding System, Part I: Introduction   79

1. Preattentive (low-level) vision extracts picture primitives based on general-purpose image processing criteria independent of the scene under analysis. It acts in parallel on the entire image as a rapid (< 50 ms) scanning system to detect variations in simple visual properties. It is known that the human visual system employs at least four spatial scales of analysis (Wilson & Bergen, 1979). Marr calls the output of the low-level vision first stage *primal sketch* or *preliminary map* (Marr, 1982).

2. Attentive (high-level) vision operates as a careful scanning system employing a focus of attention mechanism. Scene subsets, corresponding to a narrow aperture of attention, are looked at in sequence and each step is examined quickly (20–80 ms).

Finally, it is worth mentioning that, according to Marr, "vision goes symbolic almost immediately, right at the level of zero-crossing (primal sketch)... without loss of information" (Marr, 1982) (p. 343). In practice, Marr suggests the following.

a. **The output of preattentive vision (primal sketch) is symbolic. This is tantamount to saying that:**
   - **vision goes symbolic within the preattentive vision phase,**
   - **the primal sketch is a preliminary semantic map whose symbolic labels belong to a finite and discrete set of 3-D object-classes or concepts in the real (3-D) world.**

b. **The symbolic output of preattentive vision (refer to point (a) above) is lossless, i.e., when the input image is reconstructed from its semantic description, then small, but genuine image details (high spatial frequency image components) must be well preserved.**

It is also noteworthy that, in contradiction with his own intuition about what functional properties characterize a biological vision system, the CV system implemented by Marr is unable to accomplish either of the two aforementioned goals (a) and (b). For example, the Marr pre-attentive vision module consists of a contour detector (zero-crossing) whose output is a sub-symbolic primal sketch. This is not at all surprising. It accounts in general for the customary distinction between a model and the algorithm used to identify it (Baraldi et al., 2010a; Baraldi, 2011a) (also refer to Part I Section 2.6) and, in particular, for the seminal nature of the conceptual work by Marr followed by his early dramatic death.

2.4 A few comments about the inherent ill-posedness of inductive MAL from either labeled or unlabeled data

Inductive machine learning from either labeled or unlabeled data (see Fig. 1) has been central to MAL research from the beginning. In particular, "induction amounts to forming generalizations from particular true facts. This is an inherently difficult (ill-posed) problem and its solution requires a priori knowledge in addition to data" (Cherkassky & Mulier, 2006) (p. 39), to make the ill-posed inductive learning-from-data problem better posed (see Part I Section 2.1). Unfortunately, although acknowledged by a significant portion of existing literature, the inherent ill-posedness of inductive MAL from either labeled or unlabeled data appears ignored or neglected by the majority of scientists and practitioners involved with MAL common practice.

2.4.1 Inherently ill-posed unlabeled data learning

Unlabeled (unsupervised) data learning is the ability to find discrete patterns or sub-symbolic labeled data structures in an input stream of unlabeled data vectors. Well-known examples of discrete sub-symbolic data structures distinguishable in a stream of unlabeled data vectors are: (a) discrete sub-symbolic clusters (e.g., cluster 1, cluster 2, etc.) in a finite unlabeled data set belonging to a multi-dimensional measurement space and (b) discrete sub-symbolic (2-D) image segments (e.g., segment 1, segment 2, etc) found in a 2-D one-band (e.g., panchromatic) or multi-band (chromatic) image domain (see Fig. 1).

Inherently ill-posed unlabeled data clustering and image segmentation are further discussed below.

2.4.1.1 Inherently ill-posed unlabeled data clustering

Since the goal of clustering is to group the data at hand rather than to provide an accurate characterization of unobserved (future) samples generated from the same probability distribution, then the task of clustering may fall outside the framework of predictive learning (Cherkassky & Mulier, 2006). In spite of this, clustering analysis often employs unsupervised data learning approaches originally developed for vector quantization (such as the well-known k-means unsupervised data learning algorithm belonging to the family of the crisp competitive minimum-distance-to-means algorithms (Baraldi & Blonda, 1999a; Baraldi & Blonda, 1999b)), which is a predictive learning problem, see Fig. 1 (Cherkassky & Mulier, 2006).

Unlabeled data clustering is an inherently ill-posed data mapping problem. In fact, the goal of clustering is to separate a finite unlabeled dataset at hand into a finite and discrete set of "natural", hidden data structures on the basis of an often subjectively chosen measure of similarity/dissimilarity, i.e., a similarity measure chosen subjectively based on its ability to create "interesting" clusters (Backer & Jain, 1981; Baraldi & Alpaydin, 2002a; Baraldi & Alpaydin, 2002b; Cherkassky & Mulier, 2006; Fritzke, 1997). Thus, the subjective (ill-posed) nature of the nonpredictive data clustering problem precludes an absolute judgment as to the relative effectiveness of all clustering techniques (Backer & Jain, 1981). In spite of this, the inherent ill-posedness of unlabeled data clustering problems is not clearly stated in existing literature where, as a consequence, dozens of papers proposing alternative clustering algorithms are published every year (perhaps in search of a "final" best clustering algorithm which cannot exist...) (Xu & Wunsch II, 2005).

Crisp (hard) competitive minimum-distance-to-means algorithms, such as the k-means data quantization approach, try to minimize a sum-of-squares error function (Cherkassky & Mulier, 2006; Bishop, 1995). To reduce the risk of being trapped in a local minimum of the error function, soft-to-hard rather than hard competitive clustering algorithms have been conceived (Baraldi & Blonda, 1999a; Baraldi & Blonda, 1999b). In addition, it is well known that both crisp and fuzzy k-means data clustering algorithms cannot perform well with non-convex types of data, i.e., they are effective if and only if data clusters are hyperspherical (Duda et al., 2001). To overcome this problem, a k-means unsupervised data learning algorithm capable of defining automatically the number of clusters splits a non-convex data cluster, say, a data cluster shaped like a banana, into several hyperspheres. Thus, these hyperspheres should be linked to map the banana-like data cluster. To perform non-convex unlabeled data mapping, topologically preserving data clustering algorithms have been developed (Baraldi & Alpaydin, 2002a; Baraldi & Alpaydin, 2002b; Fritzke, 1997; Martinetz & Schulten, 1994).

In terms of degree of automation, which decreases monotonically with the number of system-free parameters to be user-defined, it is noteworthy that, to make the inherently ill-posed unsupervised data clustering problem better posed, every unsupervised data clustering algorithm requires at least one free parameter to be user-defined or fixed by the application developer based on heuristics. For example, it appears paradoxical that the well-known k-means vector quantizer, typically employed for unlabeled data clustering (refer to previous paragraphs), requires the user to pre-define the unknown number of unlabeled data clusters to be found in the finite unlabeled data set at hand.

In terms of computation time, unlabeled data clustering (either batch or on-line learning) is iterative (sub-optimal) in nature, therefore it is time-consuming with respect to prior knowledge-based one-pass data mapping algorithms (e.g., pattern-matching techniques).

In terms of effectiveness and robustness to changes in the input dataset, on-line (stochastic, sequential) learning unlabeled data clustering algorithms are typically subjected to local minima, e.g., they are sensitive to the order of presentation of the input data sequence. To enhance their robustness to changes in the order of presentation of the input sequence, semi-batch unlabeled data clustering algorithms have been developed (Wilson & Martinez, 2000).

2.4.1.2 Inherently ill-posed (2-D) image region extraction/contour detection

In literature, a so-called Low-Level Vision Expert (LLVE) (Matsuyama & Shang-Shouq Hwang, 1990) includes a battery of low-level sub-symbolic (non-semantic) general-purpose domain-independent inductive-learning (fine-to-coarse, bottom-up, driven-without-knowledge, see Part I Section 2.1) inherently ill-posed image processing (unlabeled data-driven) algorithms working at the signal level. This set of low-level image processing algorithms may comprise (Matsuyama & Shang-Shouq Hwang, 1990): edge-preserving noise filtering (Acton & Landis, 1997; Perona & Malik, 1990), either intensity- or color-based region/edge detection (Baraldi & Parmiggiani, 1996a; Canny, 1986), texture-based region/edge detection (Jain & Healey, 1998), region growing (Baraldi & Parmiggiani, 1996b), region extraction from not-close contours (Baraldi & Parmiggiani, 1995), etc.

In a (2-D) image domain, **region extraction is the dual problem of edge detection** and they are both inherently ill-posed visual tasks. In the rest of this paper, for simplicity's sake, in line with (Matsuyama & Shang-Shouq Hwang, 1990), all the aforementioned image processing operators are called "**segmentation**" algorithms. As output, an image segmentation algorithm generates *image features*, namely *points* and *regions* (also called segments, [2-D] objects, parcel or blobs (Carson et al., 1997; Lindeberg, 1993; Yang & Wang, 2007), also refer to Part I Section 2.3) or, vice versa, *region boundaries*, i.e., *edges*, provided with no semantic meaning. In general, a sub-symbolic image segment is: (1) made of connected pixels considered homogeneous in color and/or texture based on: (i) a subjective measure of similarity/dissimilarity and (ii) a subjective decision rule (e.g., thresholding), and (2) provided with a non-semantic label equivalent to a numerical segment-based identifier (integer value).

The inherent ill-posedness of any image segmentation algorithm is due to both systematic and accidental errors. The so-called *intrinsic insufficiency* of image segments is due to occlusion problems and dimensionality reduction (Matsuyama & Shang-Shouq Hwang, 1990) (refer to Part I Section 2.3). In addition, image segments are always affected by a so-called *artificial insufficiency* (Matsuyama & Shang-Shouq Hwang, 1990) due to the image segmentation algorithm at hand. This latter source of segmentation errors is related to the well-known *uncertainty principle* according to which, for any contextual (neighborhood) property, we cannot simultaneously measure that property while obtaining accurate localization (Corcoran & Winstanley, 2007; Petrou & Sevilla, 2006).

In practical contexts the inherent ill-posedness of any image segmentation algorithm implies the following.

(a) In real-world image segmentation problems (other than toy problems), it is inevitable for erroneous segments to be detected while genuine segments are omitted (Matsuyama & Shang-Shouq Hwang, 1990) (p. 18).

(b) Any image segmentation algorithm must rely on user-defined segmentation-free parameters based on subjective (heuristic, empirical) criteria on a site-specific basis (see Part I Section 2.1). As a consequence, any image segmentation algorithm can be considered difficult to use, i.e., its degree of automation is low, while its robustness to changes in the input data set and changes in input parameters are both low.

To overcome these shortcomings many researchers in the field of cognitive psychology believe that object segmentation cannot be achieved in a completely bottom-up manner, which is tantamount to saying that segmentation and classification are strongly coupled (Corcoran & Winstanley, 2007; Corcoran et al., 2010; Vecera & Farah, 1997). In particular, Vecera and Farah proved that the process of human visual segmentation can be strongly influenced by top-down human (subjective) factors such as prior knowledge of the image at hand in addition to desires and expectations of an external observer (Vecera & Farah, 1997).

To date, the inherent ill-posedness of any image region/boundary detection algorithm is acknowledged by a relevant portion of the CV and RS communities (Burr & Morrone, 1992; Corcoran & Winstanley, 2007; Corcoran et al., 2010; Delves et al., 1992; Hay & Castilla, 2006; Matsuyama & Shang-Shouq Hwang, 1990; Petrou & Sevilla, 2006; Vecera & Farah, 1997). For example, Castilla *et al.* observe that (Castilla et al., 2008): " Image understanding is a complex cognitive process for which we may still lack key concepts. In particular, most image segmentation methods have been developed heuristically without a deeper examination of the semantic implications of the segmentation process." Well-known image segmentation algorithms, including eCognition® by Definiens AG (Definiens Imaging GmbH, 2004), "... are conceptually inconsistent with the object-oriented approach (OOA)... an underlying hypothesis of any segmentation method is that there is a correspondence between radiometric similarity in the image and semantic similarity in the imaged landscape. Thus, it is expected that image objects (segments) coincide with landscape objects (patches)." Unfortunately, the same Size-Constrained Region Merging (SCRM) algorithm proposed by Castilla *et al.* makes no exception to their criticism since its "correspondence between radiometric similarity and semantic similarity is not straightforward" (Castilla et al., 2008).

To summarize, according to Castilla *et al.* the conceptual framework of OBIA requires generation of symbolic image segments as output. This is the same claim made by cognitive psychology (see this text above) (Corcoran & Winstanley, 2007; Corcoran et al., 2010; Vecera & Farah, 1997). This also agrees with Marr's statement: "vision goes symbolic immediately, right at the level of zero-crossing (primal sketch)... without loss of information" (Marr, 1982)

(p. 343), refer to Part I Section 2.3. As a consequence, if this conjecture holds, then existing commercial image segmentation algorithms, whose claim is to be at the basis of the GEOBIA success (Definiens Imaging GmbH, 2004; Esch et al., 2008), are actually in contrast with the true conceptual framework of GEOBIA, which requires detection of semantic image segments (e.g., landscape objects or patches).

Unfortunately, in spite of the aforementioned contributions found in existing literature, most members of the CV and RS communities, including Diamant (Diamant, 2005; Diamant, 2008; Diamant, 2010a; Diamant, 2010b) (refer to Part I Section 2.5), appear to ignore the inherently ill-posed (subjective) nature of the image segmentation (region extraction/ contour detection) problem. As a consequence, literally dozens of "novel" segmentation (region extraction/contour detection) algorithms are published each year (Zamperoni, 1996). For example, due to the availability of a commercial GEOBIA software developed by a German company (Definiens Imaging GmbH, 2004; Esch et al., 2008), OBIA approaches are currently considered the state-of-the-art in both scientific and commercial RS image mapping applications (Castilla et al., 2008; Hay & Castilla, 2006).

In commercial GEOBIA systems, to reduce the number of empirical segmentation parameters (Esch et al., 2008), a multi-scale (hierarchical) iterative segmentation first stage is employed (Definiens Imaging GmbH, 2004). As output, a hierarchical segmentation algorithm generates multi-scale segmentation solutions in the hope that the target image will appear correctly segmented at some scale. However, quantitative multi-scale assessment of segmentation quality indices requires ground truth data at each scale which are impossible or impractical to obtain in RS common practice (Corcoran & Winstanley, 2007). Therefore, the "best" segmentation map must be selected by the user on an *a posteriori* basis from the available set of multi-scale segmentation solutions according to heuristic, subjective and/or qualitative criteria analogous to those employed in the selection of prior segmentation parameters. In practice, exploitation of a hierarchical segmentation algorithm does not make a driven-without-knowledge segmentation first stage easier to use. In addition, hierarchical segmentation algorithms are computationally intensive and require large memory occupation.

The conclusion is that, to date, in spite of its commercial success, GEOBIA remains affected by a lack of general methodological consensus and research (Hay & Castilla, 2006). Scientific disagreement on the conceptual framework of GEOBIA finds its origin in the well-known information gap existing between physical information (sensations) and semantic information (percepts) (Matsuyama & Shang-Shouq Hwang, 1990) (see Part I Section 2.2.2 and Part I Section 2.3). Since GEOBIA appears unable to generate semantic image segments (e.g., landscape objects) in the pre-attentive vision phase, it appears unsuitable for filling the information gap between raster sub-symbolic imagery and vector symbolic geospatial information (typically dealt with by geographic information systems, GIS).

2.4.2 Labeled data learning for classification and function approximation

Labeled (supervised) data learning approaches deal with either classification or function approximation (regression) problems whose output variables are discrete semantic and continuous non-semantic respectively, see Fig. 1 (Alpaydin, 2010; Bishop, 1995; Cherkassky & Mulier, 2006; Mather, 1994; Mitchell, 1997).

In classification problems where the available training data set is assumed to be fully reliable (which may not always be the case (Bruzzone & Persello, 2009)), the goal of a classifier capable of learning from labeled data is to achieve a perfect fit of the training data set (to reduce to zero the training error) and, at the same time, make good semantic predictions for new (previously unobserved) inputs (to reduce to zero the testing error). An adaptive classifier can be trained in various ways, namely, on-line (sequential learning (Bishop, 1995), stochastic learning (Cherkassky & Mulier, 2006), when a large or infinite input data sequence is available and/or real-time adaptation is required), batch (it requires the storage of a complete and finite training data set (Bishop, 1995)) and semi-batch (Wilson & Martinez, 2000). In addition, there are many statistical classifiers. The most widely used statistical classifiers are the plug-in parametric maximum likelihood (ML) classifier, the non-parametric Multi-Layer Perceptron (MLP) and Radial Basis Function (RBF) networks, kernel methods (also called memory-based, which require the storage of a complete data set (Mitchell, 1997)) such as the SVM and the k-nearest neighbor (K-NN) algorithm, the naive Bayes classifier, adaptive (statistical) decision-trees such as the Classification And Regression Tree (CART), adaptive rule-based systems, mixture of experts (Jordan & Jacobs, 1994), etc. (Alpaydin, 2010; Bishop, 1995; Cherkassky & Mulier, 2006; Duda et al., 2001; Mitchell, 1997).

Classifier performance depends greatly on the characteristics of the labeled data set to be classified (Baraldi et al., 2006b). In other words, there is no single classifier that works best on all given problems; this is also referred to as the "no free lunch" theorem. In practical contexts, classification model selection, i.e., determining a suitable classifier for a given problem, is still more an art than a science.

In reinforcement learning the agent is rewarded for good responses and punished for bad ones. These can be analyzed in terms of decision theory, using concepts such as utility (Cherkassky & Mulier, 2006).

Function regression (curve fitting) takes a finite set of numerical continuous input-output pair samples and attempts to discover an unknown continuous (smooth) deterministic function which, together with added Gaussian noise, would generate those target outputs from the inputs (Bishop, 1995). The goal of function approximation is not to learn an exact representation (interpolation) of the training data, but rather to build a statistical model of the physical process that generates the training labeled data. This statistical model ought to be capable of the best trade-off between: (a) achieving a good fit of the training data (to keep low the bias term of a sum-of-squares error function) and (b) obtaining a reasonably smooth function that is not over-fitted to the training data (to keep the variance term of a sum-of-squares error function low). This is important if the self-organizing (adaptive) function approximation system is to exhibit good generalization, i.e., to make good numerical predictions for new (previously unobserved) inputs (Bishop, 1995).

To summarize, to properly deal with discrete semantic or continuous non-semantic output values, labeled (supervised) data learning systems feature different functional hypotheses and properties. For example:

- they adopt different cost functions, namely, the cross-entropy error function for adaptive classifiers versus the sum-of-squares error for function approximation approaches (Bishop, 1995) (p. 230).

Vision Goes Symbolic Without Loss of Information Within the Preattentive Vision Phase: The Need to Shift the Learning Paradigm from Machine-Learning (from Examples) to Machine-Teaching (by Rules) at the First Stage of a Two-Stage Hybrid Remote Sensing Image Understanding System, Part I: Introduction

85

- When the training labeled data set is assumed to be fully reliable the goal of adaptive classifiers is to reduce to zero both training and testing errors (e.g., if the training error is equal to zero then a classifier is called consistent (Baraldi & Alpaydin, 2002b; Mitchell, 1997)). Vice versa, reducing to zero the bias term in function regression is not recommended because it would imply over-fitting to the training data assumed to be inherently affected by Gaussian noise (which is not the case for exact interpolators) (Bishop, 1995).

2.5 Diamant's image segmentation and contour detection algorithms as proofs of his concepts

As proofs of his concepts (see Part I Section 2.2.3) Diamant presents an image segmentation algorithm and a contour detection algorithm which are summarized below.

2.5.1 Multi-scale image segmentation algorithm

In (Diamant, 2005), a multi-scale image segmentation algorithm is presented and applied to a toy problem, namely, a panchromatic (one-band) image of 640 × 480 pixels in size. The proposed segmentation algorithm is as follows.

1. Low-pass (smoothing) dyadic (sub-sampling by a factor of 2) image decomposition (down-scaling). Image decomposition levels are identified with integer numbers $l = 0,...,$ L, L+1, where level 0 identifies the input image at full spatial resolution. Value L > 0 is set to 4, thus the maximum down-scale level is L+1 = 5. A simple dyadic multi-scale panchromatic (one-band) image decomposition and averaging operator is applied as follows.

$$g^{l+1}(x,y) = [g^l(2x,2y) + g^l(2x + 1,2y) + g^l(2x + 1,2y + 1) +$$

$$+ g^l(2x,2y + 1)]/4, \quad l = 0, ..., L > 0, \tag{1-1}$$

where $g^{l+1}(x,y)$ is the gray-level value of a (down-scaled parent) pixel at the (x,y) coordinate position in a higher $(l+1)$-level image while $g^l(2x,2y)$ and its three nearest neighbors listed in Eq. (1-1) are the corresponding (up-scaled children) pixels within an image array at the lower level l.

2. Single-scale image segmentation algorithm run at the top (coarsest) (L+1)-level of the decomposition pyramid. Diamant claims that since the image size at the top level of the pyramid is significantly reduced and a severe data averaging is attained, any well-known segmentation methodology would suffice. Diamant's proprietary segmentation technique firstly outlines image boundaries (contours) (see Part I Section 2.4.1.2). Secondly, contiguous pixels of "similar" appearance (based on an unknown similarity measure and decision rule) within non-closed contours are aggregated in spatially connected segments (this is apparently a region growing from non-closed contours approach, e.g., refer to (Baraldi & Parmiggiani, 1995)). Thirdly, the segment-based mean intensity image, called characteristic intensity, is computed (this is a piecewise constant image approximation of the input image generated by replacing every pixel with the mean value of the segment where that pixel is located).

3.   (Coarse-to-fine spatial resolution) mean image and segmentation map up-scaling. At each level l = L + 1, ..., 1, with step -1, the mean image and the segmentation map are expanded to the size of the image at the nearest lower level (l-1) (at finer spatial resolution). The expansion rule is simple and the same for both up-scaling operations: the value of each parent pixel at level l is assigned to its four children at level (l-1). Diamant claims that since image regions feature a low inter-segment intensity variability, the majority of newly assigned pixels are determined in a sufficiently correct manner. Only pixels lying on object boundaries or seeds of newly emerging objects can significantly deviate from their up-scaled assigned value. Taking the corresponding l-level of the down-scaled image as a reference, these pixels can easily (!?) be detected and subjected to a refinement cycle. Here they are allowed to adjust themselves to the "proper" nearest neighbors, which certainly belong to one of the previously labeled regions or to the newly emerging ones. Unlike the lossless image decomposition/reconstruction procedure provided by Burt and Adelson's Gaussian/Laplacian pyramid (Burt & Adelson, 1983), in the Diamant case the exact reconstruction of an image is not required. In Diamant's opinion "only (?!) in special cases - medical, scientific, military, fine-art, and a couple (!?) of other applications - the reconstruction fidelity of the original image can be critically important" (Diamant, 2005), which is to say it is critical in all quantitative rather than qualitative CV applications! For example, RS image understanding applications require small, but genuine image details, say, roads, to be well preserved, which is tantamount to saying that RS image applications are among the "couple (!?) of other applications" where high fidelity in multi-scale encoding (decomposition)/decoding (reconstruction) is required.

A critical analysis of the Diamant image segmentation algorithm can be found in Part II Section 3.1.

2.5.2 Single-scale image contour detection algorithm

In (Diamant, 2005) Diamant presents a single-scale image contour detection algorithm and applies it to a toy problem, namely, a panchromatic image 256 × 256 pixels in size. This contour detector provides a measure of local information, $I_{loc}(x,y)$, as a product of two terms.

$$I_{loc}(x,y) = I_{int}(x,y) \times I_{top}(x,y) \qquad (1\text{-}2)$$

where (x,y) are the central pixel coordinates in a (2-D) image array, factor $I_{int}(x,y)$ is the intensity change component and factor $I_{top}(x,y)$ is considered a measure of topological confidence (uncertainty). In Eq. (1-2) term $I_{int}(x,y)$ is estimated as follows.

$$I_{int}(x,y) = \frac{1}{8}\sum_{n=1}^{8}\left|g_c\left(x,y\right) - g_n\left(x,y\right)\right| \ge 0. \qquad (1\text{-}3)$$

Thus, in Eq. (1-2) the first term $I_{int}(x,y)$ is estimated as the mean absolute difference between the central pixel gray value, $g_c(x,y)$, and the gray levels of its 8-adjacency neighbors, $g_n(x, y)$, n = 1, ..., 8.

In Eq. (1-2) the second term $I_{top}(x,y)$ is computed in two steps. Firstly, an expression for a pixel's interrelationship with its surrounding is defined as follows.

Vision Goes Symbolic Without Loss of Information Within the Preattentive Vision Phase: The Need to Shift the Learning
Paradigm from Machine-Learning (from Examples) to Machine-Teaching (by Rules) at the First Stage of a Two-Stage
Hybrid Remote Sensing Image Understanding System, Part I: Introduction   87

$$\text{status}(x,y) = 8g_c(x, y) - \sum_{n=1}^{8} g_n(x,y).$$ (1-4)

It is worthy of note that status(x, y) is equivalent to a contrast value computed by an isotropic *mexican-hat* operator centered on pixel (x, y). The shortest status(x, y) description (encoding) would be in a binary form, for example, 0 if status is negative, and 1 otherwise. Status(x, y) is evaluated for every pixel (x, y) in an image and mapped into a binary status map of the same size as the input image. Secondly, the spatial (topological) interactions of a pixel with its 8-adjacency neighbors can be estimated using the binary status map:

$$I_{top}(x,y) = p(1 - p) = (m/8)[(8 - m)/8], m \in \{0, 8\},$$ (1-5)

where p is the probability that the central pixel and its surrounding ones share the same status, such that $m \in \{0, 8\}$ is the number of 8-adjacency pixels that share the same status with the central pixel in the 2-D array position (x, y). Any $I_{top}(x, y)$ value is computed for every pixel (x, y) and saved in a special image of the size of the input image.

Diamant considers peaks (local extrema) in $I_{loc}(x,y)$ = Eq. (1-2) = $I_{int}(x,y) \times I_{top}(x,y)$ = Eq. (1-3) × Eq. (1-5) as signs of a visible edge present at a given location. However, establishing a proper threshold for local extrema has always been a hard and sophisticated matter. To overcome this difficulty, Diamant proposes to gather a cumulative histogram of I_{loc} values. At first, a number of equal intervals (bins) is selected and a histogram (first-order statistic) of the I_{loc} image is constructed in sequence for every histogram bin as follows: if the pixel-based I_{loc} value is greater than or equal to the bin's lower bound, then this bin counter is increased by one. As a result, the first bin represents the cardinality of all I_{loc} values > 0. It is now explicitly visible what part of the whole "image information content" is carried out by I_{loc} values equal to or greater than a particular bin lower bound. This can be used as a (subjective!) threshold for appropriate image point assignment (marking). In such a way, a set of different information content-related thresholds can be established, which can address diversified task-related requirements. For example, the most prominent image points are marked in dark gray, carrying more than 50% of the whole information content. Less important image parts can be marked in half-gray, carrying between 50 and 70% of information content, and the lowest importance image parts are marked in light gray, carrying 70 to 85% residuals of the information content. The proposed image point marking technique can be effectively used to create more enhanced low-level information content descriptors. For example, based on the status image generated from Eq. (1-4), an edge-localization image can be displayed where dark-gray is assigned to the lower intensity sides of the edges and light-gray to the higher intensity edge sides (Diamant, 2005).

A critical analysis of the Diamant image contour detection algorithm can be found in Part II Section 3.3.

2.6 Four levels of understanding of an RS-IUS

It is important to remember that there are four levels of analysis (understanding) of any information processing device, including RS-IUSs. They are listed below (Baraldi et al., 2010b; Baraldi, 2011a; Marr, 1982).

i.   Computational theory (system architecture). According to Marr, the linchpin of success in attempting to solve the CV problem is that of addressing the computational theory rather than algorithms or implementations (Marr, 1982). In other words, if the vision device architecture is inadequate, even sophisticated algorithms can produce low-quality outputs. On the contrary, improvement in the vision system architecture might achieve twice the benefit with half the effort (which is an adaptation of the original words by Wang (Fangju Wang, 1990)). For example, a two-stage stratified hierarchical hybrid RS-IUS architecture (see Part II Fig. 3) has been proposed in recent literature (Baraldi et al., 2006a; Baraldi et al., 2010a; Baraldi et al., 2010b; Baraldi et al., 2010c; Baraldi, 2011a; Baraldi, 2011b), as an alternative to the current state-of-the-art two-stage GEOBIA architecture, hereafter referred to as two-stage segment-based hybrid RS-IUS architecture (see Part II Fig. 2).

ii.   Knowledge/information representation. According to Wang, "if knowledge representation is poor, even sophisticated algorithms can produce inferior outputs. On the contrary, improvement in representation might achieve twice the benefit with half the effort" (Fangju Wang, 1990). For example, in (Baraldi et al., 2010c; Baraldi, 2011b) a crisp-to-fuzzy SIAM™ transition has been accomplished to model class mixtures.

iii. Algorithm design. This level deals with the design of the algorithm selected to fill each of the data processing modules comprised in the system architecture (refer to point (i) above). According to (Page-Jones, 1988), structured system design is "everything but code".

iv.   Implementation. This level deals with the source code generation for every algorithm designed at point (iii) above.

2.7 Quality Assurance Framework for EO (QA4EO)

Delivered by the Working Group on Calibration and Validation (WGCV) of the Committee of Earth Observations (CEOS), the space arm of the Group on Earth Observations (GEO) (GEO, 2005; GEO, 2008b), the QA4EO guidelines (GEO/CEOSS, 2008) consider mandatory the following actions: (i) calibration and validation (Cal/Val) activities from sensor build to end-of-life and (ii) every sensor-derived data product must be provided with metrological/statistically-based quality indicators (QIs) featuring a degree of uncertainty in measurement. Unfortunately, in RS common practice, these international guidelines are often ignored by scientists, practitioners and whole institutions (Baraldi, 2009).

2.7.1 Calibration and validation (Cal/Val) activities from sensor build to end-of-life

QA4EO considers mandatory an appropriate coordinated program of Cal/Val activities throughout all stages of a spaceborne mission, from sensor build to end-of-life (GEO/CEOSS, 2008). This ensures the harmonization and interoperability of multi-source observational data and derived products required by international programs such as the on-going GEOSS and GMES projects (GEO, 2008b; GEO, 2005) (refer to Part I Section 1).

In spite of the QA4EO recommendations and although it is regarded as common knowledge in the RS community, *radiometric calibration*, i.e., the transformation of dimensionless digital numbers (DNs) into a physical unit of measure related to a community-agreed radiometric scale, is often neglected in literature and surprisingly ignored by scientists, practitioners and institutions involved with RS common practice including large-scale spaceborne image mosaicking and mapping (Baraldi et al., 2006a; Baraldi, 2009; Baraldi et al., 2010a; Baraldi et al., 2010b; Baraldi, 2011a).

A relevant extension of the QA4EO recommendation for radiometric calibration of multi-source EO data is the following.

"Radiometric calibration not only ensures the harmonisation and interoperability of multi-source observational data according to the QA4EO guidelines, but is a necessary, although insufficient, condition for automating the quantitative analysis of EO data" (Baraldi et al., 2006a; Baraldi et al., 2010a; Baraldi et al., 2010b; Baraldi, 2011a) in RS data understanding problems other than toy problems at small data scale and coarse semantic granularity. By definition, a data processing system is *automatic* when it requires no user-defined parameter to run, therefore its user-friendliness cannot be surpassed (refer to Part I Section 2.8).

This necessary condition for automatic EO data understanding agrees with common sense, summarized by the expression: "garbage in means garbage out". In the terminology of MAL and CV, the radiometric calibration constraint augments the degree of prior knowledge of a RS-IUS required to complement the intrinsic insufficiency (ill-posedness) of (2-D) image features, i.e., radiometric calibration makes the inherently ill-posed CV problem better posed (Baraldi et al., 2010a; Baraldi, 2011a; Matsuyama & Shang-Shouq Hwang, 1990).

To summarize, in disagreement with the QA4EO guidelines, most existing scientific and commercial RS-IUSs, such as those listed in Table 1, do not require RS images to be radiometrically calibrated and validated. As a consequence, according to the aforementioned necessary condition for automating the quantitative analysis of EO data, these RS-IUSs are semi-automatic and/or site-specific (since one scene may represent, say, apples, while any other scene, even if contiguous or overlapping, may represent, say, oranges), refer to Table 1. Secondly, Table 1 shows that unlike SIAM™, the ERDAS Atmospheric Correction for satellite imagery (ATCOR3) (Richter, 2006) requires as input an MS image radiometrically calibrated into surface reflectance values exclusively. This implies that the ERDAS ATCOR3 software considers mandatory the inherently ill-posed and difficult-to-solve MS image atmospheric correction pre-processing stage which requires user intervention to make it better posed (Baraldi, 2011a). Thus, unlike SIAM™, the ERDAS ATCOR3 satisfies the necessary condition for automating the quantitative analysis of EO data, but is insufficient to provide a RS image classification problem with an automatic workflow requiring no user-defined empirical parameter to be based on heuristic criteria.

2.7.2 Quality Indicators (QIs) with a degree of uncertainty

In addition to considering mandatory an appropriate coordinated program of Cal/Val activities throughout all stages of a spaceborne mission, from sensor build to end-of-life (see Section 2.7.1), the QA4EO guidelines require that every sensor-derived data product generated across a satellite-based measurement system's processing chain be provided with metrological/ statistically-based QIs featuring a degree of uncertainty in measurement (GEO/CEOSS, 2008). Unfortunately, in RS common practice, as well as in existing literature,

Commercial RS-IUSs	Sub-symbolic (asemantic) versus symbolic (semantic) information primitives, namely, pixels / (2-D) objects (regions, segments) / strata	Radiometric calibration (RAD. CAL.) requirement according to the international QA4EO guidelines
PCI Geomatics GeomaticaX	Sub-symbolic pixels	NO RAD. CAL. Þ semi-automatic and site-specific
eCognition Server by Definiens AG	Unsupervised data learning sub-symbolic objects	NO RAD. CAL. Þ semi-automatic and site-specific
Pixel- and Segment-based versions of the Environment for Visualizing Images (ENVI) by ITT VIS	Either sub-symbolic pixels or unsupervised data learning sub-symbolic objects	NO RAD. CAL. Þ semi-automatic and site-specific
ERDAS IMAGING Objective	Supervised data learning symbolic objects	NO RAD. CAL. Þ semi-automatic and site-specific
ERDAS Atmospheric Correction-3 (ATCOR3) (Richter, 2006)	Sub-symbolic pixels	Consistent with the QA4EO recommendations: surface reflectance, SURF Þ inherently ill-posed atmospheric correction first stage Þ semi-automatic and site-specific.
Novel two-stage stratified hierarchical RS-IUS employing SIAM™ as its preliminary classification first stage	Prior knowledge-based symbolic pixels ∈ symbolic objects ∈ symbolic strata	Consistent with the QA4EO recommendations: top-of-atmosphere (TOA) reflectance (TOARF) or surface reflectance (SURF) values, with TOARF ⊇ SURF ⇒ atmospheric correction is optional. Automatic and robust to changes in RS optical imagery acquired across time, space and sensors.

Table 1. Existing commercial RS-IUSs and their degree of match with the international QA4EO guidelines.

these international guidelines are often ignored by scientists, practitioners and whole institutions (Baraldi, 2009). For example, most works published in RS literature assess and compare spaceborne image classification algorithms in terms of mapping accuracy exclusively, which corresponds to only one of several operational QIs of a RS-IUS (refer to Part I Section 2.8). Moreover, these classification accuracy estimates are rarely provided with a degree of uncertainty in measurement. This violates well-known laws of sample statistics (Congalton & Green, 1999; Foody, 2002; Jain et al., 2000), together with common sense envisaged under the international guidelines of the QA4EO (GEO/CEOSS, 2008).

Vision Goes Symbolic Without Loss of Information Within the Preattentive Vision Phase: The Need to Shift the Learning
Paradigm from Machine-Learning (from Examples) to Machine-Teaching (by Rules) at the First Stage of a Two-Stage
Hybrid Remote Sensing Image Understanding System, Part I: Introduction   91

It is well known, but often forgotten in common practice that any evaluation measure is inherently non-injective (Baraldi, 2011a). For example, in classification map accuracy assessment and comparison, different classification maps may produce the same confusion matrix while different confusion matrices may generate the same confusion matrix accuracy measure, such as overall accuracy. These observations suggest that *no single universally acceptable measure of quality, but instead a variety of quality indices, should be employed in practice* (Congalton & Green, 1999; Foody, 2002). To date, this general conclusion is neither obvious nor community-agreed. For example, this conclusion implies that when a test image and a reference (original) image pair is given, common attempts to identify a unique (universal) reliable image quality index, such as the relative dimensionless global error ERGAS proposed in (Wald et al., 1997), the universal image quality index Q (Wang & Bovik, 2002), the global image quality measure Q4 (Alparone et al., 2004), and the quality index with no reference QNR (Alparone et al., 2006), are inherently undermined as contradictions in terms.

In recent years the issue of uncertainty in spatial data has become increasingly recognized by the RS and geographic information systems (GIS) communities (Friedl et al., 2001). Spatial uncertainty analysis investigates sources of inaccuracies in geospatial data acquisition and understanding and investigates error propagation through a RS (2-D) image processing chain. For example, post-classification change detection between two classification maps of overall accuracy $OA_1 \in [0, 1]$ and $OA_2 \in [0, 1]$, respectively, features a change detection OA (COA) such that $COA \leq (OA_1 \times OA_2)$ (Lunetta & Elvidge, 1999). For example, Friedl *et al.* identify three primary sources of errors in spatial information generated from RS imagery (Friedl et al., 2001).

1. Errors introduced through the image acquisition process (e.g., spectral and spatial image distorsion).
2. Errors produced by the application of image processing techniques, namely, (a) image pre-processing algorithms (e.g., atmospheric correction, geometric correction, radiometric calibration) and (b) image understanding techniques (e.g., spatial and semantic accuracies in classification mapping).
3. Errors associated with interactions between the instrument time, spatial and spectral resolution and the physical nature and scale of an ecological process on the ground (e.g., pixels affected by class mixture).

2.8 Operational Quality Indicators (QIs) of an RS-IUS

In operational contexts a RS-IUS is defined as a low performer if at least one among several operational QIs scores low. Typical operational qualities of a RS-IUS encompass the following (Baraldi et al., 2010a; Baraldi et al., 2010b; Baraldi, 2011a).

i. Degree of automation. For example, a data processing system is *automatic* when it requires no user-defined parameter to run, therefore its user-friendliness cannot be surpassed. When a data processing system requires neither user-defined parameters nor reference data samples to run, it is termed *"fully automatic"* (Qiyao Yu & Clausi, 2007).

ii. Effectiveness, e.g., classification accuracy and spatial accuracy (Baraldi et al., 2005; Persello & Bruzzone, 2010).

iii. Efficiency, e.g., computation time, memory occupation.
iv. Economy (costs). Related to manpower and computing power. For example, open source solutions are welcome to reduce costs of software licenses. Supervised data learning approaches (e.g., SVMs, OBIA systems, etc.) require reference training samples which are typically scene-specific, expensive, tedious, difficult or impossible to collect.
v. Robustness to changes in the input data set, e.g., changes due to noise in the data.
vi. Robustness to changes in input parameters, if any exist.
vii. Maintainability / scalability / re-usability to keep up with changes in users' needs and sensor properties.
viii. Timeliness, defined as the time span between data acquisition and product delivery to the end user. It increases monotonically with manpower, e.g., the manpower required to collect site-specific training samples.

The aforementioned list of operational QIs is neither irrelevant nor obvious. For example, a low score in operational QIs may explain why the literally hundreds of so-called novel low-level (sub-symbolic) and high-level (symbolic) image processing algorithms presented each year in scientific literature typically have a negligible impact on commercial RS image processing software (Zamperoni, 1996). This conjecture is consistent with the fact that most works published in RS literature assess and compare spaceborne image classification algorithms in terms of mapping accuracy exclusively, which corresponds to the sole operational performance indicator (ii) listed above. Moreover, these classification accuracy estimates are rarely provided with a degree of uncertainty in measurement. This violates well-known laws of sample statistics (Congalton & Green, 1999; Foody, 2002; Jain et al., 2000), together with common sense envisaged under the international guidelines of the QA4EO (see Part I Section 2.7.2) (GEO/CEOSS, 2008).

3. Conclusions

The goal of this work is to revise, integrate and enrich previous analyses found in related papers about recent developments in the design and implementation of an operational automatic multi-sensor multi-resolution near real-time two-stage hybrid stratified hierarchical RS-IUS (Baraldi et al., 2006a; Baraldi et al., 2010a; Baraldi et al., 2010b; Baraldi, 2011a).

For publication reasons this work is split into Part I and Part II. In Part I Section 2, related works, concepts and definitions are revised to provide this paper with a significant survey value and make it self-contained. In Part II Section 2, the survey of past works is completed. The original contribution of this work can be found in Part II Section 3 to Part II Section 7.

4. Acknowledgments

This material is partly based upon work supported by the National Aeronautics and Space Administration under Grant/Contract/Agreement No. NNX07AV19G issued through the Earth Science Division of the Science Mission Directorate. The research leading to these results has also received funding from the European Union Seventh Framework Programme FP7/2007-2013 under grant agreement n° 263435. This author wishes to thank the Editorial Board for its competence and willingness to help.

Vision Goes Symbolic Without Loss of Information Within the Preattentive Vision Phase: The Need to Shift the Learning
Paradigm from Machine-Learning (from Examples) to Machine-Teaching (by Rules) at the First Stage of a Two-Stage
Hybrid Remote Sensing Image Understanding System, Part I: Introduction   93

5. References

Acton, S.T. & Landis, J. (1997). Multispectral anisotropic diffusion. *Int. J. Remote Sensing*, Vol. 18, pp. 2877-2886.

Alparone, L.; Baronti, S.; Garzelli, A. & Nencini, F. (2004). A global quality measurement of pan-sharpened multispectral imagery. *IEEE Geosci. Remote Sensing Letters*, Vol. 1, No. 4, pp. 313-317.

Alparone, L.; Aiazzi, B.; Baronti, S.; Garzelli, A. & Nencini, F. (2006). A new method for MS+Pan image fusion assessment without reference, *IEEE International Geoscience and Remote Sensing Symposium* (IEEE IGARSS 2006), Denver, Colorado, Jul. 31– Aug. 4. 2006.

Alpaydin, E. (2010). *Introduction to Machine Learning (Adaptive Computation and Machine Learning*. The MIT Press, Cambridge, MT.

Baatz, M.; Hoffmann, C.; Willhauck, G. Progressing from object-based to object-oriented image analysis. In Object-Based Image Analysis–Spatial Concepts for Knowledge-driven Remote Sensing Applications; Blaschke, T., Lang, S., Hay, G.J., Eds.; Springer-Verlag: New York, NY, 2008, Chapter 1.4, pp. 29- 42.

Backer, E. & Jain, A. K. (1981). A clustering performance measure based on fuzzy set decomposition. *IEEE Trans. Pattern Anal. Mach. Intell.*, Vol. PAMI-3, No. 1, pp. 66– 75.

Baraldi, A. & Parmiggiani, F. (1995). A refined Gamma MAP SAR speckle filter with improved geometrical adaptivity. *IEEE Trans. Geosci. Remote Sensing*, Vol. 33, No. 5, pp. 1245-1257.

Baraldi, A. & Parmiggiani, F. (1996a). Combined detection of intensity and chromatic contours in color images. *Optical Engineering*, Vol. 35, No. 5, pp. 1413-1439.

Baraldi, A. & Parmiggiani, F. (1996b). Single linkage region growing algorithms based on the vector degree of match. *IEEE Trans. Geosci. Remote Sensing*, Vol. 34, No. 1, pp. 137- 148.

Baraldi, A. & Blonda, P. (1999a). A survey of fuzzy clustering algorithms for pattern recognition: Part I. *IEEE Trans. Systems, Man and Cybernetics - Part B: Cybernetics*, Vol. 29, No. 6, pp. 778-785.

Baraldi, A. & Blonda, P. (1999b). A survey of fuzzy clustering algorithms for pattern recognition: Part II. *IEEE Trans. Systems, Man and Cybernetics - Part B: Cybernetics*, Vol. 29, No. 6, pp. 786-801.

Baraldi, A. & Alpaydin, E. (2002a). Constructive feedforward ART clustering networks– Part I. *IEEE Trans. Neural Netw.*, Vol. 13, No. 3, pp. 645–661.

Baraldi, A. & Alpaydin, E. (2002b). Constructive feedforward ART clustering networks– Part II. *IEEE Trans. Neural Netw.*, Vol. 13, No. 3, pp. 662–677.

Baraldi, A.; Bruzzone, L. & Blonda, P. (2005). Quality assessment of classification and cluster maps without ground truth knowledge. *IEEE Trans. Geosci. Remote Sensing*, Vol. 43, No. 4, pp. 857-873.

Baraldi, A.; Puzzolo, V.; Blonda, P.; Bruzzone, L. & Tarantino, C. (2006a). Automatic spectral rule-based preliminary mapping of calibrated Landsat TM and ETM+ images, *IEEE Trans. Geosci. Remote Sensing*. Vol. 44, No. 9, pp. 2563-2586.

Baraldi, A.; Bruzzone, L.; Blonda, P. & Carlin, L. (2006b). Badly-posed classification of remotely sensed images - An experimental comparison of existing data labeling systems, *IEEE Trans. Geosci. Remote Sensing*, Vol. 44, No. 1, pp. 214-235.

Baraldi, A. (2009). Impact of radiometric calibration and specifications of spaceborne optical imaging sensors on the development of operational automatic remote sensing image understanding systems. *IEEE Journal of Selected Topics in Applied Earth Observations and Remote Sensing*, Vol. 2, No. 2, pp. 104-134.

Baraldi, A.; Durieux, L.; Simonetti, D.; Conchedda, G.; Holecz, F. & Blonda, P. (2010a). Automatic spectral rule-based preliminary classification of radiometrically calibrated SPOT-4/-5/IRS, AVHRR/MSG, AATSR, IKONOS/QuickBird/ OrbView/GeoEye and DMC/SPOT-1/-2 imagery – Part I: System design and implementation. *IEEE Trans. Geosci. Remote Sensing*, Vol. 48, No. 3, pp. 1299 - 1325.

Baraldi, A.; Durieux, L.; Simonetti, D.; Conchedda, G.; Holecz, F. & Blonda, P. (2010b). Automatic spectral rule-based preliminary classification of radiometrically calibrated SPOT-4/-5/IRS, AVHRR/MSG, AATSR, IKONOS/QuickBird/ OrbView/GeoEye and DMC/SPOT-1/-2 imagery – Part II: Classification accuracy assessment. *IEEE Trans. Geosci. Remote Sensing*, Vol. 48, No. 3, pp. 1326 - 1354.

Baraldi, A.; Wassenaar, T. & Kay, S. (2010c). Operational performance of an automatic preliminary spectral rule-based decision-tree classifier of spaceborne very high resolution optical images. *IEEE Trans. Geosci. Remote Sensing*, Vol. 48, No. 9, pp. pp. 3482 - 3502.

Baraldi, A. Beyond Geographic Object-Based and Object-Oriented Image Analysis (GEOBIA/GEOOIA): Levels of understanding and degrees of novelty of an operational automatic two-stage stratified hierarchical hybrid remote sensing image understanding system, Remote Sens., submitted for consideration for publication, rs-10905, 2011.

Baraldi, A. (2011b). Fuzzification of a crisp near real-time operational automatic spectral rule-based decision-tree preliminary classifier of multi-source multi-spectral remotely-sensed images, *IEEE Trans. Geosci. Remote Sensing*, accepted for publication, July 2011.

Bishop, C. M. (1995). *Neural Networks for Pattern Recognition*. Clarendon Press, Oxford, United Kingdom.

Bruzzone, L. & Carlin, L. (2006). A multilevel context-based system for classification of very high spatial resolution images. IEEE *Trans. Geosci. Remote Sens.*, Vol. 44, No. 9, pp. 2587–2600.

Bruzzone, L. & Persello, C. (2009). A novel context-sensitive semisupervised SVM classifier robust to mislabeled training samples. *IEEE Trans. Geosci. Remote Sensing*, Vol. 47, No. 7, pp. 2142-2154.

Burr, D. C. & Morrone, M. C. (1992). A nonlinear model of feature detection, In: *Nonlinear Vision: Determination of Neural Receptive Fields, Functions, and Networks*, R. B. Pinter & N. Bahram, (Eds.), pp. 309-327, CRC Press, Boca Raton, Florida.

Burt, P. & Adelson, E. (1983). The Laplacian pyramid as a compact image code. *IEEE Trans. Communications*, Vol. COM-31, No. 4, pp. 532-540.

Canny, J. (1986). A computational approach to edge detection. *IEEE Trans. Pattern Anal. Machine Intell.*, Vol. 8, pp. 679-714.

Carson, C.; Belongie, S.; Greenspan, H. & Malik, J. (1997). Region-Based Image Querying, *Proc. Int'l Workshop Content-Based Access of Image and Video libraries*, San Juan , Puerto Rico, June 20, 1997.

Vision Goes Symbolic Without Loss of Information Within the Preattentive Vision Phase: The Need to Shift the Learning
Paradigm from Machine-Learning (from Examples) to Machine-Teaching (by Rules) at the First Stage of a Two-Stage
Hybrid Remote Sensing Image Understanding System, Part I: Introduction   95

Castilla, G., Hay, G. J. & Ruiz-Gallardo, J. R. (2008). Size-constrained Region Merging
    (SCRM): An automated delineation tool for assisted photointerpretation.
    Photogramm. Eng. Remote Sens., Vol. 74, No. 4 (Apr. 2008), pp. 409-429.
Chengquan Huang; Kuan Song; Sunghee Kim; Townshend, J.; Davis, P.; Masek, J. G. &
    Goward, S. N. (2008). Use of a dark object concept and support vector machines to
    automate forest cover change analysis. *Remote Sensing of Environment*, Vol. 112, pp.
    970-985.
Cherkassky , V. & Mulier, F. (2006). *Learning From Data: Concepts, Theory, and Methods.* Wiley,
    New York.D'Elia, S. (2009). European Space Agency (ESA), personal
    communication.
Congalton, R. G. & Green, K. (1999). *Assessing the Accuracy of Remotely Sensed Data.* Lewis
    Publishers, Boca Raton, Florida.
Corcoran, P. & Winstanley, A. (2007). Using texture to tackle the problem of scale in
    landcover classification, In: *Object-Based Image Analysis - Spatial concepts for
    knowledge-driven remote sensing applications*, T. Blaschke, S. Lang & G. Hay, (Eds.),
    pp. 113-132, Springer Lecture Notes in Geoinformation and Cartography.
Corcoran, P.; Winstanley, A. & Mooney, P. (2010). Segmentation performance evaluation for
    object-based remotely sensed image analysis. *Int. J. Remote Sensing*, Vol. 31, No. 3,
    pp. 617-645.
Definiens Imaging GmbH. (2004). *eCognition* User Guide 4.
Delves, L.; Wilkinson, R.; Oliver, C. & White, R. (1992). `Comparing the performance of SAR
    image segmentation algorithms,. *Int. J. Remote Sensing*, Vol. 13, No. 11, pp. 2121-
    2149.
Diamant, E. (2005). Searching for image information content, its discovery, extraction, and
    representation. *Journal of Electronic Imaging*, Vol. 14, No. 1, pp. 1-11.
Diamant, E. (2008). I'm Sorry to Say, But Your Understanding of Image Processing
    Fundamentals Is Absolutely Wrong, In: *Brain, Vision and AI*, C. Rossi, (Ed.), Chapter
    5, pp. 95-110, In-Tech Publishing.
Diamant, E. (2010a). Not only a lack of a right definition: Arguments for a shift in
    information-processing paradigm. 17.04.2011, Available from:
    http://arxiv.org/ftp/arxiv/papers/1009/1009.0077.pdf
Diamant, E. (2010b). Machine Learning: When and Where the Horses Went Astray?, In:
    Machine Learning, Yagang Zhang, (Ed.), pp. 1-18, In-Tech Publishing. 17.04.2011,
    Available from: http://arxiv.org/abs/0911.1386
Duda, R.; Hart, P. & Stork, D. (2001). *Pattern Classification.* Wiley, New York.
Esch, T.; Thiel, M.; Bock, M.; Roth, A. & Dech, S. (2008). Improvement of image
    segmentation accuracy based on multiscale optimization procedure. *IEEE Geosci.
    Remote Sensing Letters*, Vol. 5, No. 3 (Jul. 2008), pp. 463-467.
European Space Agency (ESA). (2008). GMES Observing the Earth, 17.04.2011, Available
    from: http://www.esa.int/esaLP/SEMOBSO4KKF_LPgines_0.html
Fangju Wang. (1990). Fuzzy supervised classification of remote sensing images. *IEEE Trans.
    Geosci. Remote Sensing*, Vol. 28, No. 2, pp. 194-201.
Foody, G. M. (2002). Status of land cover classification accuracy assessment. *Remote Sensing
    of Environment*, Vol. 80, pp. 185-201.
Friedl, M.A.; McGwire, K.C. & McIver, D. K. (2001). An overview of uncertainty in optical
    remotely sensed data for ecological applications, In: *Spatial Uncertainty in Ecology:*

Implications for Remote Sensing and GIS Applications, C.T. Hunsaker, M.F. Goodchild, M.A. Friedl & T.J. Case, (Eds), pp. 258–283, Springer, New York.

Fritzke, B. (1997). *Some competitive learning methods* (Draft document), 17.04.2011, Available from: http://www.neuroinformatik.ruhr-unibochum.de/ini/VDM/research/gsn/ DemoGNG.

GEO/CEOSS. (2008). A Quality Assurance Framework for Earth Observation, Version 2.0, 17.04.2011, Available from:
http://calvalportal.ceos.org/CalValPortal/showQA4EO.do?section=qa4eoIntro

Global Monitoring for Environment and Security (GMES) (2011). 17.04.2011, Available from: http://www.gmes.info

Gouras, P. (1991). Color vision, In: *Principles of Neural Science*, E. Kandel & J. Schwartz, (Eds.), pp. 467-479, Appleton and Lange, Norwalk, Connecticut.

Group on Earth Observations (GEO). (2005). The Global Earth Observation System of Systems (GEOSS) 10-Year Implementation Plan, 17.04.2011, Available from: http://www.earthobservations.org/docs/10-Year%20Implementation%20Plan.pdf

Group on Earth Observations (GEO). (2008a). GEO announces free and unrestricted access to full Landsat archive, 17.04.2011, Available from:
www.fabricadebani.ro/userfiles/GEO_press_release.doc

Group on Earth Observations (GEO). (2008b). GEO 2007-2009 Work Plan: Toward Convergence, 17.04.2011, Available from: http://earthobservations.org

Gutman, G. *et al.*, (Eds.). (2004). *Land Change Science*, Kluwer Academic Publishers, Dordrecht, The Netherlands.

Hay, G. J. & Castilla, G. (2006). Object-based image analysis: Strengths, weaknesses, opportunities and threats (SWOT), *Proc. 1st Int. Conf. Object-based Image Analysis* (OBIA), 2006. 17.04.2011, Available from:
www.commission4.isprs.org/obia06/Papers/01_Opening%20Session/OBIA2006_Hay_Castilla.pdf

Hudelot, C.; Atif, J. & Bloch, I. (2008). Fuzzy spatial relation ontology for image interpretation. *Fuzzy Sets and Systems Archive*, Vol. 159 , No. 15, pp. 1929-1951.

Jain, A. K.; Duin, R. & Mao, J. (2000). Statistical pattern recognition: A review. *IEEE Trans. Pattern. Anal. Machine Intell.*, Vol. 22, No. 1, pp. 4-37.

Jain, A. & Healey, G. (1998). A multiscale representation including opponent color features for texture recognition. *IEEE Trans. Image Proc.*, Vol. 7, No. 1, pp. 124-128.

Jordan, M. & Jacobs, R. (1994). Hierarchical mixtures of experts and the EM algorithm. *Neural Computation*, Vol. 6, pp.181–214.

Kandel, E. R. (1991). Perception of motion, depth and form, In: *Principles of Neural Science*, E. Kandel & J. Schwartz, (Eds.), pp. 441-466, Appleton and Lange, Norwalk, Connecticut.

Lawley, J. (2003). Self-organising complex-adaptive systems, In: *Large Group Metaphor Process* (at the Findhorn Community), 17.04.2011, Available from:
http://www.cleanlanguage.co.uk/articles/articles/216/1/Self-Organising-Systems-Findhorn/Page1.html

Legg, S. & Hutter, M. (2007). Universal intelligence: A definition of machine intelligence. 17.04.2011, Available from: http://arxiv.org/abs/0706.3639.Marr, D. (1982). *Vision*. Freeman and C., New York.

Vision Goes Symbolic Without Loss of Information Within the Preattentive Vision Phase: The Need to Shift the Learning
Paradigm from Machine-Learning (from Examples) to Machine-Teaching (by Rules) at the First Stage of a Two-Stage
Hybrid Remote Sensing Image Understanding System, Part I: Introduction   97

Lindeberg, T. (1993). Detecting salient blob-like image structures and their scales with a
    scale-space primal sketch: A method for focus-of-attention. *Int. J. Computer Vision*,
    Vol. 11, No. 3, pp. 283-318.

Lunetta, E. & Elvidge, C. (1999). *Remote Sensing Change Detection: Environmental Monitoring
    Methods and Applications*, Taylor and Francis, London, United Kingdom.

Martinetz, T. & Schulten, K. (1994). Topology representing networks. *Neural Networks*, Vol.
    7, No. 3, pp. 507-522.

Mason, C. & Kandel, E. R.. (1991). Central visual pathways, In: *Principles of Neural Science*, E.
    Kandel & J. Schwartz, (Eds.), pp. 420-439, Appleton and Lange, Norwalk,
    Connecticut.

Mather, P. (1994). *Computer Processing of Remotely-Sensed Images – An Introduction*. John Wiley
    & Sons, Chichester, GB.

Matsuyama, T. & Shang-Shouq Hwang, V. (1990). *SIGMA – A Knowledge-based Aerial Image
    Understanding System*. Plenum Press, New York.

Mitchell, T. (1997). *Machine Learning*, McGraw-Hill, New York.

Page-Jones, M. (1988). *The Practical Guide to Structured Systems Design*, Prentice-Hall,
    Englewood Cliffs, New Jersey.

Pakzad, K.; Bückner, J. & Growe, S. (1999). Knowledge Based Moorland Interpretation using
    a Hybrid System for Image Analysis, *Proc. International Society for Photogrammetry
    and Remote Sensing* (ISPRS) Conf., Munich, Germany, Sept. 8-10, 1999, 17.04.2011,
    Available from: http://www.tnt.uni-hannover.de/papers/view.php?ind=1999&
    ord=Authors&mod=ASC

Pekkarinen, A.; Reithmaier, L. & Strobl, P. (2009). Pan-european forest/non-forest mapping
    with Landsat ETM+ and CORINE Land Cover 2000 data. *ISPRS J. Photogrammetry
    & Remote Sens.*, Vol. 64, No. 2 (March 2009), pp. 171-183.

Perona, P. & Malik, J. (1990). Scale-space and edge detection using anisotropic diffusion.
    IEEE Trans. Pattern Anal. Machine Intell., Vol. 12, No. 7, pp. 629-639.

Persello, C. & Bruzzone, L. (2010). A novel protocol for accuracy assessment in classification
    of very high resolution imagery. *IEEE Trans. Geosci. Remote Sensing*, Vol. 48, No. 3,
    pp. 1232-1244.

Petrou, M. & Sevilla, P. (2006). *Image processing: Dealing with Texture*. John Wiley & Sons,
    Chichester, Great Britain.

Qiyao Yu & Clausi, D. A. (2007). SAR sea-ice image analysis based on iterative region
    growing using semantics. *IEEE Trans. Geosci. Remote Sensing*, Vol. 45, no. 12, pp.
    3919-3931.

Richter, R. (2006). Atmospheric / Topographic correction for satellite imagery – ATCOR-2/3
    User Guide, Version 6.2. 17.04.2011, Available from:
    http://www.geog.umontreal.ca/donnees/geo6333/atcor23_manual.pdf

R. Laurini and D. Thompson, Fundamentals of Spatial Information Systems, Academic
    Press, London, 1992.

Sart, F.; Inglada, J.; Landry, R., & Pultz, T (2001).Risk management using remote sensing
    data: Moving from scientific to operational applications, *Proc. SBSR Workshop*,
    Brasil, Apr. 23-27, 2001, 17.04.2011, Available from:
    www.treemail.nl/download/sarti01.pdf

Shunlin Liang. (2004). *Quantitative Remote Sensing of Land Surfaces*. J. Wiley and Sons,
    Hoboken, New Jersey.

Sjahputera, O.; Davis, C.H.; Claywell, B.; Hudson, N.J.; Keller, J.M.; Vincent, M.G.; Li, Y.; Klaric, M. & Shyu, C.R. (2008). GeoCDX: An automated change detection and exploitation system for high resolution satellite imagery, *Proc. Int. Geoscience and Remote Sensing Symposium* (IGARSS), Boston (MT), July 6-11, 2008, paper no. FR4.101.1.

T.F. Cootes and C.J.Taylor, Imaging Science and Biomedical Engineering, Draft Report, University of Manchester, Manchester, March 8, 2004, Available from: http://www.isbe.man.ac.uk/~bim/Models/app_models.pdf.

Tapsall, B.; Milenov, P. & Tasdemir, K. (2010). Analysis of RapidEye imagery for annual land cover mapping as an aid to European Union (EU) Common agricultural policy, W. Wagner, B. Székely, (Eds.): *ISPRS TC VII Symposium – 100 Years ISPRS*, Vienna, Austria, July 5–7, 2010, IAPRS, Vol. XXXVIII, Part 7B. 17.04.2011, Available from: http://mars.jrc.it/mars/content/download/1648/8982/file/P4-5_Milenov _Tapsall_BG_RapidEye_Project_JRC_ver2.pdf

USGS & NASA. (2011). Web-enabled Landsat data (WELD) Project. 17.04.2011, Available from: http://landsat.usgs.gov/WELD.php

Vecera, S. P. & Farah, M. J. (1997). Is visual image segmentation a bottom-up or an interactive process?. *Perception & Psychophysics*, Vol. 59, pp. 1280–1296.

Wald, L.; Ranchin, T. & Mangolini, M. (1997). Fusion of satellite images of different spatial resolutions: Assessing the quality of resulting images. *Photogramm. Eng. Remote Sens.*, Vol. 63, No. 6, pp. 691-699.

Wang, Z. & Bovik, A. C. (2002). A universal image quality index. *IEEE Signal Proc. Letters*, Vol. 9, No. 3, pp. 81-84.

Wilson, H. R. & Bergen, J. R. (1979). A four mechanism model for threshold spatial vision. *Vision Res.*, Vol. 19, pp. 19-32.

Wilson, D. R. & Martinez, T. R. (2000). The Inefficiency of batch training for large training sets. *IEEE-INNS-ENNS International Joint Conference on Neural Networks* (IJCNN'00), 2000, vol. 2, pp.2113.Xu, R. & Wunsch II, D. (2005). Survey of clustering algorithms. *IEEE Trans. Neural Netw.*, Vol. 16, No. 3, pp. 645-678.

Yang, J. & Wang, R. S. (2007). Classified road detection from satellite images based on perceptual organization. *Int. J. Remote Sensing*, Vol. 28, No. 20, pp. 4653-4669.

Zamperoni, P. (1996). Plus ça va, moins ça va. *Pattern Recognition Letters*, Vol. 17, No. 7, (1996), pp. 671-677.

Zins, C. (2007). Conceptual approaches for defining data, information, and knowledge. *Journal of the American Society for Information Science and Technology*, Vol. 58, No. 4, pp. 479-493.

Vision Goes Symbolic Without Loss of Information Within the Preattentive Vision Phase: The Need to Shift the Learning Paradigm from Machine-Learning (from Examples) to Machine-Teaching (by Rules) at the First Stage of a Two-Stage Hybrid Remote Sensing Image Understanding System, Part II: Novel Developments and Conclusions

Andrea Baraldi

Department of Geography, University of Maryland,
College Park, Maryland,
USA

1. Introduction

The goal of this work is to revise, integrate and enrich previous analyses found in related papers about recent developments in the design and implementation of an operational automatic multi-sensor multi-resolution near real-time two-stage hybrid stratified hierarchical remote sensing (RS) image understanding system (RS-IUS) (Baraldi et al., 2006; Baraldi et al., 2010a; Baraldi et al., 2010b; Baraldi, 2011a).

For publication reasons this work consists of two companion papers, Part I and Part II respectively. In Part I related papers, concepts and definitions are revised from existing literature to provide this work with a significant survey value and make it self-contained. The survey of past works is completed in Part II Section 2, where differences at the architectural level between different families of existing RS-IUSs, namely, multi-agent hybrid RS-IUSs, two-stage segment-based RS-IUSs and two-stage stratified hierarchical hybrid RS-IUSs, are highlighted.

The original contribution of Part II is to propose novel definitions of objective continuous sub-symbolic sensory data, continuous physical information, subjective discrete semi-symbolic data structure, discrete semantic-square (semantic^2) information (which is naturally generated from the simultaneous combination of three components: (I) an objective continuous sensory data set, (II) an external subjective supervisor (observer) and (III) his/her own subjective prior ontology equivalent to a model of the (3-D) world existing before looking at the objective sensory data at hand) and prior knowledge base.

In practical contexts the aforementioned original definitions imply the following.

a. It is impossible to *extract* semantic[2] information from objective continuous sensory data because the latter, *per se*, are provided with no semantics at all.

b. It is possible to *correlate* discrete semantic[2] information to objective continuous sensory data. Unfortunately, correlation between continuous sensory data and a finite and discrete set of categorical variables, corresponding to independent random variables generating separable data structures (data aggregations, data clusters, data objects), is low in real-world RS image mapping problems at large data scale or fine semantic granularity, other than toy problems at small data scale and coarse semantic granularity.

Some practical conclusions of potential interest to the RS, computer vision (CV), artificial intelligence (AI) and machine learning (MAL) communities stem from these speculations. Firstly, in operational contexts (e.g., RS image classification problems at national/ continental/ global scale), other than toy problems (e.g., RS image mapping at coarse spatial resolution and local/regional scale), inductive classifiers capable of learning from a finite labeled data set are considered structurally inadequate to correlate (rather than extract, see this text above) discrete semantic[2] information with objective sensory data provided, *per se*, with no semantics at all.

Secondly, to increase the operational quality indicators (QIs) of existing two-stage hybrid RS-IUSs (namely, degree of automation, accuracy, efficiency, robustness to changes in input parameters, robustness to changes in the input data set, scalability, timeliness and economy), any first-stage inductive MAL-from-examples approach should be replaced by a deductive Machine Teaching (MAT)-by-rules capable of generating a preliminary classification first stage where small, but genuine image details are well preserved (Baraldi et al., 2006; Baraldi et al., 2010a; Baraldi et al., 2010b; Baraldi, 2011a).

Thirdly, in RS-IUSs, MAL-from-data algorithms, either labeled (supervised) or unlabeled (unsupervised), either context-insensitive (e.g., pixel-based) or context-sensitive (e.g., 2-D object-based), should be adapted to work on a driven-by-knowledge stratified (semantic masked, layered) basis and moved to the second stage of a novel two-stage stratified hierarchical hybrid RS-IUS architecture recently proposed in RS literature (Baraldi et al., 2006a; Baraldi et al., 2010a; Baraldi et al., 2010b; Baraldi et al., 2010c; Baraldi, 2011a; Baraldi, 2011b).

As a proof of these concepts, the operational automatic multi-sensor multi-resolution near real-time Satellite Image Automatic Mapper™ (SIAM™), recently presented in RS literature[1] (Baraldi et al., 2006; Baraldi et al., 2010a; Baraldi et al., 2010b; Baraldi et al., 2010c; Baraldi, 2011a; Baraldi, 2011b), is adopted as first stage.

The rest of Part II of this work is organized as follows. Part II Section 3 discusses theoretical inconsistencies and algorithmic drawbacks found in Diamant's works (discussed in Part I Section 2.2 and Part I Section 2.5). Revised/novel definitions of objective continuous sensory data, continuous physical information, discrete semantic[2] information and prior knowledge are provided in Part II Section 4. In Part II Section 5 practical consequences of the novel definitions provided in Part II Section 4 are considered for CV, AI and MAL applications. Part II Section 6 presents the operational automatic multi-sensor multi-resolution near real-time SIAM™ as a proof of the original concepts proposed in this work. Conclusions are reported in Part II Section 7.

---

[1] SIAM™ - Patent pending - © Andrea Baraldi & University of Maryland.

Vision Goes Symbolic Without Loss of Information Within the Preattentive Vision Phase: The Need to Shift the Learning
Paradigm from Machine-Learning (from Examples) to Machine-Teaching (by Rules) at the First Stage of a Two-Stage
Hybrid Remote Sensing Image Understanding System, Part II: Novel Developments and Conclusions   101

2. Related works (continued): Taxonomy of hybrid RS-IUS architectures

As reported in Part I Section 2.1, there is a new trend of research and development in both
CV (Cootes & Taylor, 2004) and RS literature (Matsuyama & Shang-Shouq Hwang, 1990;
Shunlin Liang, 2004) to outperform existing scientific and commercial image understanding
systems. This novel trend focuses on the development of quantitative hybrid models for
retrieving sub-symbolic continuous variables (e.g., LAI) and symbolic categorical discrete
variables (e.g., land cover composition) from multi-spectral (MS) imagery. By definition,
hybrid models combine both statistical and physical models to take advantage of the unique
features of each and overcome their shortcomings (see Part I Section 2.1). The study of
hybrid quantitative models is also called AI systems integration. In this section, the
taxonomy of hybrid RS-IUSs is summarized in line with (Baraldi et al., 2010a). It consists of:

- multi-agent hybrid RS-IUSs,
- two-stage segment-based RS-IUSs, whose conceptual foundation is well known in RS
  literature as as geographic (2-D) object-based image analysis (GEOBIA), including a so-
  called iterative geographic OO image analysis (GEOOIA) approach (Baatz et al., 2008).
  and
- two-stage stratified hierarchical hybrid RS-IUSs employing SIAM™ as preliminary
  classification first stage.

2.1 Multi-agent hybrid RS-IUSs

In existing literature multi-agent hybrid RS-IUSs provide application-specific combinations
of inductive and deductive inference mechanisms (Matsuyama & Shang-Shouq Hwang,
1990). A traditional multi-agent hybrid RS-IUS architecture comprises the following
modules (see Fig. 1).

1. (3-D) Scene domain knowledge, also called *world model* (Matsuyama & Shang-Shouq
    Hwang, 1990). It is represented as a semantic network consisting of classes of objects as
    nodes and relationships between classes as arcs between nodes (refer to Part I Section
    2.2.2).
2. A Low-Level Vision Expert (LLVE, refer to Part I Section 2.4.1.2) (Matsuyama & Shang-
    Shouq Hwang, 1990). In general, an LLVE can be applied either image-wide or within a
    local image area specified by a Specialized Object Model Selection Expert (SOMSE, see
    this text below) (Mather, 1994). LLVE includes a battery of low-level sub-symbolic (non-
    semantic) general-purpose domain-independent inductive-learning (fine-to-coarse,
    bottom-up) driven-without-knowledge inherently ill-posed image processing
    algorithms called *image segmentation* for simplicity's sake (also refer to Part I Section
    2.4.1.2) (Matsuyama & Shang-Shouq Hwang, 1990). As output, the image segmentation
    first stage provides image features, namely points and regions (segments, [2-D] objects,
    parcel or blobs (Carson et al., 1997; Lindeberg, 1993; Yang & Wang, 2007), see Part I
    Section 2.3) or, vice versa, region boundaries, i.e., edges, provided with no semantic
    meaning (see Part I Section 2.4.1.2).
3. A high-level interpretation second stage employing a combination of top-down (model-
    driven) and bottom-up (data-driven) inference mechanisms to establish the
    correspondence between sub-symbolic (2-D) image features extracted from the image
    domain and symbolic (3-D) object models stored in the world model to construct
    plausible structural (semantic) description(s) of the depicted scene (refer to Part I Section

2.3). The combination of top-down with bottom-up inference strategies achieves two operational advantages: (a) provides better conditions for an otherwise ill-posed driven-without-knowledge segmentation first stage (refer to Part I Section 2.3) and (b) allows restriction of intensive processing to a small portion of the image data (Matsuyama & Shang-Shouq Hwang, 1990), analogously to a focus of visual attention in pre-attentive biological vision (Mason & Kandel, 1991; Gouras, 1991; Kandel, 1991). The high-level processing second stage comprises (Matsuyama & Shang-Shouq Hwang, 1990): (I) a Spatial Reasoning Expert (SRE) whose aim is to trigger the instantiation, within a candidate local area, of plausible generic (3-D) object models found in the available world model, e.g., house, and (II) a SOMSE (refer to this text above) which uses domain-dependent knowledge about specific applications to: (i) prune the search space of specialized (3-D) object models (e.g., rectangular house, L-shaped house, etc.) linked by A-KIND-OF relations to the generic target (3-D) object model (e.g., house) provided by SRE; (ii) transform the 3-D appearance properties of the specialized (3-D) object model into a selected set of 2-D appearance properties based on the imaging sensor model; (iii) transform a target spatial relation in fuzzy terms (e.g., in front of) provided by SRE into a local area based on a trial-and-error heuristic search with no concrete theoretical basis and (iv) provide a consistency examination between quantitative absolute image features collected by LLVE in a local area and the target 2-D appearance constraints. In other words, the 2-D appearance properties must be satisfied by image features extracted by LLVE from a local area. Since the image structure in a local area is very simple compared with that of the entire image, image feature extraction performed by an object model-driven and locational constrained LLVE can be very efficient and reliable compared with that performed by the same LLVE run image-wide at the first stage (Matsuyama & Shang-Shouq Hwang, 1990) (p. 41).

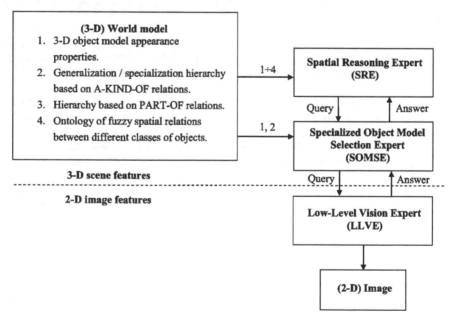

Fig. 1. Multi-agent hybrid systems for RS image understanding (derived from Figure 2.1 in (Matsuyama & Shang-Shouq Hwang, 1990), p. 36).

	Legenda. Y: Yes, N: No, C: Complete, I: Incomplete (radiometric calibration offset parameters are set to zero), (E)TM: (Enhanced) Thematic Mapper, B: Blue, G: Green, R: Red, NIR: Near Infra-Red, MIR: Medium IR, TIR: Thermal IR, SR: Spatial Resolution, Pan: Panchromatic. ■■■ columns: visible channels typical of water and haze. ■ column: NIR band typical of vegetation. ■ columns: MIR channels characteristics of bare soils. ■ column: TIR channel.										
SIAM™ system of systems		B – (E)TM1, 0.45-0.52 (μm)	G – (E)TM2, 0.52-0.60 (μm)	R – (E)TM3, 0.63-0.69 (μm)	NIR – (E)TM4, 0.76-0.90 (μm)	MIR1 – (E)TM5, 1.55-1.75 (μm)	MIR2 – (E)TM7, 2.08-2.35 (μm)	TIR – (E)TM6, 10.4-12.5 (μm)	SR (m)	Rad. Cal. Y/N, C/I	Pan SR (m)	Notes
L-SIAM™ (95/47/18 Sp. Cat.)	Landsat-4/-5 TM	×	×	×	×	×	×	×	30	Y-C		Refer to Table I in (Baraldi et al., 2006a).
	Landsat-7 ETM+	×	×	×	×	×	×	×	30	Y-C	15	Same as above.
	MODIS	×	×	×	×	×	×	×	250, 500, 1000	Y-C		Same as above.
	ASTER		×	×	×	×	×	×	15-30	Y-C		Same as above.
	CBERS-2B	×	×	×	×	×	×	×		N		
S-SIAM™ (68/40/15 Sp. Cat.)	SPOT-4 HRVIR		×	×	×	×			20	Y-I	10	Refer to Table II in (Baraldi et al., 2006a).
	SPOT-5 HRG		×	×	×	×			10	Y-I	2.5 - 5	Same as above.
	SPOT-4/-5 VMI		×	×	×	×			1100	Y-I		Same as above.
	IRS-1C/-1D LISS-III		×	×	×	×			23.5	Y-I		
	IRS-P6 LISS-III		×	×	×	×			23.5	Y-I		
	IRS-P6 AWiFS		×	×	×	×			56	Y-I		
AV-SIAM™ (82/42/16 Sp. Cat.)	NOAA AVHRR			×	×	×		×	1100	Y		Refer to Table II in (Baraldi et al., 2006a).
	MSG		×	×	×			×	3000	Y		Same as above.
AA-SIAM™ (82/42/16Sp. Cat.)	ENVISAT AATSR		×	×	×	×		×	1000	Y		Same as above.
	ERS-2 ATSR-2		×	×	×	×		×	1000	Y		
I-SIAM™ (52/28/12Sp. Cat.)	IKONOS-2	×	×	×	×				4	Y	1	
	QuickBird-2	×	×	×	×				2.4	Y	0.61	
	WorldView-2	×	×	×	×				2.0	Y	0.5	
	GeoEye-1	×	×	×	×				1.64	Y	0.41	
	OrbView-3	×	×	×	×				4	Y	1	
	RapidEye-1 to -5	×	×	×	×				6.5	Y-I		
	ALOS AVNIR-2	×	×	×	×				10	Y		
	KOMPSAT-2	×	×	×	×				4	N	1	
	TopSat	×	×	×	×				5	N	2.5	
	FORMOSAT-2	×	×	×	×				8	N	2	
D-SIAM™ (52/28/12Sp. Cat.)	Landsat-1/-2/-3/-4/-5 MSS		×	×	×				79	Y		
	IRS-P6 LISS-IV		×	×	×				5.8	Y-I		
	SPOT-1/-2/-3 HRV		×	×	×				20	Y-I	10	
	DMC		×	×	×				22-32	N		

Table 1. SIAM™ system of systems. List of spaceborne optical imaging sensors eligible for use as input.

Multi-agent hybrid systems typically suffer from two main limitations.

- In addition to the intrinsic insufficiency of image features, e.g., due to occlusion and dimensionality reduction (refer to Part I Section 2.3), these systems are affected by the so-called artificial insufficiency caused by the inherent ill-posedness of the image segmentation problem (Matsuyama & Shang-Shouq Hwang, 1990) (see Part I Section 2.4.1.2). This means that in RS common practice any first-stage image segmentation algorithm is simultaneously affected by both omission and commission segmentation errors. Although the inherent ill-posedness of image segmentation is acknowledged by a reasonable portion of existing literature (Burr & Morrone, 1992; Corcoran et al., 2010; Corcoran & Winstanley, 2007; Delves et al., 1992; Hay & Castilla, 2006; Matsuyama & Shang-Shouq Hwang, 1990; Petrou & Sevilla, 2006; Vecera & Farah, 1997), this is often forgotten by a large segment of the RS community where literally dozens of "novel" segmentation algorithms are published each year (Zamperoni, 1996) (refer to Part I Section 2.4.1.2).
- Semantic nets lack flexibility and scalability to cope with changes in sensor characteristics and users' changing needs, i.e., they are unsuitable for commercial RS image processing software toolboxes and remain limited to scientific applications.

To overcome these limitations, an alternative two-stage stratified hierarchical hybrid RS-IUS architecture, such as that shown in Fig. 3, was proposed in recent literature (Baraldi et al., 2006; Baraldi et al., 2010a; Baraldi et al., 2010b; Baraldi, 2011a; Baraldi, 2011b; Baraldi et al., 2010c).

2.2 Two-stage segment-based RS-IUSs

Two-stage segment-based RS-IUSs comprise an inductive driven-without-knowledge image segmentation first stage and a second-stage object-based classifier, see Fig. 2. The latter can be implemented based on deductive or inductive inference mechanisms, say, as a prior knowledge-based non-adaptive decision-tree or a supervised data learning classifier (e.g., a Support Vector Machine, SVM (Bruzzone & Carlin, 2006)).

Due to the availability of a commercial GEOBIA software developed by a German company (Definiens Imaging GmbH, 2004; Esch et al., 2008), two-stage segment-based RS-IUSs have recently gained widespread popularity and are currently considered the state-of-the-art in both scientific and commercial RS image mapping application domains (Mather, 1994; Pekkarinen, Reithmaier & Strobl, 2009). In practice, under the guise of 'flexibility' current commercial 2-D object-based software provides overly complicated options to choose from (Hay & Castilla, 2006). This means that with their increasing diffusion commercial two-stage segment-based RS-IUSs show an increasing lack of productivity (Tapsall et al., 2010), consensus and research (Castilla et al., 2008; Hay & Castilla, 2006) (refer to Part I Section 2.4.1.2).

2.3 Two-stage stratified hierarchical hybrid RS-IUS employing SIAM™ as its preliminary classification first stage

Accounting for the customary distinction between a model and the algorithm used to identify it (Baraldi et al., 2010a; Baraldi, 2011a), an original two-stage stratified hierarchical hybrid RS-IUS architecture (see Fig. 3) was identified starting from several RS-IUS

Vision Goes Symbolic Without Loss of Information Within the Preattentive Vision Phase: The Need to Shift the Learning
Paradigm from Machine-Learning (from Examples) to Machine-Teaching (by Rules) at the First Stage of a Two-Stage
Hybrid Remote Sensing Image Understanding System, Part II: Novel Developments and Conclusions   105

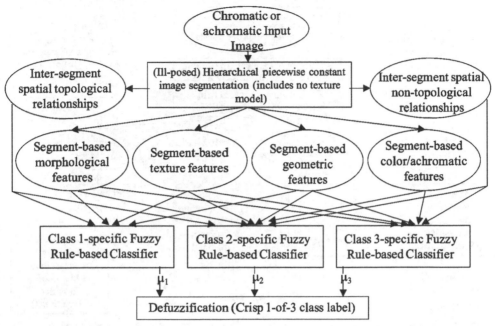

Fig. 2. Two-stage segment-based hybrid RS–IUS architecture adopted, for example, by the eCognition commercial software toolbox (Definiens Imaging GmbH, 2004). Preliminary image simplification is pursued by means of an (ill-posed hierarchical) image segmentation approach which generates as output a segmented (discrete) map, either single-scale or multi-scale. Worthy of note is that first-stage output sub-symbolic informational primitives, namely, labeled segments (2-D objects, parcels), e.g., segment 1, segment 2, etc., are provided with no semantic meaning.

implementations proposed by Shackelford and Davis in recent years (Shackelford & Davis, 2003a; Shackelford & Davis, 2003b). This novel RS-IUS architecture comprises the following phases (Baraldi et al., 2006; Baraldi et al., 2010a; Baraldi et al., 2010b; Baraldi et al., 2010c; Baraldi, 2011a; Baraldi, 2011b).

a.   A radiometric calibration pre-processing stage, where DNs are transformed into top-of-atmosphere reflectance (TOARF) or surface reflectance (SURF) values, with TOARF $\supseteq$ SURF, the latter being an ideal (atmospheric noise-free) case of the former. This radiometric calibration constraint not only ensures the harmonization and interoperability of multi-source observational data in line with the Quality Assurance Framework for EO (QA4EO) guidelines (GEO/CEOSS, 2008), but is considered a necessary, although not sufficient, condition for input Earth observation (EO) imagery to be automatically interpreted (see Part I Section 2.7.1). It is worth mentioning that a RS-IUS suitable for mapping TOARF values into surface categories makes the inherently ill-posed (therefore, difficult to solve) atmospheric correction problem an optional MS image pre-processing stage unlike competing classification approaches employing surface reflectance spectra, such as the ERDAS ATCOR3 (Richter, 2006) (see Part I Section 2.7.1).

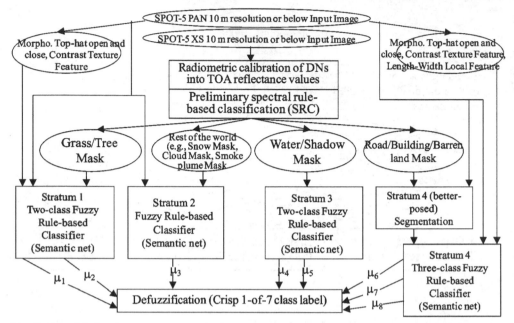

Fig. 3. Novel hybrid two-stage stratified hierarchical RS-IUS architecture. This data flow
diagram (DFD) shows processing blocks as rectangles and sensor derived data products as
circles. In this example, a SPOT-5 MS image is adopted as input. The panchromatic (PAN)
image can be generated from the MS image. The MS image is input to the preliminary
classification first stage and, if useful, to second-stage class-specific classification modules.
The PAN image is exclusively employed as input to second-stage stratified class-specific
context-sensitive classification modules, where color information is dealt with by
stratification. For example, stratified texture detection is computed in the PAN image
domain, which reduces computation time.

b.   A first-stage application-independent per-pixel (non-contextual) top-down (prior
      knowledge-based, see Part I Section 2.1) preliminary classifier in the Marr sense (Marr,
      1982).
c.   A second-stage battery of stratified hierarchical context-sensitive application-dependent
      modules for class-specific feature extraction and classification.

In (Baraldi et al., 2006; Baraldi et al., 2010a; Baraldi et al., 2010b; Baraldi et al., 2010c; Baraldi,
2011a; Baraldi, 2011b), the abovementioned first-stage pixel-based preliminary classifier was
designed and implemented as an original operational automatic near-real-time per-pixel
multi-source multi-resolution application-independent SIAM™. To employ as input a
radiometrically calibrated MS image acquired by almost any of the ongoing or future
planned satellite optical missions, SIAM™ is designed as an integrated system of systems. It
comprises a "master" 7-band Landsat-like SIAM™ (L-SIAM™) together with five down-
scaled ("slave", derived) versions of L-SIAM™ whose input is a MS image featuring a
spectral resolution that overlaps with, but is inferior to, Landsat's. To summarize, SIAM™
combines six sub-systems (refer to Table 1).

i. A "master" 7-band L-SIAM™ capable of detecting 95/ 47/ 18 spectral categories at fine/ intermediate/ coarse semantic granularity (see Fig. 4). The legend of the preliminary classification map generated by L-SIAM™ at fine semantic granularity is shown in Table 2.

ii. A four-band Satellite Pour l'Observation de la Terre (SPOT)-like SIAM™ (S-SIAM™), which detects 68/ 40/ 15 spectral categories at fine/ intermediate/ coarse semantic granularity (see Fig. 5).

iii. A four-band National Oceanic and Atmospheric Administration (NOAA) Advanced Very High Resolution Radiometer (AVHRR)-like SIAM™ (AV-SIAM™), which detects 82/ 42/ 16 spectral categories at fine/ intermediate/ coarse semantic granularity.

iv. A five-band ENVISAT Advanced Along-Track Scanning Radiometer (AATSR)-like SIAM™ (AA-SIAM™), which detects 82/ 42/ 16 spectral categories at fine/ intermediate/ coarse semantic granularity.

v. A four-band IKONOS-like SIAM™ (I-SIAM™), which detects 52/ 28/ 12 spectral categories at fine/ intermediate/ coarse semantic granularity (see Fig. 6). The legend of the preliminary classification map generated by I-SIAM™ at fine semantic granularity is shown in Table 3.

vi. A three-band Disaster Monitoring Constellation (DMC)-like SIAM™ (D-SIAM™), which detects 52/28/12 spectral categories at fine/intermediate/coarse semantic granularity.

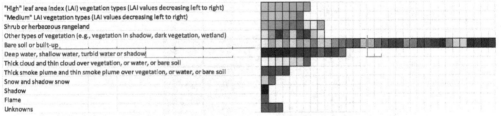

Table 2. Preliminary classification map legend adopted by L-SIAM™ at fine semantic granularity. Pseudo-colors of the 95 spectral categories are gathered based on their spectral end member (e.g., bare soil or built-up) or parent spectral category (e.g., "high" LAI vegetation types). The pseudo-color of a spectral category is chosen as to mimic natural colors of pixels belonging to that spectral category.

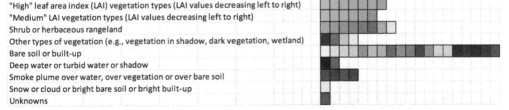

Table 3. Preliminary classification map legend adopted by I-SIAM™ at fine semantic granularity. Pseudo-colors of the 52 spectral categories are gathered based on their spectral end member (e.g., bare soil or built-up) or parent spectral category (e.g., "high" LAI vegetation types). The pseudo-color of a spectral category is chosen as to mimic natural colors of pixels belonging to that spectral category.

Fig. 4 to Fig. 6 show qualitatively that, in disagreement with a common opinion in the RS community where GEOBIA is considered indispensable for spaceborne VHR image understanding (Bruzzone & Carlin, 2006; Bruzzone & Persello, 2009; Persello & Bruzzone, 2010), the pixel-based SIAM™ is very successful in the automatic mapping of RS imagery, including VHR images (Baraldi et al., 2006; Baraldi et al., 2010a; Baraldi et al., 2010b; Baraldi et al., 2010c; Baraldi, 2011a; Baraldi, 2011b). This means that SIAM™ is not affected by the well-known salt-and-pepper classification noise effect which traditionally affects ordinary pixel-based classifiers (e.g., maximum-likelihood classifiers (Cherkassky and Mulier, 2006)), which is tantamount to saying that SIAM™ is successful in modeling the within-spectral-category variance.

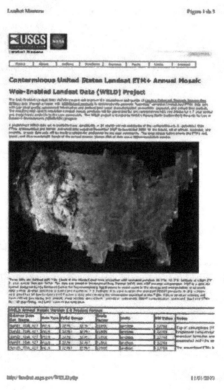

Fig. 4(a). Web-Enabled Landsat Data (WELD) Project (USGS & NASA, 2011). This is a joint NASA and USGS project providing seamless consistent mosaics of fused Landsat-7 Enhanced TM Plus (ETM+) and MODIS data radiometrically calibrated into top-of-atmosphere reflectance (TOARF) and surface reflectance. These mosaics are made freely available to the user community. Each consists of 663 fixed location tiles. Spatial resolution: 30 m. Area coverage: Continental USA and Alaska. Period coverage: 7-year. Product time coverage: weekly, monthly, seasonal and annual composites.

Vision Goes Symbolic Without Loss of Information Within the Preattentive Vision Phase: The Need to Shift the Learning
Paradigm from Machine-Learning (from Examples) to Machine-Teaching (by Rules) at the First Stage of a Two-Stage
Hybrid Remote Sensing Image Understanding System, Part II: Novel Developments and Conclusions   109

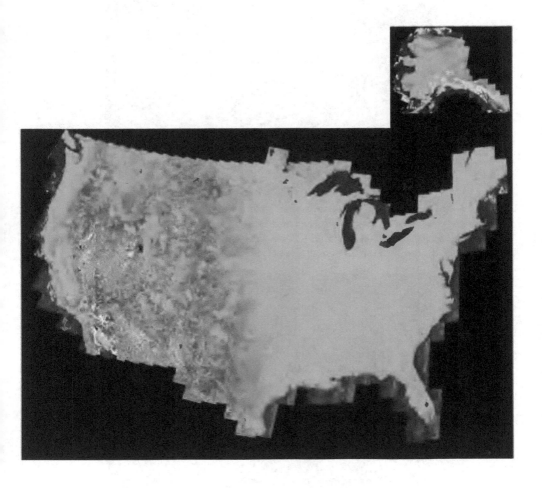

Fig. 4(b). Including the map of Alaska at the top right. Preliminary classification map automatically generated by L-SIAM™ from the 2008 annual WELD mosaic shown in Fig. 4(a). Output spectral categories are depicted in pseudo colors. Map legend: refer to Table 2. To generate this map at national scale L-SIAM™ was run overnight by L. Boschetti (Univ. of Maryland) in Dec. 2010. To the best of this author's knowledge, this is the first example of such a high-level product automatically generated at both the NASA and USGS.

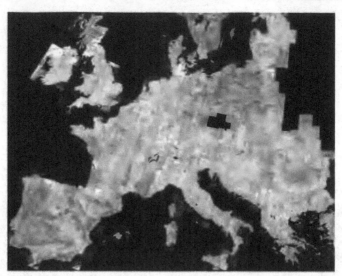

Fig. 5(a). 4-band GMES-IMAGE2006 Coverage 1 mosaic, consisting of approximately two thousand 4-band IRS-P6 LISS-III, SPOT-4, and SPOT-5 images, mostly acquired during the year 2006, depicted in false colors: Red – Band 4 (Short Wave InfraRed, SWIR), Green – Band 3 (Near IR, NIR), Blue – Band 1 (Visible Green). Down-scaled spatial resolution: 25 m.

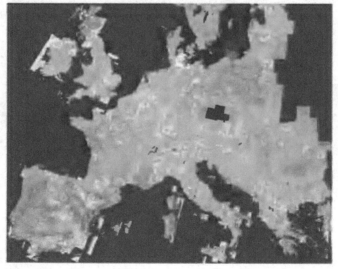

Fig. 5(b). Preliminary classification map automatically generated by S-SIAM™ from the mosaic shown in Fig. 5(a). Output spectral categories are depicted in pseudo colors. A map legend similar to Table 2 is adopted: water and shadow areas are in blue, clouds in white, snow and ice in light blue, vegetation types in different shades of green, rangeland types in different shades of light green, barren land types in different shades of brown and grey. To the best of this author's knowledge, this is the first example of such a high-level product automatically generated at the European Commission – Joint Research Center (EC-JRC).

Vision Goes Symbolic Without Loss of Information Within the Preattentive Vision Phase: The Need to Shift the Learning
Paradigm from Machine-Learning (from Examples) to Machine-Teaching (by Rules) at the First Stage of a Two-Stage
Hybrid Remote Sensing Image Understanding System, Part II: Novel Developments and Conclusions   111

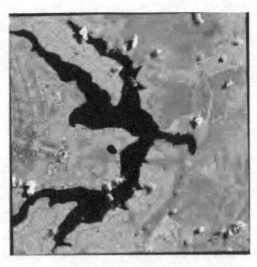

Fig. 6(a). QuickBird-2 image, 2.4 m spatial resolution, acquisition date 2010-03-16, radiometrically calibrated into TOARF values, depicted in false colors (R: 3, G: 4, B: 1). Default image histogram stretching: ENVI linear stretching 2%.

Fig. 6(b). Automatic Q-SIAM™ preliminary mapping of the QB-2 image shown in Fig. 6(a). Spectral categories are depicted in pseudo colors. Map legend: see Table 3. It is noteworthy that, within the Q-SIAM™ mutually exclusive and completely exhaustive classification scheme, cloud detection is *per se* an interesting operational product with relevant commercial applications and, to the best of these authors' knowledge, without alternative solutions in either commercial or scientific RS-IUSs.

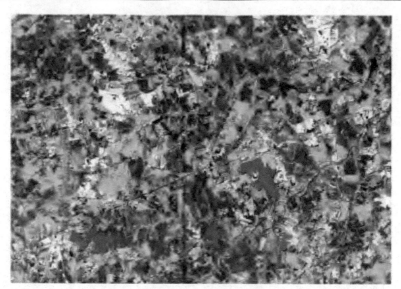

Fig. 7(a). Zoomed area of a Landsat 7 ETM+ image of Virginia, USA (path: 16, row: 34, acquisition date: 2002-09-13), depicted in false colors (R: band ETM5, G: band ETM4, B: band ETM1), 30 m resolution, calibrated into TOARF values.

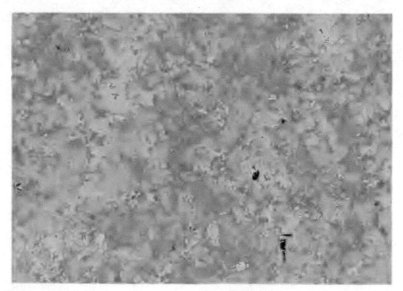

Fig. 7(b). 2nd-stage stratified vegetated land cover classification map generated in series with the L-SIAM™ first stage from Fig. 7(a). This 2nd-stage map consists of 19 vegetated/non-vegetated land cover classes, depicted in pseudo-colors, including: crop field or grassland, broad-leaf forest, needle-leaf forest and non-vegetated pixels (in black). Input features are: spectral layers generated by L-SIAM™, (achromatic) brightness and multi-scale isotropic texture features extracted from the brightness image.

Vision Goes Symbolic Without Loss of Information Within the Preattentive Vision Phase: The Need to Shift the Learning
Paradigm from Machine-Learning (from Examples) to Machine-Teaching (by Rules) at the First Stage of a Two-Stage
Hybrid Remote Sensing Image Understanding System, Part II: Novel Developments and Conclusions   113

To the best of this author's knowledge no unifying automatic multi-sensor multi-resolution near real-time RS image classification platform alternative to SIAM™ can be found in existing literature. This is tantamount to saying that SIAM™ provides the first operational example of an automatic multi-sensor multi-resolution near real-time EO system of systems envisaged under on-going international research programs such as the Global EO System of Systems (GEOSS) conceived by the Group on Earth Observations (GEO) (GEO, 2005; GEO, 2008a) and the Global Monitoring for the Environment and Security (GMES), which is an initiative led by the European Union (EU) in partnership with the European Space Agency (ESA) (ESA, 2008; GMES, 2011) (see Part I Section 1).

Fig. 7 shows an example of an automatic 2nd-stage stratified rule-based vegetated land cover classification system in series with the L-SIAM™ first stage. The two-stage automatic classifier employing L-SIAM™ as preliminary classification first stage (refer to Fig. 3) is input with a 7-band Landsat image radiometrically calibrated into TOARF values, shown in Fig. 7(a). The 2nd-stage stratified rule-based vegetated land cover classification system in series with the L-SIAM™ first stage employs as input features: spectral-based layers (strata, generated by L-SIAM™ at first stage), (achromatic) brightness and multi-scale isotropic texture extracted from the brightness image. The 2nd-stage classifier provides as output a classification map consisting of 19 vegetated/non-vegetated land cover classes, depicted in pseudo-colors, including: crop field or grassland, broad-leaf forest, needle-leaf forest and non-vegetated pixels (in black), see Fig. 7(b).

3. Inconsistencies and limitations of the Diamant computational theory and algorithms

An original analysis of the Diamant definitions reported in Part I Section 2.2.3 and Diamant's image segmentation and contour detection algorithms summarized in Part I Section 2.5 is provided below.

3.1 Comments on the Diamant definitions of data, information and knowledge

According to this author, the Diamant definitions reported in Part I Section 2.2.3 are affected by three major drawbacks.

i.   Diamant states that "information elicitation (extraction) does not require incorporation of any high-level knowledge" (Diamant, 2010a; Diamant, 2010b), which is tantamount to saying that detection of non-semantic primary data structures (data objects), e.g., (2-D) image segments, in an unlabeled data set, e.g., a (2-D) image, does not require incorporation of any high-level (prior) knowledge. Based on this statement it is possible to conclude that despite his theoretical anti-conformism, namely, his willingness to replace the MAL-from-examples paradigm with the MAT-by-rules approach, Diamant is a conformist in practice. In fact, the Diamant image contour detection and image segmentation algorithms (see Part I Section 2.5) fit existing CV system architectures well established in literature, such as, respectively, the Marr CV system architecture, conceived in the 1980s and comprising a zero-crossings (contour detection) primal sketch, and RS-IUSs where an image segmentation first stage is adopted in agreement with the GEOBIA approach (see Part I Section 2.4.1.2). In other words, there is a clear contradiction in terms between the Diamant claim of replacing the MAL-from-examples with a MAT-by-rules paradigm and his practical proofs of concept, consisting of image segmentation and contour detection algorithms 100% consistent with the same MAL-from-examples paradigm he intends to overcome.

ii.   If the Diamant CV system coincides with a Marr CV system or an GEOBIA approach (refer to paragraph (i) above), then, in practical contexts, its operational QIs (see Part I Section 2.8) are expected to score as low as Marr's or OBIA's (refer to Part I Section 1, Part I Section 2.4.1.2 and Part II Section 2). At the level of understanding of an information processing system known as computational theory (system architecture, see Part I Section 2.6), GEOBIA scores low in operational contexts because, according to the present author, it goes symbolic as late as possible, namely, at the output of its second and last stage (see Fig. 2). This is in contrast with an important intuition by Marr stating that "vision goes symbolic almost immediately, right at the level of zero-crossings (first-stage primal sketch)... without loss of information" (Marr, 1982) (p. 343) (see Part I Section 2.3).

iii. To recover from the gap existing between Diamant's theoretical anti-conformism, but practical conformism (refer to paragraphs (i) and (ii) above), it is sufficient to observe that statements such as "information elicitation (aggregation) does not require incorporation of any high-level knowledge" (Diamant, 2010a; Diamant, 2010b), are in clear contradiction with a relevant section of existing literature (see Part I Section 2.4.1.2). In particular, Diamant considers primary data structures, equivalent to non-semantic data objects (e.g., image segments), as "natural data structures which reflect some similarities among neighboring elements in the data. Therefore, defining them is certainly a well-grounded procedure that does not raise any objection, because objective (physical) laws underpin such a procedure" (Diamant, 2010a) (see Part I Section 2.2.3.2). In other words, "physical information, being a natural property of the data, can be extracted instantly from the data, and any special rules for such task accomplishment are not needed" (Diamant, 2010a). Unfortunately, no well-grounded (well-posed) inductive learning-from-unlabeled-data approach exists (see Part I Section 2.1). For example, both unlabeled data clustering and (2-D) image segmentation algorithms are inherently ill-posed (see Part I Section 2.4.1). By adopting the Diamant terminology it is possible to state that detection of "discernable" data structures is not at all a physical problem of objective nature: it is rather a typical semantic problem of a qualitative (subjective) nature, where prior knowledge (provided by an external supervisor) must come into play to make the inherently ill-posed inductive learning-from-data problem better posed, although subjective (see Part I Section 2.1). This is tantamount to saying that the conceptual foundation of GEOBIA, i.e., the relationship between inherently ill-posed sub-symbolic (2-D) image segments and symbolic (3-D) landscape objects, remains affected by a lack of general consensus and research (Hay & Castilla, 2006) (see Part I Section 2.4.1.2).

To conclude, Diamant appears to have totally misunderstood one of two facts about the MAL-from-examples paradigm. These two facts hold true for MAL from unlabeled data and MAL from labeled data algorithms, respectively, as described below.

a.   MAL from unlabeled (unsupervised) data (see Part I Section 2.1 and Part I Section 2.4.1). Any machine learning from unlabeled data approach (e.g., unlabeled data clustering, image segmentation) is inherently ill-posed and requires prior knowledge to become better posed. It means that any attempt to extract non-semantic primary data structures (data objects), e.g., image segments and unlabeled data clusters, from an unlabeled data set (e.g., an image) without incorporation of high-level knowledge provided by an

Vision Goes Symbolic Without Loss of Information Within the Preattentive Vision Phase: The Need to Shift the Learning
Paradigm from Machine-Learning (from Examples) to Machine-Teaching (by Rules) at the First Stage of a Two-Stage
Hybrid Remote Sensing Image Understanding System, Part II: Novel Developments and Conclusions 115

external supervisor is a fatal misconception, committed by Diamant himself, stemming from the fallacies (inherent ill-posedness) of the MAL-from-examples paradigm.

b. MAL from labeled (supervised) data (see Part I Section 2.1 and Part I Section 2.4.2). It is true that, in Diamant's words, "knowledge about the rules that underpin (semantic) secondary (data) structures formation (from primary data structures considered as non-semantic and driven-without-knowledge) is a property of human observers (or their artificial counterparts) and not an inherent property of the data... (therefore) attempts to extract semantics from data are a fatal misconception stemming from the fallacies of the data-processing paradigm..." (Diamant, 2010a). This quote implies that no semantic information can be extracted from objective sensory data, but a correlation function can be established between semantic concepts and objective data for toy data understanding problems exclusively (refer to Part I Section 1 and Part I Section 2.1).

3.2 Comments on the Diamant image segmentation algorithm

In practical terms, the image segmentation algorithm proposed by Diamant can be subjected to the following criticisms.

- Not enough information is provided for the implementation to be reproduced. In practice the Diamant image segmentation algorithm cannot be duplicated and, therefore, cannot be tested by others.
- Diamant does not provide his image segmentation algorithm with QIs such as those listed in Part I Section 2.8. For example, based on Diamant's paper it is impossible to assess the following operational QIs.
  - Degree of automation. The following questions remain unanswered. What is the number of the image segmentation-free parameters to be user-defined? Have these user-defined parameters a physical meaning? What is their range of change?
  - Robustness to changes in input parameters to be user-defined.
  - Robustness to changes in the input data set acquired across time, space and sensors. In his paper (Diamant, 2005) Diamant applies his image segmentation algorithm to a single toy problem whose input data set consists of a panchromatic image 640×480 pixels in size. What about color images? What about satellite imagery? What about synthetic images of known visual properties?
  - Scalability. For example, does this image segmentation algorithm apply to data sets of different spatial scales, e.g., mosaics of hundreds of satellite images to generate classification maps at global scale where small but genuine image details (e.g., one pixel-wide roads) must be well preserved? I am afraid it does not... Does it apply to different sensors and users?
  - Efficiency in computation time and memory occupation.
  - Accuracy in terms of spatial quality of the segment boundaries (Baraldi et al., 2005; Persello & Bruzzone, 2010).

The conclusion is that based on existing literature the overall quality of the Diamant image segmentation algorithm remains unknown, which is often the case with the dozens of alternative image segmentation algorithms published in RS and CV literature each year (refer to Part I Section 2.4.1.2). Perhaps it is also due to these implementation shortcomings that so many researchers and practitioners ignored or criticized Diamant's methodological speculations.

- The Diamant image segmentation algorithm is not quantitatively compared (see Part I Section 2.8) against at least one alternative approach in a test image set consisting of both real and synthetic images (Baraldi et al., 2010c).
- The image segmentation algorithm proposed in (Diamant, 2005) is not technically sound.
  - In (Diamant, 2005) Diamant writes "segmentation/classification" and then "spatially connected regional groups (of pixels)" as "clusters" rather than segments, blobs or regions (see Part I Section 2.3). It is well known that (2-D) image segmentation, labeled (supervised) data classification and unlabeled (unsupervised) data clustering are completely different inductive learning-from-data problems (see Part I Section 2.4). Mixing these terms is a relevant conceptual mistake.
  - It is well known that image region extraction is the dual task of edge detection, in fact they are both inherently ill-posed inductive learning-from-unlabeled data problems (see Part I Section 2.4.1.2). In (Diamant, 2005), quite surprisingly Diamant acknowledges the ill-posedness of edge detection, but appears to ignore the inherent ill-posedness (subjective nature) of image region extraction acknowledged by a relevant portion of existing literature (see Part I Section 2.4.1.2). In fact, he states: "the efficiency of (my own) unsupervised top-down directed region-based (learning from unlabeled data) image segmentation is hard to disprove today" (Diamant, 2005). For example, by replacing pixels belonging to the same segment with their segment-based mean value (often called mean image), Diamant's image segmentation algorithm provides as output a piecewise constant approximation of the input image. Of course, researchers and practitioners interested in texture segmentation would find the Diamant piecewise constant image segmentation of little utility. In fact, the Diamant image segmentation algorithm incorporates no texture model. In practice, it detects texture elements (textons) rather than textures (made of textons) in the image. This accounts for the subjective nature of the image segmentation problem which is apparently ignored by Diamant.
- Breaking points and failure modes of the implemented algorithm are not documented in the paper.
- Conclusions are not properly supported by results contained in the manuscript. Indeed claims such as "the efficiency of (my own) unsupervised top-down directed region-based (learning from unlabeled data) image segmentation is hard to disprove today" (Diamant, 2005) are completely unjustified in both theoretical and practical terms (see previous comments).

To summarize, the Diamant image segmentation algorithm appears as "yet another image segmentation algorithm" (Baraldi et al., 2010a) based on heuristics whose superiority against alternative approaches is completely unproved. In other words, the image segmentation algorithm proposed by Diamant cannot be considered as adequate proof of his concepts (see Part I Section 2.2.3.2).

3.3 Comments on the Diamant contour detector

In practical terms, the contour detection algorithm proposed by Diamant can be subjected to the following criticisms.

Vision Goes Symbolic Without Loss of Information Within the Preattentive Vision Phase: The Need to Shift the Learning
Paradigm from Machine-Learning (from Examples) to Machine-Teaching (by Rules) at the First Stage of a Two-Stage
Hybrid Remote Sensing Image Understanding System, Part II: Novel Developments and Conclusions   117

- Status = Eq. (1-4), is nothing new, but a well-known isotropic zero dc-value mexican-hat operator for contrast detection (Canny, 1986; Burt, & Adelson,1983; Marr, 1982; Jain & Healey, 1998).
- Intensity information, I_{int} = Eq. (1-3), is another contrast value. However, it does not feature zero dc-value. This means the following.
  - Correlation between I_{int} = Eq. (1-3) and status = Eq. (1-4) can be relevant, i.e., I_{loc} = Eq. (1-2) = Eq. (1-3) × Eq. (1-5) is the product of two correlated contrast values where one-of-two is absolute valued.
  - Term I_{int} = Eq. (1-3) is not consistent with the psychophysical phenomenon of the Mach bands: where a luminance (radiance, intensity) ramp meets a plateau, there are spikes of brightness (perceived luminance), whereas there are none in the luminance profile. This is the sole case of continuity in the luminance profile capable of generating spikes of brightness (Baraldi & Parmiggiani, 1996a).
- The Diamant contour detection is single scale. On the contrary, it is known that the human visual system employs at least four spatial scales of analysis (Wilson & Bergen, 1979) (see Part I Section 2.3).
- The Diamant contour detector is not quantitatively compared (see Section 2.7) against at least one alternative approach in a test image set consisting of both real and synthetic images (Baraldi et al., 2010c).

To summarize, the Diamant contour detector appears to be neither new nor biologically plausible. It can be considered as "yet another contour detector" (Baraldi et al., 2010a) based on heuristics whose superiority against alternative approaches is completely unproved. In other words, the contour detector proposed by Diamant cannot be considered as adequate proof of his concepts (see Part I Section 2.2.3.2).

4. Revised/novel definitions of objective continuous sub-symbolic sensory data, continuous physical information, subjective discrete semi-symbolic data structure, discrete semantic-square (semantic^2) information and prior knowledge base

As a revision of Diamant's works (Diamant, 2005; Diamant, 2008; Diamant, 2010a; Diamant, 2010b), a new set of definitions of: (i) sub-symbolic objective primary data element in an objective sensory data set, (ii) semi-symbolic subjective secondary data structure, (iii) objective physical information, (iv) subjective semantic-square (semantic^2) information and (v) subjective prior knowledge base (ontology or model of the 3-D world) provided by an external subjective supervisor (human, God or equivalent machine).

4.1 Levels of aggregation of objective continuous sub-symbolic sensory data

There are five fine-to-coarse possible levels of aggregation of objective continuous sub-symbolic sensory data. These levels of aggregation are either sub-symbolic (non-semantic), semi-symbolic or symbolic. Semi-concepts are defined as stable concepts (percepts, classes of 3-D objects in the world) whose semantic meaning is adopted at the bottom level (layer 0) of an ontology (see Part I Section 2.2.2). The semantic information of semi-concepts (e.g., in a RS image, land cover semi-concepts are spectral categories such as *water or shadow, snow or ice, bare soil or built-up, vegetation*, etc.) is superior to that of objective data, whose semantic information is null, but equal or inferior (i.e., not superior) to that of concepts belonging to higher levels of abstraction (aggregation) in the ontology at hand (e.g., in a RS image classification taxonomy such as the International Global Biosphere Programme (IGBP) land cover classification scheme (FAO, 2000), target (3-D) land cover classes are *water bodies, snow or ice, barren, urban and built-up, needle-leaf forest, broad-leaf forest, mixed forest, shrubland, grassland, cropland,* etc.) (Baraldi et al., 2006; Baraldi et al., 2010a; Baraldi et al., 2010b; Baraldi et al., 2010c; Baraldi, 2011a; Baraldi, 2011b). An ontology is a hierarchical abstract representation (model) of the (3-D) world. For example, well-known examples of RS data classification taxonomies are the aforementioned IGBP land cover classification scheme (FAO, 2000), the Co-ordination of Information on the Environment (CORINE) (European Commission Joint Research Center, 2005), the U.S. Geological Survey (USGS) classification hierarchy (Lillesand & Kiefer, 1994) and the Food and Agriculture Organization of the United Nations (FAO) Land Cover Classification System (LCCS) (Di Gregorio & Jansen, 2000; Herold et al., 2006). An ontology can be modeled as a semantic network consisting of a hierarchical class taxonomy, represented as an inverted tree whose leaves are at the bottom layer 0, plus relationships between classes as arcs between nodes (refer to Part I Section 2.2.2).

The five fine-to-coarse possible levels of aggregation of objective sub-symbolic sensory data are listed below.

1. **An unlabeled objective continuous (quantitative) sub-symbolic (non-semantic) sensory** *scalar data element.* For example, a one-band pixel value in an image, a character in a vocabulary, etc. This is a scalar (simple, atomic, elementary, primitive) fact (measurement, sign, symbol, character, element) resulting from an observation (examination, inspection, monitoring, measurement) of the (3-D) world.

2. **An unlabeled objective continuous sub-symbolic** *primary data vector / primary data n-tuple / primary data element*, where $n \geq 1$ is the vector dimensionality. Each primary data n-tuple consists of $n \geq 1$ scalar data elements, e.g., a multi-spectral pixel value in an image, a word in a dictionary, etc. In the rest of this paper, if an unlabeled objective data set consisting of primary data elements is discrete and finite (e.g., an image as a 2-D data array), then its cardinality is identified as p (e.g., an image consists of p pixels). In this case primary data elements may be identified by integer numbers, e.g., a pixel is identified by a (row, column) coordinate pair in a (2-D) image domain. A set of sub-symbolic primary data elements (e.g., an image) can be described according to a given mathematical vocabulary/language. For example, a 2-D array of pixels (image) can be encoded as a 2-D spatial frequency function by means of a 2-D fast Fourier transform (FFT).

3. **A finite set** (e.g., a (2-D) image array) **of** p **unlabeled objective continuous sub-symbolic primary data elements** (e.g., pixels), with $p \in \{1, \infty\}$. To be described in physical terms, a set of objective sub-symbolic primary data elements requires a mathematical vocabulary/language, e.g., a 2-D FFT of a (2-D) image. This is related to the concept of continuous physical information in an objective sensory data set (refer to this text below).

4. **A labeled subjective discrete semi-symbolic** *secondary data structure / secondary data object.* It consists of one or more primary data elements of a given objective data set grouped together (based on any possible subjective aggregation criterion) and labeled

as one semi-symbolic secondary data structure. Each label belongs to a discrete and finite set of semi-concepts. The semantic meaning of semi-concepts (e.g., *vegetation*) is superior to zero (like that of unlabeled primary data elements) and not superior (i.e., equal or inferior) to that of concepts in the real (3-D) world. A discrete and finite quantitative data set consisting of p unlabeled objective primary data elements (e.g., a multi-spectral image consisting of p pixels, refer to point 3. above) always consists of a discrete and finite set of semi-symbolic secondary data structures whose cardinality is identified hereafter as s, such that inequality ($s \le p$) always holds. It is noteworthy that if equality ($s == p$) holds, this does not correspond to a trivial case since secondary data structures are semi-symbolic while primary data elements are sub-symbolic. To the best of this author's knowledge, it is at the level of subjective semi-symbolic secondary data structures that the view of the present author starts diverging from *all* existing CV algorithms and implementations, including GEOBIA-based RS-IUSs and Diamant's image segmentation and contour detection algorithms. This degree of novelty is consistent with well-known evidence collected in CV and MAL domains. For example:

- A large section of the scientific community acknowledges that detection of data structures in an unlabeled objective data set, such as the detection of unlabeled data clusters and unlabeled (2-D) image segments (see Part I Section 2.4.1), is an inherently subjective (which is tantamount to saying semantic, since words subjective and semantic are synonyms, refer to Part I Section 2.1) ill-posed problem, therefore difficult to solve, which requires prior (semantic) knowledge to become better posed (tractable) (refer to Part I Section 2.1).

- According to Marr, "vision goes symbolic immediately, right at the level of zero-crossing (primal sketch)... without loss of information" (Marr, 1982) (p. 343) (refer to Part I Section 2.3). Secondary semi-symbolic data structures (e.g., image segments labeled as *vegetation*) can be described (encoded) according to a given pair of one mathematical and one natural vocabulary/language to account for, respectively, their objective (quantitative) and subjective (semantic, qualitative) properties. For example, semi-symbolic image segments can be described by a segment description table whose columns consist of: (a) a segment-specific semantic label belonging to a discrete and finite set of semi-concepts (refer to this text above) and (b) segment-specific quantitative descriptors such as (Matsuyama & Shang-Shouq Hwang, 1990): (i) locational properties (e.g., minimum enclosing rectangle), (ii) photometric properties (e.g., mean, standard deviation, etc.), (iii) geometric/shape properties (e.g., area, perimeter, compactness, straightness of boundaries, elongatedness, rectangularity, number of vertices, etc.), (iv) texture properties, (v) morphological properties, (vi) spatial non-topological relationships between objects (e.g., distance, angle/orientation, etc.), (vii) spatial topological relationships between objects (e.g., adjacency, inclusion), etc. (Baraldi et al., 2010a).

In practice, the following definition holds.

Discrete semi-symbolic secondary data structure = Continuous sub-symbolic primary data element(s) + discrete semi-symbolic label belonging to a discrete and finite set of semi-concepts (e.g., in RS image understanding, possible semi-concepts are spectral categories equivalent to land cover class sets consisting of one or more land cover classes; examples of spectral categories are *vegetation, water or shadow, bare soil or built-*

up, etc. (Baraldi et al., 2006; Baraldi et al., 2010a; Baraldi et al., 2010b; Baraldi, 2011a; Baraldi, 2011b; Baraldi et al., 2010c)).

This also means that the set of discrete semi-symbolic secondary data structures incorporates the continuous objective sensory data set.

5. **A finite set of ($s \leq p$) labeled secondary subjective semi-symbolic data structures, which include the objective sensory data set** (refer to point 4. above), with $s \in \{1, p\}$. In this author's terminology, it is called *preliminary classification map* or *primal sketch*. These terms are:

    -   in line with the CV system proposed by Marr at the level of computational theory (see Part I Section 2.6) when he states: "vision goes symbolic almost immediately, right at the level of zero-crossings (primal sketch)... without loss of information" (Marr, 1982) (p. 343) (refer to Part I Section 2.3)

    -   In contrast with the CV system proposed by Marr at the level of algorithm design and implementation (see Part I Section 2.5), where the term primal sketch identifies the non-symbolic output of a zero-crossings algorithm, which is an instance of the unlabeled data learning class of image edge detectors/region extractors (Marr, 1982).

    It is noteworthy that in a (2-D) preliminary classification map domain, a labeled semi-symbolic segment may be defined as a spatially connected set of secondary semi-symbolic data structures featuring the same label, say, connected pixels featuring label *vegetation*. Therefore, in a (2-D) preliminary classification map domain, semi-symbolic pixels belong to semi-symbolic image segments which belong to semi-symbolic image strata (layers) defined as image-wide sets of semi-symbolic segments featuring the same semi-symbolic label. In other words, in the preliminary classification map domain, three spatial types co-exist: **semi-symbolic pixels in semi-symbolic image segments in semi-symbolic image strata**. This would end the bad-faith antagonism between unlabeled pixels versus labeled non-symbolic segments (e.g., segment 1, segment 2, etc.) which affects traditional pixel-based versus object-based RS-IUSs and CV systems (refer to Part I Table 1). A labeled subjective semi-symbolic quantitative data set can be described (encoded) according to a given pair of one mathematical and one natural vocabulary/language capable of accounting for both the quantitative and semantic (qualitative, subjective) nature of labeled subjective semi-symbolic secondary data structures (refer to point 4. above).

4.2 Continuous physical information

Continuous physical (quantitative, objective, sensory) information. This is a hierarchical (i.e., multi-scale, including one-scale as a special case) description (representation), namely, down-scale encoding (decomposition), up-scale decoding (reconstruction) or one-scale transcoding (from one data format to another at the same hierarchical level), of the physical objective data set based on a given mathematical non-natural vocabulary/language. This hierarchical description/ representation of the objective sensory data set can be either lossless or lossy, depending on the exact/non-exact reconstruction (decoding) of the original data set from its representation (encoding). For example, an FFT of a time-signal is a one-scale transcodification of the signal from the time to the frequency domain. A well-known example of down-scale encoding/up-scale decoding is the Gaussian-Laplacian image pyramid (Burt & Adelson, 1983). It means that physical information stems from the combination of an objective data set with a mathematical non-natural vocabulary/language. To summarize the concept of physical information, we can write the following definition.

Continuous objective data set + (arbitrary) multi-scale down-scale encoding, up-scale decoding or one-scale transcoding/description/data format = hierarchical physical information encompassing down-scale/ fine-to-coarse resolution/ compression/ encoding, up-scale/ coarse-to-fine resolution/ decompression/ decoding, and/or one-scale transcodification (from one data format to another at the same hierarchical level), either lossless or lossy.

4.3 Discrete semantic-square information

Discrete semantic-square (semantic[2]) (where semantic is a synonym of categorical, symbolic, subjective, abstract, qualitative, vague, but persistent, stable, see Part I Section 2.1) **information (concepts, percepts) stems from the semantic[2] labeling of an objective data set performed by an external subjective supervisor** (human, God or equivalent machine) **provided with a subjective hierarchical prior knowledge base** (ontology or model of the (3-D) world, equivalent to an inverted tree with leaves at the bottom level 0, see Part I Section 2.2.2). **Semantic[2] labeling occurs when a subjective supervisor (first source of subjectivity), provided with his/her own subjective ontology (second source of subjectivity), observes and scrutinizes the objective data set, consisting of p sub-symbolic primary data elements (refer to point 3. in Section 4.1), to achieve the following.**

    a.   At the bottom level 0 of the inverted tree (ontology, see Part I Section 2.2.2), a semi-symbolic label, belonging to a discrete and finite set of semi-concepts (e.g., in a RS image, spectral categories are *vegetation, water or shadow, bare soil or built-up*, etc.), is assigned to each sub-symbolic primary data element (e.g., each pixel in a RS image) of a set of p sub-symbolic primary data elements to form a finite and discrete set of s semi-symbolic secondary data elements, with $s \leq p$ (refer to point 5. in Section 4.1).

    b.   At hierarchical levels ≥ 1 of the inverted tree (see Part I Section 2.2.2), a hierarchical symbolic label is assigned to the set of s semi-symbolic secondary data elements based on symbolic reasoning (Matsuyama & Shang-Shouq Hwang, 1990).

This definition of semantic[2] labeling disagrees at the level of the aforementioned point a. with the traditional definition of semantic labeling provided by MAL, which encompasses existing CV systems (e.g., Diamant's (Diamant, 2005)) and RS-IUSs (e.g., (Definiens Imaging GmbH, 2004; Matsuyama & Shang-Shouq Hwang, 1990)). In fact, point a. above states that **semantic[2] information stems naturally (automatically, instantaneously) from the simultaneous interaction of three necessary and sufficient components.**

    i.   An objective sensory data set (consisting of facts, measures, etc.) described in terms of continuous physical information (representation, description) based on a mathematical vocabulary/language.

    ii.   A subjective supervisor/actor (human, God or equivalent machine). He/she acts as the first source of subjectivity in the labeling (mapping) process. To be considered as such, a supervisor must be the carrier of a prior semantic knowledge base (ontology). He/she acts as follows:

- observes the objective data set and
- interprets/scrutinizes the objective data set to match (label) data with his/her own ontology.

iii. A subjective hierarchical (multi-scale) prior ontology which exists before looking at the data. Since it deals with semantic information, a prior knowledge base is subjective by definition (since subjective and semantic are synonyms, refer to Part I Section 2.1). In practice, this ontology acts as the second source of subjectivity in the labeling (mapping) process. According to Diamant this hierarchical ontology is equivalent to a narrative story or tale which requires a natural language to comprise, in a top-down representation: the story title, index, sections, paragraphs, sentences and words. It is graphically represented and implemented as a semantic net or inverted tree whose leaves are at the bottom level 0 where physical information is incorporated (refer to this text above and Part I Section 2.2.2).

The aforementioned points i.-iii. imply that **objective sensory data,** *per se,* **do not possess any semantic[2] information, but physical information exclusively.** Rather, **semantic[2] information incorporates objective data as one-of-three components.** This also means that nobody should disagree with Diamant when he repeats over and over that sensory data do not possess semantic information, therefore semantic information cannot be *extracted* from sensory data (Diamant, 2010a). On the contrary, Diamant's statement should not be considered original at all because it has been perfectly acknowledged in philosophy for hundreds of years, as well as in psychophysical studies of perception (Matsuyama & Shang-Shouq Hwang, 1990) and MAL in the last 50 years (Cherkassky & Mulier, 2006). This concept is summarized below.

- Philosophy and psychophysical studies of perception. The statement that sensory data do not possess semantic information is tantamount to saying there is an information gap between physical information and semantic information, which is the well-known information gap between (sensory and varying) sensations and (vague, but stable) perceptions. In practice, "we are always seeing objects we have never seen before at the sensation level, while we perceive familiar objects everywhere at the perception level" (Matsuyama & Shang-Shouq Hwang, 1990) (see Part I Section 1 and Part I Section 2.2.2).
- MAL. In unlabeled data learning algorithms (e.g., unlabeled data clustering), no semantics is detected as output (e.g., unlabeled data cluster 1, unlabeled data cluster 2), see Fig. 1. In labeled data learning algorithms for classification applications (see Part I,Fig. 1), no semantic information is extracted from a finite set of training data pairs consisting of an (objective data vector, subjective discrete label), but a correlation function can be estimated between continuous sensory data and a discrete and finite set of subjective labels (refer to Part I Section 2.1 and Part I Section 2.4.2).

The foregoing comments also mean that Diamant is right, although vague, when he states that "semantics is a property of a human observer" (Diamant, 2010a). **To state this more precisely, since semantic[2] information naturally (automatically, instantaneously) stems from the interaction of three necessary and sufficient components i.-iii. (see above in this text), then semantic[2] information cannot be separated from any of its three components.** For example, let us think of a piano (symbolic data structure) whose objective presence (fact) requires the simultaneous presence of a subjective human actor (or equivalent machine) to generate whatever sound (semantic information). The sound (generated semantic information) is neither in the piano, nor in the piano player, nor in his/her prior knowledge of what a piano is all about, but in the instantaneous combination of these three factors. This also means that **semantic[2] information** quite obviously **changes with the objective data set, the subjective human supervisor and his/her own subjective ontology.** In particular (refer to this text above), **semantic[2] information means there are two subjective actors in the semantic labeling of objective sensory data, namely, the subjective external observer and scrutinizers (or equivalent machine) and his/her own ontology or semantic (abstract) model of the world.** In fact, it is well known that all humans do not adopt the same ontology and two humans who adopt the same ontology do not apply this ontology the same way through time in interpreting a given observation. For example, two players will never generate the same music when playing the same musical score on the same piano. Not even the same player will ever generate the same music when playing twice the same musical score on the same piano. To summarize these concepts we can write the following definition.

Objective sensory data set + subjective supervisor provided, as such, with a subjective prior hierarchical knowledge base (ontology) = hierarchical semantic[2] (subjective[2]) information, which includes physical information at the bottom level 0 of the inverted tree which deals with the semantic granularity of semi-concepts assigned to semi-symbolic secondary data structures.

4.4 Subjective hierarchical (multi-scale) prior knowledge base

Subjective hierarchical (multi-scale) prior knowledge base (ontology, model of the (3-D) world) equivalent to a semantic net or inverted tree with leaves at the bottom level 0 where physical information is incorporated. Refer to this text above.

4.5 Intelligence

Intelligence (cognition) is the system's ability to aggregate bottom-up (from-data-to-concepts) and disassemble top-down (from-concepts-to-data) semantic information (which incorporates physical information) across the hierarchical levels of a subjective prior knowledge base.

4.6 Information processing system

An information processing system, cognitive system or intelligent system transforms an input sensory data set into an output instantiation of a story in natural language whose hierarchical structure is provided by an ontology or inverted tree retained in the system's memory before looking at the sensory data.

To summarize, the aforementioned novel definitions sketch a RS-IUS where information goes symbolic during the pre-attentive vision phase to generate a semi-symbolic primal sketch (preliminary classification map). This is in line with the CV system proposed by Marr at the level of computational theory (see Part I Section 2.6) when he states: "vision goes symbolic almost immediately, right at the level of zero-crossings (primal sketch)" (Marr, 1982), p. 343 (see Part I Section 2.3). However, it differs from the CV system proposed by Marr at the level of primal sketch implementation (see Part I Section 2.6) consisting of a sub-symbolic zero-crossing algorithm (Marr, 1982). In addition, the novel RS-IUS sketched above differs at the level of both computational theory and algorithm design and implementation from existing CV systems such as GEOBIA systems (Definiens Imaging GmbH, 2004; Esch et al., 2008), including Diamant's (Diamant, 2005; Diamant, 2008; Diamant, 2010a; Diamant, 2010b), where an unlabeled data learning (driven-without-knowledge) algorithm is adopted at the first stage.

5. Practical consequences of the proposed definitions on CV, AI and MAL system design and implementation strategies

Practical consequences of the definitions proposed in Part II Section 4 on CV, AI and MAL system design and implementation strategies are several, more detailed, better posed and, therefore, far more relevant than Diamant's (Diamant, 2010a). Thus, they should benefit from more favorable consideration by the scientific community.

1.   Definitions provided in Part II Section 4 are consistent with the Marr statement: "vision goes symbolic almost immediately, right at the level of zero-crossings (primal sketch)... without loss of information" (Marr, 1982) (p. 343) (refer to Part I Section 2.3). This is tantamount to saying that exploitation of the deductive subjective prior knowledge-based inference paradigm must regard the preattentive visual phase whose output, the so-called *primal sketch* (Marr, 1982), must be as follows:
     * semantic in nature (see Part I Section 2.3), therefore it is called *preliminary classification map*;
     * capable of preserving small, but genuine image details (high spatial frequency image components). This requirement is inconsistent with existing image segmentation algorithms which are inherently affected by the *uncertainty principle* according to which, for any contextual (neighborhood) property, we cannot simultaneously measure that property while obtaining accurate localization (Corcoran & Winstanley, 2007; Petrou & Sevilla, 2006) (see Part I Section 2.4.1.2).
     Although he stated that vision goes symbolic right at the output of the preattentive vision phase, which has to affect the architectural level of understanding of a CV system (see Part I Section 2.6), Marr selected a sub-symbolic edge detection (zero-crossing) algorithmic for primal sketch generation (Marr, 1982). By embracing the Marr computational theory rather than his algorithmic solutions, the present author concludes that, as output, **the preattentive visual phase no longer generates sub-symbolic image primitives, namely, non-semantic points and edges or, vice versa, image regions (which is what was implemented by Marr (Marr, 1982)), but semi-symbolic secondary data structures, namely, semi-symbolic pixels in semi-symbolic segments in semi-symbolic strata** (see Part II Section 4) (Baraldi et al., 2006; Baraldi et al., 2010a; Baraldi et al., 2010b; Baraldi et al., 2010c; Baraldi, 2011a; Baraldi, 2011b).

2.   **It is impossible to *extract* semantic[2] information from objective continuous sensory data because the latter, *per se*, are provided with no semantics at all.** This is the well-known information gap between semantic[2] information and physical information (refer to Part I Section 2.2.2 and Part I Section 2.3).

3.   **Although it is impossible to *extract* semantic[2] information from objective continuous sensory data, it is possible to *correlate* discrete semantic[2] information to objective**

Vision Goes Symbolic Without Loss of Information Within the Preattentive Vision Phase: The Need to Shift the Learning
Paradigm from Machine-Learning (from Examples) to Machine-Teaching (by Rules) at the First Stage of a Two-Stage
Hybrid Remote Sensing Image Understanding System, Part II: Novel Developments and Conclusions   125

continuous sensory data. This conclusion is by no means novel as it is well known in literature. For example, Shunlin Liang summarizes this concept in a few words: statistical pattern recognition systems are based on correlation relationships between objective sensory (e.g., RS) data and either continuous (e.g., LAI) or categorical (e.g., land surface) variables (see Part I Section 2.1) (Shunlin Liang, 2004). **Unfortunately, low or no correlation can be found between continuous sensory data and a finite and discrete set of categorical variables, corresponding to independent random variables generating "distinguishable" data structures (data aggregations, data clusters) in real-world data mapping problems at large data scale or fine semantic granularity, other than toy problems at small data scale and coarse semantic granularity.** This low correlation effect is due to the combination of two factors.

- According to the *central limit theorem,* the distribution of the sample average of g independent and identically distributed (iid) random variables (corresponding to, say, g categorical variables) approaches the normal distribution, featuring no "distinguishable" data sub-structure, as the sample size g increases. In other words, the separability of "distinguishable" data structures in a given objective sensory data set belonging to a given measurement space is monotonically non-increasing with (i.e., it decreases with or remains equal to) the finite number of discrete semantic concepts involved with the cognitive (classification) problem at hand.

- Within-class variability (vice versa, inter-class separability) is monotonically non-decreasing (i.e., it increases or remains equal) (vice versa, non-increasing) with the magnitude of the sample set per categorical variable when this variable-specific sample set size is "large" according to large-sample statistics (although large sample is a synonym for 'asymptotic' rather than a reference to an actual sample magnitude, a sample set cardinality of 30÷50 samples per random variable is typically considered sufficiently large that, according to a special case of the central limit theorem, the distribution of many sample statistics becomes approximately normal). For example, in (Chengquan Huang et al., 2008), where an SVM training and classification model selection strategies are applied to every image in a RS image mosaic at global scale to separate forest from non-forest pixels, a so-called training data automation (TDA) procedure identifies a forest peak in a one-band first-order statistic (histogram) of a local image window. The size of this local image window must be fine-tuned based on heuristics because inter-class spectral separability between classes forest and non-forest (vice versa, within-class variability) decreases (vice versa, increases) monotonically with the local window size above a certain (empirical) threshold (minimum window size, below which the collected sample is not statistically significant).

4. As an extension of points 2. and 3. above, **unlabeled (unsupervised) data learning algorithms,** namely, driven-without-knowledge image segmentation algorithms and unlabeled data clustering algorithms (see Part I Section 2.4.1), **should be considered highly inappropriate** (like using a fork for cutting food: unless the food is particularly soft, it will never work) **when the objective sensory data acquisition occurs in the domain of real-world data mapping problems at large data scale or fine semantic granularity** (where the separability of "distinguishable" data structures in a given objective sensory data set belonging to a given measurement space is expected to be low), other than toy problems at small data scale and coarse semantic granularity.

Does this mean the relevant effort spent by the MAL community to develop driven-without-knowledge image segmentation algorithms (Castilla et al., 2008) or, say, self-organizing topology-preserving unlabeled data clustering algorithms (Fritzke, 1997; Martinetz & Schulten, 1994), has been worthless? Fortunately, not. It rather means the following.

i.   The main application domain of, say, self-organizing topology-preserving unlabeled data clustering algorithms should remain the modeling of stationary and non-stationary distributions, see Part I Fig. 1.

ii. When an unlabeled (unsupervised) data learning algorithm, either a driven-without-knowledge image segmentation algorithm or an unlabeled data clustering algorithm (see Part I Section 2.4.1), is adopted as the first stage of a two-stage hybrid cognitive system, CV system or RS-IUS, it should be considered highly inappropriate. In particular:

I   It should be replaced by a deductive MAT-by-rules approach where community-agreed prior knowledge is conveyed to generate as output a lossless semi-symbolic product (consisting of semi-concepts). For example, in a RS-IUS, the MAT-by-rules first stage should generate a preliminary classification map (see Part II Section 4) where small, but genuine image details are well preserved (refer to this text above).

II   If useful, it should be:

a.   adapted to work on a driven-by-knowledge stratified (semantic masked) basis and

c.   next, moved to the second stage of a two-stage stratified hierarchical hybrid cognitive system. For example, a two-stage stratified hierarchical hybrid RS-IUS architecture has been proposed in recent literature, see Fig. 3 (Baraldi et al., 2006; Baraldi et al., 2010a; Baraldi et al., 2010b; Baraldi et al., 2010c; Baraldi, 2011a; Baraldi, 2011b).

5.   As an extension of points 2. and 3. above, **labeled (supervised) data learning classifiers** (see Part I Section 2.4.2) **should be considered highly inappropriate** (like using a fork for cutting food; unless the food is particularly soft, it will never work) **in real-world data mapping problems at large data scale or fine semantic granularity** (where within-class variability is monotonically non-decreasing (i.e., it increases or remains equal) with the cardinality of the objective sensory data set), other than toy problems at small data scale and coarse semantic granularity. This conclusion is by no means novel. Rather, it is well known in literature. For example, Shunlin Liang summarizes this concept in few words: statistical model are usually site-specific (see Part I Section 2.1) (Shunlin Liang, 2004). Does this mean the relevant effort spent by the MAL community to develop supervised data learning classifiers has been worthless? Fortunately, no. It rather means the following.

i.   The main application domain of supervised data learning algorithms should be considered function regression where input and output variables are continuous non-semantic, see Fig. 1.

ii. When a supervised data learning classifier (see Part I Section 2.4.2) is adopted as the first stage of a two-stage hybrid cognitive system, CV system or RS-IUS, it should be considered highly inappropriate. An experimental proof of this concept is that supervised MAL algorithms (say, SVMs), either context-insensitive (e.g.,

Vision Goes Symbolic Without Loss of Information Within the Preattentive Vision Phase: The Need to Shift the Learning
Paradigm from Machine-Learning (from Examples) to Machine-Teaching (by Rules) at the First Stage of a Two-Stage
Hybrid Remote Sensing Image Understanding System, Part II: Novel Developments and Conclusions   127

pixel-based) or context-sensitive (Bruzzone & Carlin, 2006; Bruzzone & Persello, 2009; Persello & Bruzzone, 2010), considered successful in terms of operational QIs (refer to Part I Section 2.7.2) at local/regional scale, become impracticable in mapping RS image mosaics consisting of hundreds of images at national/continental/global scale (Chengquan Huang et al., 2008). In these real world problems the cost, timeliness, quality and availability of adequate reference (training) data sets derived from field sites, existing maps and tabular data are currently considered the most limiting factors on RS data product generation and validation (Gutman et al., 2004). In particular, the first-stage supervised data learning classifier of a two-stage hybrid RS-IUS should be:

I   replaced by a deductive MAT-by-rule approach where community-agreed prior knowledge is conveyed to generate a preliminary classification map (see Part II Section 4) where small, but genuine image details are well preserved (refer to this text above);

II   if useful, it should be:

a.   adapted to work on a driven-by-knowledge stratified (semantic masked) basis and

d.   next, moved to the second stage of a two-stage stratified hierarchical hybrid RS-IUS architecture proposed in recent literature, see Fig. 3 (Baraldi et al., 2006; Baraldi et al., 2010a; Baraldi et al., 2010b; Baraldi, 2011a; Baraldi, 2011b; Baraldi et al., 2010c).

6. SIAM™ as a proof of the efficacy of the required shift of learning paradigm from MAL-from-examples to MAT-by-rules at the first stage of two-stage hybrid RS-IUSs

To the best of this author's knowledge SIAM™ provides the first experimental proof of the efficacy of the required switch of learning paradigm from MAL-from-examples to MAT-by-rules at the first stage of a two-stage hybrid RS-IUS architecture (refer to Part II Section 2.3), see Table 4. SIAM™ is an operational (good-to-go, press-and-go, turnkey) software button (executable). In particular, SIAM™ is automatic, efficient, scalable, accurate and robust to changes in the input data acquired across time, space and sensors. For example, the automatic SIAM™ is consistent and accurate across sensors at the national/ continental/ global scale (refer to Part II Section 2.3) (Baraldi et al., 2006; Baraldi et al., 2010a; Baraldi et al., 2010b; Baraldi et al., 2010c; Baraldi, 2011a; Baraldi, 2011b), whereas semi-automatic inductive data learning neural network approaches, such as SVMs, require to be re-trained (supervised) image-wide (Chengquan Huang et al., 2008).

SIAM™ belongs to the family of physical models that follow the physical laws of the real (3-D) world to represent an abstract of the reality (see Part I Section 2.1) (Shunlin Liang, 2004). In particular, SIAM™ follows the physical laws of spaceborne optical imaging devices to provide a two-stage hybrid RS-IUS with a first-stage deductive prior knowledge-based inference mechanism. Unfortunately, it takes a long time for human experts to learn physical laws of the real (3-D) world and tune physical models based on human intuition, domain expertise and evidence from data observations (Mather, 1994; Shunlin Liang, 2004). For example, the development of the SIAM™ dates back to the year 2002 (Baraldi, 2011a).

Quality Indicators (QIs)	State-of-the-art RS-IUSs	SIAM™
Degree of automation: (a) number, physical meaning and range of variation of user-defined parameters, (b)collection of the required training data set, if any.	VL, L	VH (fully automatic, it cannot be surpassed)
Effectiveness : (a) semantic accuracy and (b) spatial accuracy.		VH
Semantic information level	Land cover class (e.g., *deciduous forest*)	
Efficiency: (a) computation time and (b) memory occupation.	VL, L in training (hours per images)	VH (5 m to 30 s per Landsat image in a laptop)
Robustness to changes in input image	VL (specific training per image)	VH
Robustness to changes in input parameters	VL	VH (it cannot be surpassed)
Scalability to changes in the sensor's specifications or user's needs.	VL	VH (it works with any existing spaceborne sensor)
Timeliness (from data acquisition to high-level product generation, increases with manpower and computing power).	VH (e.g., the collection of reference samples is a difficult and expensive task)	VL, i.e., timeliness is reduced to almost zero
Economy (inverse of costs increasing with manpower and computing power).	VL, L, high costs in manpower and also computing power	VH, i.e., costs in manpower and computing power are reduced to almost zero

Table 4. QIs of SIAM™ versus state-of-the-art RS-IUSs' (refer to Part I Section 2.8). Legend of fuzzy sets: Very low (VL), Low (L), Medium (M), High (H), Very High (VH). Legend of colors: Red-Bad, Blue-Average, Green-Good

Part I Section 2.2.2 reported the question: is human biology as irrelevant to AI research as bird biology is to aeronautical engineering? Actually, biological vision has always represented a fundamental source of inspiration for the CV community. While SIAM™ considers its degree of biological plausibility as a value added, straightforward imitation of biological vision solutions is not always possible. This is the reason why SIAM™ cannot be considered highly plausible in biological terms although it is very useful in practice. For example, SIAM™ cannot work with panchromatic imagery whereas the human visual system is perfectly able to interpret gray-tone images.

7. Conclusions

It is well known that semantic information is not in objective sensory data, which is tantamount to saying there is a well-known information gap between semantic[2] information and physical information. This conceptual work observes that semantic[2] information is naturally (automatically, instantaneously) generated by the simultaneous interaction of a subjective external supervisor who observes and scrutinizes an objective sensory data set based on his/her own subjective prior knowledge base (ontology, model of the 3-D world). Semantic[2] information resulting from this interaction takes the intermediate form of semi-symbolic secondary data structures that incorporate physical information at the bottom level (layer 0) of an ontology represented as an inverted tree.

A shift of learning paradigm from MAL-from-examples to MAT-by-rules in the first stage of two-stage hybrid RS-IUSs is recommended. Experimental proof of this concept is provided by the operational automatic SIAM™ recently proposed in RS literature.

Vision Goes Symbolic Without Loss of Information Within the Preattentive Vision Phase: The Need to Shift the Learning
Paradigm from Machine-Learning (from Examples) to Machine-Teaching (by Rules) at the First Stage of a Two-Stage
Hybrid Remote Sensing Image Understanding System, Part II: Novel Developments and Conclusions   129

The practical conclusion of this conceptual work is twofold.

1. In line with a relevant section of existing literature (Shunlin Liang, 2004), labeled (supervised) data learning classifiers (see Part I Section 2.4.2) should be considered highly inappropriate, being affected by low operational QIs (see Part I Section 2.8), in dealing with real-world data mapping problems at large data scale (e.g., RS image mapping at national/ continental/ global scale) or fine semantic granularity, except in the case of toy problems at small data scale and coarse semantic granularity (e.g., RS image mapping at coarse spatial resolution and local/regional scale). This awareness should be divulged among the RS, CV, AI and MAL communities.

2. Any inductive MAL-from-examples algorithm, whether labeled (supervised, e.g., SVMs) or unlabeled (e.g., image segmentation, unlabeled data clustering), whether context-insensitive (e.g., pixel-based) or context-sensitive (e.g., (2-D) object-based), employed as the first stage of a two-stage hybrid cognitive system, CV system or RS-IUS, should be:

   a. replaced by a deductive MAT-by-rules approach where community-agreed prior knowledge is conveyed and,

   b. if useful, adapted to work on a driven-by-knowledge stratified (semantic masked) basis and moved to the second stage of a two-stage stratified hierarchical hybrid cognitive system. For example, a two-stage stratified hierarchical hybrid RS-IUS architecture has been proposed in recent literature, see Fig. 3 (Baraldi et al., 2006; Baraldi et al., 2010a; Baraldi et al., 2010b; Baraldi et al., 2010c; Baraldi, 2011a; Baraldi, 2011b).

This required shift of the learning paradigm from MAL-from-examples to MAT-by-rules adopted in the first stage of a two-stage hybrid RS-IUS is similar in nature to previous conceptual shifts occurring between deductive coarse-to-fine (from symbolic concepts to sub-symbolic data) AI/MAI and inductive fine-to-coarse (from sub-symbolic data to symbolic concepts) Cybernetics/MAL, see Part I Section 2.2. What is novel about the proposed shift of the learning paradigm from MAL-from-examples to MAT-by-rules at the first stage of a two-stage hybrid RS-IUS is the following.

- Its aim is to accomplish the following fundamental observation by Marr: "vision goes symbolic almost immediately, right at the level of zero-crossings (primal sketch)... without loss of information" (Marr, 1982) (p. 343) (see Part I Section 1, Part I Section 2.2.2 and Part I Section 2.3), which means that exploitation of the deductive subjective prior knowledge-based inference paradigm must regard the preattentive visual phase whose output product, known as *primal sketch*, must be: (i) semantic in nature (in disagreement with the Marr algorithmic solution of zero-crossings), therefore it is called *preliminary classification map* (see Part II Section 4) and (ii) capable of preserving small, but genuine image details, unlike existing image segmentation algorithms affected by the *uncertainty principle* (Corcoran & Winstanley, 2007; Petrou & Sevilla, 2006) (see Part I Section 2.4.1.2).

- It comes together with a novel conceptual framework consisting of explicit definitions of: (i) sub-symbolic objective primary data element in an objective sensory data set, (ii) semi-symbolic subjective secondary data structure, (iii) objective physical information, (iv) subjective semantic[2] information and (v) subjective prior knowledge base (ontology or model of the 3-D world (Matsuyama & Shang-Shouq Hwang, 1990)) provided by an external subjective supervisor (human, God or equivalent machine), refer to Part II Section 4.

- It affects exclusively the inductive learning-from-data first stage of traditional two-stage hybrid CV systems (e.g., Marr's (Marr, 1982), Diamant's (Diamant, 2005; Diamant, 2008; Diamant, 2010a; Diamant, 2010b)) or RS-IUSs, whether or not this first stage is implemented as an inductive algorithm capable of learning from either unlabeled (unsupervised) or labeled (supervised) data, whether context-insensitive (e.g., pixel-based) or context-sensitive (e.g., (2-D) object-based). If useful, these inductive data learning algorithms may be adapted to run on a driven-by-knowledge stratified (semantic masked, layered) basis and moved to the second stage of a novel two-stage stratified hierarchical hybrid RS-IUS architecture proposed in recent literature, see Fig. 3 (Baraldi et al., 2006; Baraldi et al., 2010a; Baraldi et al., 2010b; Baraldi et al., 2010c; Baraldi, 2011a; Baraldi, 2011b).

- It comes together with a novel two-stage stratified hierarchical hybrid RS-IUS architecture employing a first-stage spectral rule-based preliminary classification algorithm based on prior spectral knowledge, see Fig. 3 (Baraldi et al., 2006; Baraldi et al., 2010a; Baraldi et al., 2010b; Baraldi et al., 2010c; Baraldi, 2011a; Baraldi, 2011b).

- It comes together with an operational (namely, automatic, efficient, accurate, robust, scalable, see Part I Section 2.8) Satellite Image Automatic Mapper™ (SIAM™) implementation (software executable), equivalent to an automatic (good-to-go, press-and-go, turnkey) software button, provided as an experimental proof of the efficacy of the required shift in learning paradigm from MAL-from-examples to MAT-by-rules at the first stage of a two-stage hybrid RS-IUS architecture, see Fig. 3 (Baraldi et al., 2006; Baraldi et al., 2010a; Baraldi et al., 2010b; Baraldi et al., 2010c; Baraldi, 2011a; Baraldi, 2011b).

To summarize, to the best of this author's knowledge this is the first time a novel computational theory (RS-IUS architecture) is supported by operational (good-to-go, press-and-go, turnkey) algorithmic and implementation solutions as proofs of concept. For example, this was not the case of the Marr (Marr, 1982) or the Diamant CV systems (Diamant, 2005; Diamant, 2008; Diamant, 2010a; Diamant, 2010b), whose computational theories (see Part I Section 2.6) are both inconsistent with algorithmic solutions adopted by their authors. As a consequence, these two CV systems become two more instances of the well-known class of two-stage segment-based hybrid CV systems, also termed GEOBIA systems, traditionally affected by a lack of general consensus and research (Hay & Castilla, 2006; Matsuyama & Shang-Shouq Hwang, 1990).

The proposed conclusions of potential interest to the RS, CV, AI and MAL communities are supported by unquestionable independent sources of evidence listed below.

- Since the late 1950s, the original ambitious goals of AI/MAI and Cybernetics/MAL have been fragmented into "practical" and "manageable" problems equivalent to "a family of relatively disconnected efforts" (Diamant, 2005; Diamant, 2008; Diamant, 2010a; Diamant, 2010b).

- It is well-known in literature that inductive learning-from-examples "is an inherently difficult (ill-posed) problem and its solution requires a priori knowledge in addition to data" (Cherkassky & Mulier, 2006) (p. 39) (see Part I Section 2.1). In practical contexts this means the following.
  - Unlabeled (unsupervised) data learning algorithms, namely, unlabeled data clustering (Backer & Jain, 1981; Baraldi & Alpaydin, 2002a; Baraldi & Alpaydin, 2002b; Cherkassky & Mulier, 2006; Fritzke, 1997) and unlabeled (2-D) image segmentation algorithms (Burr & Morrone, 1992; Corcoran et al., 2010; Corcoran & Winstanley, 2007; Delves et al., 1992; Hay & Castilla, 2006; Matsuyama & Shang-Shouq Hwang, 1990; Petrou & Sevilla, 2006; Vecera & Farah, 1997), are recognized as inherently ill-posed problems subjective in nature by a relevant portion of existing literature.
  - Labeled (supervised) data learning classifiers are unable to establish correlation relationships between objective sensory (e.g., RS) data and categorical variables (e.g., land cover classes) at large data scale or fine semantic granularity. For example, in (Chengquan Huang et al., 2008) a forest/non-forest one-class SVM battery of classifiers must be re-trained and re-selected for every image in an image mosaic at global scale. Vice versa, labeled data learning classifiers are exclusively suitable for finding correlation relationships between objective sensory data and categorical variables at small data scale and coarse semantic granularity (e.g., in RS data mapping problems at coarse spatial resolution and local/regional scale). In fact, in practical RS data applications where supervised data learning algorithms are employed at large spatial scale, fine spatial resolution or fine semantic granularity (Chengquan Huang et al., 2008), the cost, timeliness, quality and availability of adequate reference (training/testing) datasets derived from field sites, existing maps and tabular data have turned out to be the most limiting factors on RS data product generation and validation (Gutman et al., 2004).
- The prior knowledge-based SIAM™ is provided with unsurpassed operational QIs (see Part I Section 2.8) in the mapping of RS image mosaics at national/ continental/ global scale (e.g., refer to Table 4).

To the best of this author's knowledge, while the proposed practical conclusions of potential interest to the RS, CV, AI and MAL communities are supported by the aforementioned independent sources of evidence, these conclusions are not contradicted by any practical achievement gained by the RS, CV, AI and MAL communities in recent years. Thus, rather than being agreed or disagreed upon, these conclusions ought to be accepted by the scientific community unless proved otherwise when the increasing rate of collection of RS data of enhanced spatial, spectral and temporal quality will no longer outpace our capability of generating (rather than extracting) semantic[2] information from RS data provided, *per se*, with no semantics at all.

8. Acknowledgments

This material is partly based upon work supported by the National Aeronautics and Space Administration under Grant/Contract/Agreement No. NNX07AV19G issued through the Earth Science Division of the Science Mission Directorate. The research leading to these results has also received funding from the European Union Seventh Framework Programme

FP7/2007-2013 under grant agreement n° 263435. This author wishes to thank the Editorial Board for its competence and willingness to help.

9. References

Baatz, M.; Hoffmann, C.; Willhauck, G. Progressing from object-based to object-oriented image analysis. In Object-Based Image Analysis–Spatial Concepts for Knowledge-driven Remote Sensing Applications; Blaschke, T., Lang, S., Hay, G.J., Eds.; Springer-Verlag: New York, NY, 2008, Chapter 1.4, pp. 29- 42.

Backer, E. & Jain, A. K. (1981). A clustering performance measure based on fuzzy set decomposition. *IEEE Trans. Pattern Anal. Mach. Intell.*, Vol. PAMI-3, No. 1, pp. 66–75.

Baraldi, A. & Parmiggiani, F. (1996). Combined detection of intensity and chromatic contours in color images. *Optical Engineering*, Vol. 35, No. 5, pp. 1413-1439.

Baraldi, A. & Alpaydin, E. (2002a). Constructive feedforward ART clustering networks– Part I. *IEEE Trans. Neural Netw.*, Vol. 13, No. 3, pp. 645–661.

Baraldi, A. & Alpaydin, E. (2002b). Constructive feedforward ART clustering networks– Part II. *IEEE Trans. Neural Netw.*, Vol. 13, No. 3, pp. 662–677.

Baraldi, A.; Bruzzone, L. & Blonda, P. (2005). Quality assessment of classification and cluster maps without ground truth knowledge. *IEEE Trans. Geosci. Remote Sensing*, Vol. 43, No. 4, pp. 857-873.

Baraldi, A.; Puzzolo, V.; Blonda, P.; Bruzzone, L. & Tarantino, C. (2006). Automatic spectral rule-based preliminary mapping of calibrated Landsat TM and ETM+ images, *IEEE Trans. Geosci. Remote Sensing.* Vol. 44, No. 9, pp. 2563-2586.

Baraldi, A.; Durieux, L.; Simonetti, D.; Conchedda, G.; Holecz, F. & Blonda, P. (2010a). Automatic spectral rule-based preliminary classification of radiometrically calibrated SPOT-4/-5/IRS, AVHRR/MSG, AATSR, IKONOS/QuickBird/ OrbView/GeoEye and DMC/SPOT-1/-2 imagery – Part I: System design and implementation. *IEEE Trans. Geosci. Remote Sensing*, Vol. 48, No. 3, pp. 1299 - 1325.

Baraldi, A.; Durieux, L.; Simonetti, D.; Conchedda, G.; Holecz, F. & Blonda, P. (2010b). Automatic spectral rule-based preliminary classification of radiometrically calibrated SPOT-4/-5/IRS, AVHRR/MSG, AATSR, IKONOS/QuickBird/ OrbView/GeoEye and DMC/SPOT-1/-2 imagery – Part II: Classification accuracy assessment. *IEEE Trans. Geosci. Remote Sensing*, Vol. 48, No. 3, pp. 1326 - 1354.

Baraldi, A.; Wassenaar, T. & Kay, S. (2010c). Operational performance of an automatic preliminary spectral rule-based decision-tree classifier of spaceborne very high resolution optical images. *IEEE Trans. Geosci. Remote Sensing*, Vol. 48, No. 9, pp. pp. 3482 - 3502.

Baraldi, A. (2011a). Levels of understanding and degrees of novelty of a two-stage remote sensing image understanding system employing the Satellite Image Automatic Mapper™ (SIAM™) as its preliminary classification first stage. *IEEE Trans. Geosci. Remote Sensing*, submitted for consideration for publication, TGRS-2011-00241.

Baraldi, A. (2011b). Fuzzification of a crisp near real-time operational automatic spectral rule-based decision-tree preliminary classifier of multi-source multi-spectral

Vision Goes Symbolic Without Loss of Information Within the Preattentive Vision Phase: The Need to Shift the Learning
Paradigm from Machine-Learning (from Examples) to Machine-Teaching (by Rules) at the First Stage of a Two-Stage
Hybrid Remote Sensing Image Understanding System, Part II: Novel Developments and Conclusions   133

remotely-sensed images, *IEEE Trans. Geosci. Remote Sensing*, accepted for publication, July 2011.

Bruzzone, L. & Carlin, L. (2006). A multilevel context-based system for classification of very high spatial resolution images. IEEE *Trans. Geosci. Remote Sens.*, Vol. 44, No. 9, pp. 2587–2600.

Bruzzone, L. & Persello, C. (2009). A novel context-sensitive semisupervised SVM classifier robust to mislabeled training samples. *IEEE Trans. Geosci. Remote Sensing*, Vol. 47, No. 7, pp. 2142-2154.

Burr, D. C. & Morrone, M. C. (1992). A nonlinear model of feature detection, In: *Nonlinear Vision: Determination of Neural Receptive Fields, Functions, and Networks*, R. B. Pinter & N. Bahram, (Eds.), pp. 309-327, CRC Press, Boca Raton, Florida.

Burt, P. & Adelson, E. (1983). The Laplacian pyramid as a compact image code. *IEEE Trans. Communications*, Vol. COM-31, No. 4, pp. 532-540.

Canny, J. (1986). A computational approach to edge detection. *IEEE Trans. Pattern Anal. Machine Intell.*, Vol. 8, pp. 679-714.

Carson, C.; Belongie, S.; Greenspan, H. & Malik, J. (1997). Region-Based Image Querying, *Proc. Int'l Workshop Content-Based Access of Image and Video libraries*, San Juan , Puerto Rico, June 20, 1997.

Castilla, G., Hay, G. J. & Ruiz-Gallardo, J. R. (2008). Size-constrained Region Merging (SCRM): An automated delineation tool for assisted photointerpretation. *Photogramm. Eng. Remote Sens.*, Vol. 74, No. 4 (Apr. 2008), pp. 409-429.

Chengquan Huang; Kuan Song; Sunghee Kim; Townshend, J.; Davis, P.; Masek, J. G. & Goward, S. N. (2008). Use of a dark object concept and support vector machines to automate forest cover change analysis. *Remote Sensing of Environment*, Vol. 112, pp. 970–985.

Cherkassky , V. & Mulier, F. (2006). *Learning From Data: Concepts, Theory, and Methods*. Wiley, New York.D'Elia, S. (2009). European Space Agency (ESA), personal communication.

Cootes, T. F. & Taylor, C. J. (2004). *Statistical Models of Appearance for Computer Vision, Imaging Science and Biomedical Engineering*, University of Manchester, 17.04.2011, Available from: www.isbe.man.ac.uk/~bim/Models/app_model.ps.gz.

Corcoran, P. & Winstanley, A. (2007). Using texture to tackle the problem of scale in landcover classification, In: *Object-Based Image Analysis - Spatial concepts for knowledge-driven remote sensing applications*, T. Blaschke, S. Lang & G. Hay, (Eds.), pp. 113-132, Springer Lecture Notes in Geoinformation and Cartography.

Corcoran, P.; Winstanley, A. & Mooney, P. (2010). Segmentation performance evaluation for object-based remotely sensed image analysis. *Int. J. Remote Sensing*, Vol. 31, No. 3, pp. 617-645.

Definiens Imaging GmbH. (2004). *eCognition* User Guide 4.

Delves, L.; Wilkinson, R.; Oliver, C. & White, R. (1992). `Comparing the performance of SAR image segmentation algorithms,. *Int. J. Remote Sensing*, Vol. 13, No. 11, pp. 2121-2149.

Diamant, E. (2005). Searching for image information content, its discovery, extraction, and representation. *Journal of Electronic Imaging*, Vol. 14, No. 1, pp. 1-11.

Diamant, E. (2008). I'm Sorry to Say, But Your Understanding of Image Processing Fundamentals Is Absolutely Wrong, In: *Brain, Vision and AI*, C. Rossi, (Ed.), Chapter 5, pp. 95-110, In-Tech Publishing.

Diamant, E. (2010a). Not only a lack of a right definition: Arguments for a shift in information-processing paradigm. 17.04.2011, Available from: http://arxiv.org/ftp/arxiv/papers/1009/1009.0077.pdf

Diamant, E. (2010b). Machine Learning: When and Where the Horses Went Astray?, In: *Machine Learning*, Yagang Zhang, (Ed.), pp. 1-18, In-Tech Publishing. 17.04.2011, Available from: http://arxiv.org/abs/0911.1386

Di Gregorio, A. & Jansen, L. (2000). *Land Cover Classification System (LCCS): Classification concepts and user manual*, FAO Corporate Document Repository, 17.04.2011, Available from: http://www.fao.org/DOCREP/003/X0596E/X0596e00.htm

Esch, T.; Thiel, M.; Bock, M.; Roth, A. & Dech, S. (2008). Improvement of image segmentation accuracy based on multiscale optimization procedure. *IEEE Geosci. Remote Sensing Letters*, Vol. 5, No. 3 (Jul. 2008), pp. 463-467.

European Space Agency (ESA). (2008). GMES Observing the Earth, 17.04.2011, Available from: http://www.esa.int/esaLP/SEMOBSO4KKF_LPgmes_0.html

Forestry Department, Food and Agriculture Organization (FAO) of the United Nations (2000). *FRA 2000 - Forest Cover Mapping & Monitoring with NOAA-AVHRR & Other Coarse Spatial Resolution Sensors*, Rome, 2000.

Fritzke, B. (1997). *Some competitive learning methods* (Draft document), 17.04.2011, Available from: http://www.neuroinformatik.ruhr-unibochum.de/ini/VDM/research/gsn/DemoGNG.

GEO/CEOSS. (2008). A Quality Assurance Framework for Earth Observation, Version 2.0, 17.04.2011, Available from: http://calvalportal.ceos.org/CalValPortal/showQA4EO.do?section=qa4eoIntro

Global Monitoring for Environment and Security (GMES) (2011). 17.04.2011, Available from: http://www.gmes.info

Gouras, P. (1991). Color vision, In: *Principles of Neural Science*, E. Kandel & J. Schwartz, (Eds.), pp. 467-479, Appleton and Lange, Norwalk, Connecticut.

Group on Earth Observations (GEO). (2005). The Global Earth Observation System of Systems (GEOSS) 10-Year Implementation Plan, 17.04.2011, Available from: http://www.earthobservations.org/docs/10-Year%20Implementation%20Plan.pdf

Group on Earth Observations (GEO). (2008). GEO 2007-2009 Work Plan: Toward Convergence, 17.04.2011, Available from: http://earthobservations.org

Gutman, G. *et al.*, (Eds.). (2004). *Land Change Science*, Kluwer Academic Publishers, Dordrecht, The Netherlands.

Hay, G. J. & Castilla, G. (2006). Object-based image analysis: Strengths, weaknesses, opportunities and threats (SWOT), *Proc. 1st Int. Conf. Object-based Image Analysis* (OBIA), 2006. 17.04.2011, Available from: www.commission4.isprs.org/obia06/Papers/01_Opening%20Session/OBIA2006_Hay_Castilla.pdf

Herold, M.; Woodcock, C.; Di Gregorio, A.; Mayaux, P.; Belward, A. S.; Latham, J. & Schmullius, C. (2006). A joint initiative for harmonization and validation of land cover datasets. *IEEE Trans. Geosci. Remote Sensing*, Vol. 44, No. 7, pp. 1719–1727.

Vision Goes Symbolic Without Loss of Information Within the Preattentive Vision Phase: The Need to Shift the Learning
Paradigm from Machine-Learning (from Examples) to Machine-Teaching (by Rules) at the First Stage of a Two-Stage
Hybrid Remote Sensing Image Understanding System, Part II: Novel Developments and Conclusions   135

Jain, A. K.; Duin, R. & Mao, J. (2000). Statistical pattern recognition: A review. *IEEE Trans. Pattern. Anal. Machine Intell.*, Vol. 22, No. 1, pp. 4-37.

Jain, A. & Healey, G. (1998). A multiscale representation including opponent color features for texture recognition. *IEEE Trans. Image Proc.*, Vol. 7, No. 1, pp. 124-128.

Lillesand, T. & Kiefer, R. (1994). *Remote Sensing and Image Interpretation*. Wiley & Sons, New York.

Lindeberg, T. (1993). Detecting salient blob-like image structures and their scales with a scale-space primal sketch: A method for focus-of-attention. *Int. J. Computer Vision*, Vol. 11, No. 3, pp. 283-318.

Martinetz, T. & Schulten, K. (1994). Topology representing networks. *Neural Networks*, Vol. 7, No. 3, pp. 507–522.

Mason, C. & Kandel, E. R.. (1991). Central visual pathways, In: *Principles of Neural Science*, E. Kandel & J. Schwartz, (Eds.), pp. 420-439, Appleton and Lange, Norwalk, Connecticut.

Mather, P. (1994). *Computer Processing of Remotely-Sensed Images – An Introduction*. John Wiley & Sons, Chichester, GB.

Matsuyama, T. & Shang-Shouq Hwang, V. (1990). *SIGMA – A Knowledge-based Aerial Image Understanding System*. Plenum Press, New York.

Pekkarinen, A.; Reithmaier, L. & Strobl, P. (2009). Pan-european forest/non-forest mapping with Landsat ETM+ and CORINE Land Cover 2000 data. *ISPRS J. Photogrammetry & Remote Sens.*, Vol. 64, No. 2 (March 2009), pp. 171-183.

Persello, C. & Bruzzone, L. (2010). A novel protocol for accuracy assessment in classification of very high resolution imagery. *IEEE Trans. Geosci. Remote Sensing*, Vol. 48, No. 3, pp. 1232-1244.

Petrou, M. & Sevilla, P. (2006). *Image processing: Dealing with Texture*. John Wiley & Sons, Chichester, Great Britain.

Richter, R. (2006). Atmospheric / Topographic correction for satellite imagery – ATCOR-2/3 User Guide, Version 6.2. 17.04.2011, Available from: http://www.geog.umontreal.ca/donnees/geo6333/atcor23_manual.pdf

Shackelford, A. K. & Davis, C. H. (2003a). A hierarchical fuzzy classification approach for high-resolution multispectral data over urban areas. *IEEE Trans. Geosci. Remote Sensing*, Vol. 41, No. 9, pp. 1920–1932.

Shackelford, A. K. & Davis, C. H. (2003b). A combined fuzzy pixel-based and object-based approach for classification of high-resolution multispectral data over urban areas. *IEEE Trans. Geosci. Remote Sensing*, Vol. 41, No. 10, pp. 2354–2363.

Shunlin Liang. (2004). *Quantitative Remote Sensing of Land Surfaces*. J. Wiley and Sons, Hoboken, New Jersey.

Tapsall, B.; Milenov, P. & Tasdemir, K. (2010). Analysis of RapidEye imagery for annual land cover mapping as an aid to European Union (EU) Common agricultural policy, W. Wagner, B. Székely, (Eds.): *ISPRS TC VII Symposium – 100 Years ISPRS*, Vienna, Austria, July 5–7, 2010, IAPRS, Vol. XXXVIII, Part 7B. 17.04.2011, Available from: http://mars.jrc.it/mars/content/download/1648/8982/file/P4-5_Milenov _Tapsall_BG_RapidEye_Project_JRC_ver2.pdf

USGS & NASA. (2011). Web-enabled Landsat data (WELD) Project. 17.04.2011, Available from: http://landsat.usgs.gov/WELD.php

Vecera, S. P. & Farah, M. J. (1997). Is visual image segmentation a bottom-up or an interactive process?. *Perception & Psychophysics*, Vol. 59, pp. 1280–1296.

Wang, Z. & Bovik, A. C. (2002). A universal image quality index. *IEEE Signal Proc. Letters*, Vol. 9, No. 3, pp. 81-84.

Wilson, H. R. & Bergen, J. R. (1979). A four mechanism model for threshold spatial vision. *Vision Res.*, Vol. 19, pp. 19-32.

Yang, J. & Wang, R. S. (2007). Classified road detection from satellite images based on perceptual organization. *Int. J. Remote Sensing*, Vol. 28, No. 20, pp. 4653-4669.

Zamperoni, P. (1996). Plus ça va, moins ça va. *Pattern Recognition Letters*, Vol. 17, No. 7, (1996), pp. 671-677.

Convex Set Approaches for Material Quantification in Hyperspectral Imagery

Juan C. Valdiviezo-N and Gonzalo Urcid
Optics Department, INAOE
Mexico

1. Introduction

Emerging as the combination of optics and spectroscopy, the development of high resolution imaging spectrometers has allowed a new perspective for the monitoring, identification and quantification of natural resources in Earth's surface, that is known today as *hyperspectral remote sensing*. An imaging spectrometer is an instrument that images the energy reflected or scattered by an object in hundred of spectral bands at different portions of the electromagnetic spectrum. Although these devices have been developed for remote sensing purposes, their applications have substantially increased in the last years because of their capabilities in materials identification, being also used in biology, medicine and related areas (Huebshman et al, 2005). In contrast to multispectral devices where each imaged spectral band covers a wide spectral range, a hyperspectral sensor has a higher spectral resolution that usually is less than 10 nm; thus, the number of spectral bands captured by the sensors represents an important difference between both technologies. Once the hyperspectral data have been appropriately calibrated taking into account the illumination factors and the atmospheric effects, the spectral information registered at each pixel of the image allows a direct identification of any imaged object based on its spectrum.

When Earth observation is the application, a hyperspectral sensor usually presents a low spatial resolution caused by either the characteristics of the instrument or the flight altitude of the aerial platform, which causes that the spatial resolution decreases as the distance from the Earth increases. Considering such a sensor having a spatial resolution in the order of meters, the spectral reflectance captured in a single pixel of the image would be comprised by the mixed reflectance spectra of different materials or objects present in that physical area. Therefore, the image data will be formed by a number of pixels whose spectral information corresponds to the mixture of the constituent materials spectra. Many authors in the literature have proposed to represent these spectral mixtures as a linear combination of constituent materials spectra with their corresponding abundances (Boardman, 1993; Keshava, 2003; Winter, 1999). This model, frequently known as the *constrained linear mixing model* (CLMM), has been the basis for some autonomous techniques oriented toward the unsupervised identification of constituent materials from hyperspectral imagery, and can be considered as a convex set representation.

This chapter presents a general overview of the techniques based on a convex set representation that have been used to identify the constituent materials from a hyperspectral scene. Besides the presentation of some classical methods used for this purpose, we are going to emphasize a recently published technique whose properties are based on lattice algebra to approximate a minimum convex set. The organization of this chapter is as follows. In Section 2 the physical foundations concerning the hyperspectral imaging process, including data characteristics and their appropriate calibration will be presented. In Section 3 we will state the necessary mathematical background to understand fundamental concepts such as minimum convex sets, affine independence, and their relation with constituent materials in hyperspectral data. Section 4 will describe some classical as well as recent techniques to achieve the autonomous endmember determination process. Section 5 will start with a brief mathematical background on lattice algebra that is necessary to understand the endmember determination method that will be described later. The section will be complemented with the presentation of two canonical lattice associative memories whose geometrical properties are used to define a convex hull from hyperspectral data. In Section 6 we will provide two application examples to illustrate the autonomous identification of natural resources from two scenes registered, respectively, over the Gulf of Mexico, and the Belstville area in Maryland (USA). Thus, the endmember identification will be realized using lattice associative memories and another novel method known as *vertex component analysis* (VCA). Finally, in section 6 we will give some pertinent comments and conclusions of this chapter.

2. Hyperspectral imaging

The development of more sophisticated imaging technologies in combination with high resolution spectrometers has given place to a new perspective in remote sensing, in which it is possible to register simultaneously the spatial and spectral information of the energy reflected from Earth's surface. These instruments, known as imaging spectrometer systems, image the Sun radiance reflected from or emitted by materials on the surface, in hundred of narrow and contiguous spectral bands usually in the reflective solar portion of the spectrum (from 0.35 to 2.5 μm). In remote sensing terminology, the region from approximately 0.35 to 1.0 μm is known as the *visible/near infrared* (VNIR) and the range from 1.0 to 2.5 μm is known as the *short wavelength infrared* (SWIR). Therefore, the resulting hyperspectral data consist of an image cube conformed by a number of radiance images that can be used to estimate the reflectance spectra of the scene. Thus, the information contained in a single pixel of a hyperspectral image can be used to compare and identify any object based on its characteristic spectrum, at a specific location of the zone of interest.

2.1 Physical foundations

There are fundamental matter-energy interaction processes that constitute the basis of the information captured by spectrometer instruments. The electromagnetic radiation coming from the sun can be modified in its direction, intensity or polarization when reaching the Earth's surface. These radiation changes depend on the physical and chemical constitution of the materials comprising the surface, and can be classified as radiation *transmission, reflection, absorption* or *emission*. When an electromagnetic wave propagating in free space reaches the frontier of a different medium, one part of its energy can be transmitted through the material and the other part can be reflected by the surface. Thus, the portion of energy that has been transmitted can be absorbed by some molecules at certain frequencies, causing an increment of the energy in their electrons and a change in the energy level. After a short time in the excitation state, the electrons return to their original state producing an emission

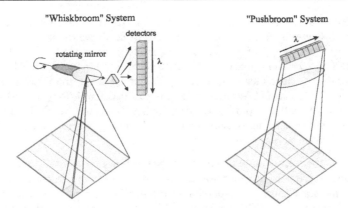

Fig. 1. Two types of scanning systems used to register a hyperspectral scene; the number of spectral bands are determined by the detectors that cover specific wavelength intervals λ.

of energy at lower frequencies. These interaction processes are used in spectroscopy for the characterization of materials in nature since they absorb or emit electromagnetic radiation at different wavelengths depending on their physical constitution. Hence, the materials covering the Earth's surface can be identified in hyperspectral data according to some absorption or emission bands present in the spectra recorded by the sensor.

For remote sensing purposes imaging spectrometers are placed onboard aerial platforms, mainly satellites or airplanes. Thus, the fundamental parts conforming hyperspectral remote sensing systems are: (1) optics to collect light, (2) a mechanism to scan the *instantaneous field of view* (IFOV) of the spectrometer over a scene, and (3) a set of spectrometers. The image acquisition process is as follows. A scanning mirror coupled to the mechanical system and the platform motion are used as part of the scanning process to collect the reflected energy coming from the surface. Furthermore, the scanning process of each line of the image can be realized using different systems. If the optics forms an image of a single point on the ground such that a line scanner scans a long line that is cross tracked to the platform motion, the scanner is called a "whiskbroom system". If the optics forms the image of a large slit such that no scan mechanism is needed other than the platform motion to form an image, the scanner is called a "pushbroom system" (see Fig. 1). Still another kind of systems use a linear variable filter over a two dimensional array of photodetectors (Jensen, 2007). After the collection of energy has been realized, the incoming light is then leaded through a set of spectrometers that splits the light into many narrow bands of energy by means of a dispersive element that can be either a grating or a prism. The energy coming from the dispersive elements is recorded by photodetectors whose sensibility responds to a specific wavelength interval, giving place to several image spectral bands.

2.1.1 Spatial and spectral resolution

In imaging spectrometers there are two basic characteristics that define the degree of resolution of the system. The *spatial resolution* is a measure of the minimum detail on the surface that can be captured for a given remote sensor. Thus, spatial resolution depends on the proper characteristics of the sensor and the flight altitude of the aerial platform. In particular, for a grating spectrograph hyperspectral imager, the spatial resolution is set by the size of the pixels of the *charge couple device* (CCD) camera in the y direction and the microscope system magnification. However, in the x direction, the resolution depends on the spectrometer slit width and the microscope system magnification (Huebshman et al, 2005). Let Δx and Δy be respectively the x and y dimensions of the CCD pixels and the magnification be M. Note that the slit width of the spectrometer w_s is always going to be larger than Δx. Then, the spatial resolution in the y direction is $2\Delta y/M$, while in the x direction is $2w_s/M$. Moreover, other common definition of spatial resolution relating the pixel size and the flight altitude refers to the physical area over the surface occupied by a single pixel. Clearly, the resolution increases as the altitude of the aerial platform decreases.

On the other hand, the *spectral resolution* refers to the number and bandwidth of spectral bands that a sensor can register. In fact, spectral resolution depends on spectrometer components which includes the slit width, the dispersion of the grating or prism, and the sensor device pixel size. For example, for a CCD pixel size of 10 square microns, the dispersion at normal operation is determined to be approximately 40 nm per mm or, equivalently, 0.4 nm per pixel.

2.2 Reflectance estimation from sensors

The light intercepted by the entrance aperture of a sensor is the quantity know as radiance. Given that the spectral reflectance is a physical quantity that is related to material properties, it is necessary to estimate the reflectance spectra[1] from radiance information captured in hyperspectral data. For this purpose, the background energy level of the Sun must be removed and the scattering and absorbing effects of the atmosphere must be compensated for. There are three main techniques that can be used in order to estimate the spectral reflectance, which can be considered as being either an image, empirical, or model based approach. An image based approach uses only data measured by the instrument, requiring that the images include regions of relatively uniform reflectance. Thus, any absorption presented in the measured reflectance of these regions will be related with one of the mentioned effects and therefore, such effects can be compensated for the complete image. Dividing each image spectrum by the flat field spectrum, the scene is converted to relative reflectance. On the other hand, empirical methods employ both remotely sensed data and field measurements of reflectance, denoted by $r(\lambda)$, to solve a linear equation of at-sensor radiance, such that,

$$L(\lambda) = br(\lambda) + c, \tag{1}$$

where $L(\lambda)$ is the radiance captured by the sensor that varies with wavelength λ, and b, c represent, respectively, multiplicative and additive terms that adjust the sensor radiance.

Model based approaches seek to represent all factors involved in the radiance acquired at a pixel by pixel basis including atmospheric perturbations. For this purpose, a simulated solar irradiance spectrum is used, then the method estimates the solar radiance in the day and hour of image acquisition and the absorption and scattering effects of the atmosphere. Hence, the solar radiance impinging on sensor L_s as a function of wavelength λ can be modeled as

$$L_s(\lambda) = \frac{1}{\pi}(Er(\lambda) + M_T)\tau_\theta + L_p, \tag{2}$$

---

[1] Recall that reflectance is defined as the ratio of the energy reflected from a material to the incident light falling on it.

where E is the irradiance on the Earth's surface, $r(\lambda)$ is the reflectance of the surface, M_T is the spectral radiant exitance at temperature T, τ_θ is the transmissivity of the atmosphere at zenith angle θ and L_p is the spectral path radiance of the atmosphere (Farrand, 2005). Solving Eq. (2) gives accurate results in reflectance estimation since it includes all factors contributing to the image acquisition process. Model based approaches are also employed to estimate atmospheric properties directly from the hyperspectral data.

2.3 Imaging spectrometers

One of the first hyperspectral instruments placed onboard an aircraft for Earth observation is the *Airborne, Visible and Infrared Imaging Spectrometer* (AVIRIS). The sensor was developed at NASA's Jet Propulsion Laboratory and it is composed by a whiskbroom scanning mirror and a linear array of 224 silicon and indium-antimonide sensors. The fine spectral resolution of the instrument, around 10 nm, allows to acquire 224 spectral bands in the spectral range from 0.4 to 2.5 μm. When the sensor is placed onboard the ER-2 aircraft, flying at an altitude of 20 km above ground level, the spatial resolution of the sensor is around 400 m^2, having a 30° total field of view, and an IFOV of 1.0 mrad. However, if the instrument is placed on an aircraft flying at an altitude of 4 km over the sea level, the spatial resolution of the sensor is about 16 m^2.

Furthermore, CHRIS is a current European imaging spectrometer that is operating in its ninth year. The instrument has a spatial resolution of 17 m in up to 62 bands. The data captured by the sensor is serving in more than 50 countries to support a wide range of applications, such as, land surface and coastal zone monitoring. Other imaging spectrometers that are in use today are the *hyperspectral digital imagery collection experiment* (HYDICE) and the image spectrometers belonging to SpecTir (SpecTir, 2009).

Besides the current hyperspectral sensors, three missions are planned to work within the next five years. Italy's ASI space agency plans to launch a medium resolution hyperspectral imaging mission, known as Prisma, in 2012. The instrument will combine a hyperspectral sensor with a panchromatic medium resolution camera, being able to acquire 235 spectral bands in the VNIR and SWIR. The German Aerospace Center (DLR) and the German Research Centre for Geosciences (GFZ) are planning to launch the EnMAP hyperspectral satellite in 2014; the sensor is designed to register Earth's surface in over 200 narrow color bands at the same time. In 2015, NASA plans to launch the *Hyperspectral Infrared Imager*, known as HyspIRI. The HyspIRI mission includes two instruments mounted on a satellite in Low Earth Orbit. The first, an imaging spectrometer, will measure from the visible to short wavelength infrared at a resolution of 10 nm. Also, a multispectral sensor will cover from 3 to 12 μm in the mid and thermal infrared. Both instruments have a spatial resolution of 60 m at nadir. Thus, HyspIRI will acquire 210 spectral bands, whose data will be used to study the world's ecosystems and provide critical information on natural disasters, such as, the processes that indicate volcanic eruption, the nutrients and water status of vegetation, deforestation, among others (Esa, 2010).

3. Mathematical background

In this section, a general mathematical background is given for several endmember search techniques briefly described in the next section. Many of these techniques developed and used for the unsupervised classification of materials in hyperspectral data have been based on convex sets theory; hence, it is necessary to define some important concepts such as minimum convex sets and endmembers together with its geometrical representation in multidimensional spaces. In the following definitions, we assume that a finite set X of n-dimensional vectors with real entries is given. Thus, using column notation we can denote this set as $X = \{x^1, \ldots, x^k\} \in \mathbb{R}^n$ where k is the number of vectors.

3.1 Convex sets and affine independence

In the theory of convex sets, a set of vectors $X = \{x^1, \ldots, x^k\} \subset \mathbb{R}^n$, also considered as points, is said to be *convex* if a straight line joining any two points resides within the set X (Lay, 2007). Being $\{a_\zeta\} \subset \mathbb{R}$ a set of scalars for all $\zeta \in K = \{1, \ldots, k\}$, a *linear combination* of vectors in X is an expression of the form $\sum_{\zeta=1}^{k} a_\zeta x^\zeta$. Then, X is said to be a linearly independent set if the unique solution to the equation $\sum_{\zeta=1}^{k} a_\zeta x^\zeta = 0$ is given by $a_\zeta = 0$ for $\zeta \in K$. Otherwise, the vectors in X are said to be linearly dependent. Furthermore, from a geometrical point of view, an *affine combination* is a linear combination of X subject to the condition $\sum_{\zeta=1}^{k} a_\zeta = 1$. If, in addition to the preceding condition, we require that $a_\zeta \geq 0 \; \forall \zeta \in K$ then the set is called a *convex combination* of vectors. The set of all convex combinations formed with elements of X is known as the *convex hull* of X, denoted as $C(X)$.

The notion of affine independence is of fundamental importance in the theory of convex sets and is defined as follows. Let $K_\eta = K \setminus \{\eta\}$ denote the index set from which index η has been deleted. If the set of vector differences, $X' = \{x^\zeta - x_\eta : \zeta \in K_\eta\}$ is linearly independent for some $\eta \in K$, it can be shown that X' is a linearly independent set $\forall \eta \in K$. Therefore, the set $X = \{x^1, \ldots, x^k\} \subset \mathbb{R}^n$ is said to be *affine independent* if and only if the set $X' = \{x^\zeta - x_\eta : \zeta \in K_\eta\} \subset \mathbb{R}^n$ is a linearly independent set for some $\eta \in K$ (Gallier, 2001). Notice that the vectors $x^1, \ldots, x^k$ are *affinely independent* if the unique solution to the simultaneous equations $\sum_{\zeta=1}^{k} a_\zeta x^\zeta = 0$ and $\sum_{\zeta=1}^{k} a_\zeta = 0$ is given by $a_\zeta = 0$ for all $\zeta \in K$. Hence, linear independence implies affine independence. It follows from this definition that any two distinct points are affinely independent, any three non-collinear points are affinely independent, and in general any m points in $\mathbb{R}^n$, with $m \leq n + 1$ are affinely independent if and only if they are not points of a common $(m - 2)$-dimensional linear subspace of $\mathbb{R}^n$. The convex hull of affinely independent points form a simplex that is the minimum convex set formed by $n + 1$ vertices. In particular, if X is affinely independent, then $C(X)$ is an m-dimensional simplex or m-simplex. Thus, a 0-simplex is simply a point, a 1-simplex is a line segment determined by two affinely independent points, a 2-simplex is a triangle determined by three affinely independent points, while a 3-simplex is a tetrahedron defined by four affinely independent points.

3.2 The constrained linear mixing model

As discussed in the previous section, a noticeable characteristic of hyperspectral images is that most of the pixels contain mixtures of the spectra of constituent materials in the scene. According to the physical interaction of light with matter, it is possible to represent such mixtures using a non-linear model if we consider that photons contribute with each molecule separately. However, in this representation the estimation of the proportions of each constituent material could be a difficult task. A more practical representation, known as the *constrained linear mixing model* (CLMM), has been used to represent the spectral mixtures at a pixel basis in hyperspectral data; the CLMM model has shown to be a good approximation for the abundance estimation of constituent materials when dealing with spectral mixtures, and it is mathematically expressed by,

$$x = \sum_{i=1}^{p} a_i s^i + r = Sa + r \tag{3}$$

$$a_i \geq 0 \; \forall i \quad \text{and} \quad \sum_{i=1}^{p} a_i = 1, \tag{4}$$

where $x \in \mathbb{R}^n$ is a spectral pixel acquired over n bands, $S = \{s^1, s^2, \ldots, s^p\}$ is an $n \times p$ matrix whose columns are the spectra of constituent materials (also known as endmembers), $a=(a_1, a_2, \ldots, a_p)^t$ is a p-dimensional vector of corresponding fractional abundances present in x and r is a noise vector (Keshava, 2003). The CLMM requires the set S of p endmembers be linearly independent and, in general, that the number of endmembers be much less than the dimensionality of the data pixel spectra ($p \ll n$).

In a geometrical representation, the CLMM described above can also be thought as a minimum convex set enclosing most of the hyperspectral data, where the p pure pixels spectra are the vertices of the corresponding simplex (see Fig. 2). Moreover, because of the spatial position of pure pixels, these vertices are technically known as *endmembers*. This way, any other spectral pixel of the image belongs to this convex set and can be completely represented by those endmembers. The last statement is the cornerstone of the geometrical based approach so frequently used to extract the constituent materials spectra from hyperspectral data. Furthermore, the estimation of fractional abundances for each endmember can be performed through the inversion of Eq. (3) subject to the imposed restrictions specified by Eq. (4). This process, known as *spectral unmixing* (or *demixing*), allows to quantify the proportion of each endmember in every image pixels. A simple and direct numerical method is provided, in the unconstrained case, by the least square estimation method expressed by

$$a = S^+ x = (S^t S)^{-1} S^t x, \tag{5}$$

where S^+ denotes de Moore-Penrose pseudoinverse matrix. This estimation exists when the S matrix is of full rank. The abundances that result from this estimation do not necessarily satisfy the constraints imposed in Eq.(4). Therefore, full additivity can be satisfied using the method of *Lagrange Multipliers*, while the non-negativity condition can be enforced by applying the *non-negative least squares numerical method* (Lawson & Hanson, 1974). It is also possible to employ a hybrid method in order to satisfy both constraints simultaneously.

4. Autonomous methods for endmember determination

Because the goal in the analysis of hyperspectral data is the quantification of materials comprising the scene, it is important to determine experimentally or even numerically the endmembers spectra. An experimental identification of these spectra implies the use of another device such as a spectroradiometer or a spectrometer to measure directly the reflectance spectra of materials belonging to the area under study; however this methodology is impractical in many situations because it requires an additional effort to collect samples from the zone of interest. A more practical methodology is to extract the same information as much as possible directly from the image data. In addition, assuming most pixels in the

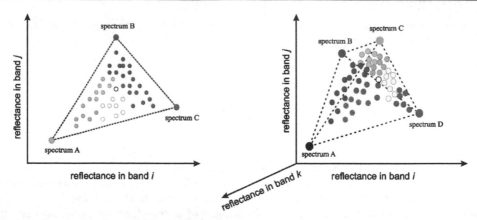

Fig. 2. Left: a 2-simplex whose vertices are three spectrally pure pixels in the image. Right: a 3-simplex defined by four spectrally pure pixels in the image defining a tetrahedron. Both simplex encloses all the spectral data.

image are conformed of spectral mixtures, then constituent materials are identified as those pixels having the spectrum of only one material. Based on this hypothesis, several authors have recently proposed and developed different methodologies used for the autonomous identification of spectrally pure pixels from the image itself. In this section we will make a review of some important techniques that have been applied for this purpose and whose methodology takes the constrained linear mixing model to represent the spectral mixtures at image pixels.

One of the earliest efforts for endmember extraction was proposed by Boardman and is known as *pixel purity index* (PPI) (Boardman, 1995). The algorithm is based on the geometry of convex sets to extract the vertices of a convex hull. Starting with a dimensionality reduction applied to the original data cube by using the minimum noise fraction transform, PPI generates a large number of random n-dimensional vectors, known as "skewers", through the dataset. Every pixel vector in the input data is projected onto each skewer, and its position is specified. The data that correspond to extreme points in the direction of a skewer are identified and placed on a list, indicating an increment in their pixel purity score. After many repeated projections, those pixels with a score above a certain threshold are determined as candidate "pure" pixels. From the resulting set of endmembers spectra, one can manually select those pixels that correspond to pure spectra. It is important to remark that the PPI algorithm was originally conceived as a guide to endmember determination since it requires to compare the determined spectra with those obtained from a spectral library in order to identify the final set of endmembers.

The *minimum volume transform* (MVT) algorithm, computes the minimum volume simplex enclosing the data (Craig, 1994). This proposal is based on the observation that scatter diagrams of multispectral remote sensing data tend to be triangular or pyramidal for the two or three band cases, respectively. Hence, they radiate away from the *dark-point*, which represents the sensor's response to an unilluminated object. Therefore, a minimum volume transform may be described as a non-orthogonal linear transformation of the multivariate data to new axes passing through the dark-point, and whose directions are chosen such that they embrace the data cloud. Thus, the determined MVT can be used to unmix images into new spatial variables showing the proportions of the different cover types present in the remotely sensed scene.

The NFIND-R algorithm is an *iterative simplex volume expansion* procedure that assumes the volume contained by an n-simplex whose vertices are specified by the purest pixels is always greater than any other volume formed by other combination of pixels (Winter, 1999). The input for the algorithm is the full data cube, which after subsequent projection is reduced in dimension. The selection of these vertices is initially realized by a random selection of a set of q vectors as endmembers candidates and then computing the volume of the simplex formed by these initial endmembers. The process continues iteratively by replacing every endmember one at a time with a pixel in the image and computing the respective volume. Hence, the pixel purity likelihood is evaluated by calculating the volume for every pixel in the place of each endmember. If the replacement results in a volume increase, then the pixel replaces the corresponding endmember. The procedure is repeated until there is no more replacement of endmembers; hence, the final spectra are considered as pure pixels and can be used as endmembers to estimate their corresponding abundances. It is important to remark that the accuracy in the method depends on the initial selection of endmembers.

The algorithm termed as *vertex component analysis* (VCA), is an unsupervised technique that relies on singular value decomposition and principal component analysis as subprocedures assuming the existence of pure pixels (Nascimento & Bioucas-Dias, 2005). In particular, VCA exploits the fact that endmembers are vertices of a simplex and that the affine transformation of a simplex is also a simplex. This algorithm iteratively projects data onto a direction orthogonal to the subspace spanned by the endmembers already determined. The new endmember spectrum is the extreme of the projection and the main loop continues until all given endmembers are exhausted.

The *minimum volume enclosing symplex* (MVES) algorithm is an autonomous technique supported on a linear programming solver that does not require the existence of pure pixels in the hyperspectral data (Chan et al, 2009). For the case when there exist pure pixels, the MVES technique leads to unique identification of endmembers. In particular, dimension reduction is accomplished by affine set fitting and Craig's unmixing criterion (Craig, 1994) is applied to formulate hyperspectral unmixing as an MVES optimization problem. The algorithm first determines the affine parameters set, solves by linear programming an initial feasibility problem with linear convex constraints, and iteratively optimizes two linear programming problems with nonconvex objective functions. Notice that the algorithm requires knowing in advance the number of endmembers to be found.

5. Lattice based approach for endmember extraction

5.1 Lattice algebra operations

The use of lattice algebra for science and engineering applications in which the usual matrix operations of addition and multiplication are replaced by corresponding lattice operations, has increased in the last years. These ideas have been applied in diverse areas, such as pattern recognition (Ritter et al, 1998), associative memories in image processing (Ritter et al, 2003; Ritter & Gader, 2006; Urcid & Valdiviezo, 2009), computational intelligence (Graña, 2008), industrial applications modeling and knowledge representation (Kaburlasos & Ritter, 2007), and hyperspectral image segmentation (Graña et al, 2009; Ritter et al, 2009; Ritter & Urcid, 2010; Valdiviezo & Urcid, 2007).

The basic numerical operations of taking the maximum or minimum of two numbers, denoted as functions $\max(x, y)$ and $\min(x, y)$, will be written as binary operators using the "join" and "meet" symbols employed in lattice theory, i.e., $x \vee y = \max(x, y)$ and $x \wedge y = \min(x, y)$. We use lattice matrix operations that are defined componentwise using the underlying structure of $\mathbb{R}_{-\infty}$ or $\mathbb{R}_{\infty}$ as semirings. For example, the maximum of two matrices X, Y of the same size $m \times n$ is defined as $(X \vee Y)_{ij} = x_{ij} \vee y_{ij}$ for $i = 1, \ldots, m$ and $j = 1, \ldots, n$. Inequalities between matrices are also verified componentwise, for example, $X \leq Y$ if and only if $x_{ij} \leq y_{ij}$. Also, the *conjugate matrix* X^* is defined as $-X^t$ where X^t denotes usual matrix transposition. Given an $m \times p$ matrix X and a $p \times n$ matix Y with entries in $\mathbb{R}$, we define a pair of dual matrix operations named as the *max-sum* and the *min-sum* denoted, respectively by $X \boxtimes Y$ and $X \boxtimes Y$ and whose i, j-th entry for $i = 1, \ldots, m$ and $j = 1, \ldots, n$, respectively, is given by $(X \boxtimes Y)_{ij} = \bigvee_{k=1}^{p}(x_{ik} + y_{kj})$ and $(X \boxtimes Y)_{ij} = \bigwedge_{k=1}^{p}(x_{ik} + y_{kj})$. For $p = 1$ these lattice matrix operations reduce to the *outer sum* of two vectors $\mathbf{x} = (x_1, \ldots, x_n)^t \in \mathbb{R}^n$ and $\mathbf{y} = (y_1, \ldots, y_m)^t \in \mathbb{R}^m$, defined by the $m \times n$ matrix

$$\mathbf{y} \times \mathbf{x}^t = \begin{pmatrix} y_1 + x_1 & \cdots & y_1 + x_n \\ \vdots & \ddots & \vdots \\ y_m + x_1 & \cdots & y_m + x_n \end{pmatrix}. \tag{6}$$

5.2 Lattice associative memories

Lattice based operations have been applied for pattern recognition problems as the computational model for a novel class of neural networks that are used as associative memories (Ritter et al, 1998). In general, let $(\mathbf{x}^1, \mathbf{y}^1), \ldots, (\mathbf{x}^k, \mathbf{y}^k)$ be k vector pairs with $\mathbf{x}^\xi = (x_1^\xi, \ldots, x_n^\xi)^t \in \mathbb{R}^n$ and $\mathbf{y}^\xi = (y_1^\xi, \ldots, y_m^\xi)^t \in \mathbb{R}^m$ for $\xi \in K$. Given a set of vector associations $\{(\mathbf{x}^\xi, \mathbf{y}^\xi) : \xi \in K\}$ we define a pair of associated matrices (X, Y), where $X = (\mathbf{x}^1, \ldots, \mathbf{x}^k)$ and $Y = (\mathbf{y}^1, \ldots, \mathbf{y}^k)$, with an association given by $(\mathbf{x}^\xi, \mathbf{y}^\xi)$ for $\xi \in K$. Thus, X is of dimension $n \times k$ with i, j-th entry x_i^j and Y is of dimension $m \times k$ with i, j-th entry y_i^j. Two $m \times n$ *lattice associative memories* able to store k vectors such that, for $\xi = 1, \ldots, k$, the memory recalls $\mathbf{y}^\xi$ when is presented the vector $\mathbf{x}^\xi$ are defined as follows: the *min-memory* W_{XY} and the *max-memory* M_{XY}, both of size $m \times n$, that store a set of associations (X, Y) are given by the expressions

$$W_{XY} = \bigwedge_{\xi=1}^{k} [\mathbf{y}^\xi \times (-\mathbf{x}^\xi)^t] \quad ; \quad w_{ij} = \bigwedge_{\xi=1}^{k} (y_i^\xi - x_j^\xi), \tag{7}$$

$$M_{XY} = \bigvee_{\xi=1}^{k} [\mathbf{y}^\xi \times (-\mathbf{x}^\xi)^t] \quad ; \quad m_{ij} = \bigvee_{\xi=1}^{k} (y_i^\xi - x_j^\xi). \tag{8}$$

The left part of Eqs. (7) and (8) are in matrix form, while the expressions to the right correspond to the i, j-th entry of *min*-W and *max*-M memories, respectively. In this case the memories are named *lattice hetero-associative* memories (LHAMs); if $X = Y$, we have a *lattice auto-associative* memory (LAAM), the case used for endmember determination. Furthermore, the main diagonals of both memories, i.e., w_{ii} and m_{ii}, consist entirely of zeros. Since $Y = X$, $X \boxtimes X^* = (X^*)^* \boxtimes X^* = (X \boxtimes X^*)^*$, then $M = W^*$. Hence, the *min-* and *max-*memories are dual to each other in the sense of matrix conjugation and $m_{ij} = -w_{ji}$.

5.3 Endmember determination from LAAMs

For a given set of vectors $X = \{x^1, \ldots, x^k\} \in \mathbb{R}^n$ and the corresponding matrix memories W_{XX} and M_{XX} computed from X, rewritten as $W = \{w^1, \ldots, w^n\}$ and $M = \{m^1, \ldots, m^n\}$ to specify their column vectors, an n-dimensional convex hull enclosing most if not all of the vectors in the given space can be derived. The points defining the convex hull will correspond to the vertices of an n-simplex and can be extracted from the columns of W and M. An important fact of the column values of LAAMs is that the relationship with the set of original data X is not direct, for example, W usually has negative values by definition. Hence, an *additive scaling* is required to relate the column values with the data set X. Thus, two scaled matrices, denoted respectively as $\overline{W}$ and $\overline{M}$, are defined for all $i = 1, \ldots, n$ according to the following expressions,

$$\overline{w}^i = u_i + w^i \quad ; \quad u_i = \bigvee_{\xi=1}^{k} x_i^\xi \quad ; \quad u = \vee_{\xi=1}^{k} x^\xi, \tag{9}$$

$$\overline{m}^i = v_i + m^i \quad ; \quad v_i = \bigwedge_{\xi=1}^{k} x_i^\xi \quad ; \quad v = \wedge_{\xi=1}^{k} x^\xi, \tag{10}$$

where u and v denotes, respectively, the *maximum* and *minimum vector bounds* of X, and whose entries are defined for all $i = 1, \ldots, n$.

Once the columns of $\overline{W}$ and $\overline{M}$ have been scaled, a fundamental result from this method is that the set of points $\overline{M} \cup \overline{W} \cup \{u, v\}$, forms a *convex polytope* $\mathfrak{B}$ with $2(n+1)$ vertices that contains X. These points must satisfy the affine independence condition and any subset of them can be used as endmembers. As it was proven in (Ritter & Urcid, 2010), the following theorems establish sufficient conditions to extract two subsets, W' and M' of affine independent vectors from the columns of both $\overline{W}$ and $\overline{M}$. The first theorem provides four equivalent conditions that furnish a computationally simple test for the affine independence of the sets $\overline{W}$ and $\overline{M}$; the symbols $\overline{w}_i$ and $\overline{m}_i$ denote the i-th row of $\overline{W}$ and $\overline{M}$, respectively; also, $c = (c, \ldots, c)$ denotes a constant vector.

Theorem 1. If $i, j \in \{1, \ldots, n\}$, then the following statements are equivalent: (1) $\overline{w}_i - \overline{w}_j = c$, (2) $\overline{w}^i = \overline{w}^j$, (3) $\overline{m}_i - \overline{m}_j = c$, and (4) $\overline{m}^i = \overline{m}^j$.

An important consequence of the Theorem 1 is that to verify that $\overline{W}$ or $\overline{M}$ is affinely independent, all that one needs to do is to check that no two vectors of $\overline{W}$ or $\overline{M}$ are identical. The next theorem provides a simple method for deriving a set of affine independent vectors from $\overline{W}$ and $\overline{M}$. In this notation, J' denotes an arbitrary non-empty subset of J.

Theorem 2. $W' \subset \overline{W}$ is *affinely independent* if and only if $\overline{w}^i \neq \overline{w}^j$ for all distinct pairs $\{i, j\} \subset J'$. Similarly, $M' \subset \overline{M}$ is *affinely independent* if and only if $\overline{m}^i \neq \overline{m}^j$ for all distinct pairs $\{i, j\} \subset J'$.

In the next section we will use this method to derive affinely independent sets from $\overline{M} \cup \overline{W} \cup \{u, v\}$ as endmembers of particular data sets X.

6. Identification of endmembers: application examples

The validity in the convex set representation for endmembers identification, discussed in the previous sections, can be illustrated through experiments using real hyperspectral data sets. In fact, the aim of the application examples is to provide enough details in the use of a novel endmember determination technique. In particular, lattice auto-associative memories, W_{XX} and M_{XX}, have shown to be an efficient procedure for the autonomous endmember determination, from which a subset of final endmembers can be selected to accomplish hyperspectral image segmentation. As a complement to the theoretical results given before, the endmembers output set from the VCA algorithm will be presented and compared with that set obtained with the LAAMs method. At the end of this section, we present composite abundance maps generated from the estimation of endmember proportions using constrained linear unmixing on each hyperspectral scene.

The following data sets were taken from the SpecTir's extensive hyperspectral baseline environmental dataset (SpecTir, 2009). The available information about the image acquisition indicates that a VNIR-SWIR hyperspectral instrument, covering a wavelength range from 0.395 to 2.45 μm, was used to collect the images. Each scene has a spectral resolution of 5 nm, with a number of 360 spectral bands. Thus, a single hyperspectral cube is conformed by 600 lines $\times$ 320 pixels $\times$ 360 bands (about 132 Mbytes). Given the high spectral resolution of a hyperspectral image, a common practice to avoid redundant information consists in a spectral dimensionality reduction of the data cube by application of a chosen technique, such as principal component analysis, minimum noise fraction transform, or adjacent band removal of highly correlated bands (Keshava, 2003). These reductions are often necessary to eliminate undesirable effects produced during the acquisition process and to diminish computational requirements. Hence, in the hyperspectral cubes used for this simulation, the number of spectral bands was reduced to 90 by making a selection of spectral bands at subintervals of 20 nm covering the same wavelength interval. This spectral reduction allows to speed up the computation times with no significant effects in the endmembers identification task.

Example 1. Gulf of Mexico wetland sample

This hyperspectral cube was registered over the *Lower Suwanee National Wildlife Refuge*, which is located in the north coast of the Gulf of Mexico belonging to the USA. The Refuge lodges one of the largest undeveloped river-delta estuarine systems in this nation. Some of the numerous wildlife species that inhabit the zone are: swallow-tailed kites, bald eagles, West Indian manatees, Gulf sturgeon, whitetailed deer, and eastern wild turkeys. Natural salt marshes, tidal flats, bottomland hardwood swamps, and pine forests provide habitat for thousands of creatures. This particular hyperspectral data set was acquired at a spatial resolution of 4 m^2, covering tidal wetlands and multiple national wildlife reserves during the period of May to June 2010. In fact, the images captured the state of vegetation at the time of flights and can be used to locate the presence or absence of hydrocarbons at the surface of vegetation. Also, since the data were acquired prior to the oil disaster (occurred in 2010), they can be compared to images from later flights to assist in the damage assessments. Figure 3 shows two color composite images of the Gulf Coast hyperspectral scene. The left part of the figure

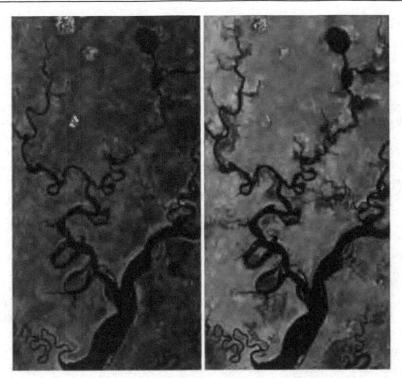

Fig. 3. Color images of the Gulf Coast hyperspectral scene used for example 1. Left: image formed by combining bands 54 (red), 34 (green) and 14 (blue). Right: combination of bands 81 (red), 217 (green), and 54 (blue).

was formed by combining bands 54 (red, $\lambda = 693$ nm), 34 (green, $\lambda = 584$ nm) and 14 (blue, $\lambda = 469.5$ nm), giving the appearance of a true color image; the right part was formed with bands 81 (red, $\lambda = 851$ nm), 217 (green, $\lambda = 1631$ nm) and 54 (blue, $\lambda = 693$ nm), whose combination using two infrared bands highlights the vegetation areas in green, orange and brown colors.

For the endmember determination process, we first form the set $X = \{x^1, \ldots, x^k\} \in \mathbb{R}^n$, where $k = 600 \times 320 = 192{,}000$ and $n = 90$, arranged with the total number of spectral vectors comprising the scene. The second step consists in the computation of the memories W_{XX} and M_{XX} from X, with Eqs. (7) and (8). Using the vectors $\mathbf{v}$ and $\mathbf{u}$ calculated from Eqs. (9) and (10), the columns of W and M are then scaled to obtain $\overline{W}$ and $\overline{M}$. In order to determine a subset of affinely independent vectors, it is necessary to prove that no two columns of $\overline{W}$ or $\overline{M}$ are equal. For the application in hand, the resulting $\overline{W}$ and $\overline{M}$ are conformed, respectively by 90 affinely independent columns and, therefore, each one provides us with 90 "candidate" endmembers. In addition, because of the additive scaling previously performed, the column vectors from $\overline{W}$ present an "upward spike" since $\overline{w}_{ii} = u_i$, and vectors from $\overline{M}$ presents a "downward spike" due to $\overline{m}_{ii} = v_i$. It is then necessary to realize a simple smoothing procedure considering the nearest one or two spectral samples next to $\overline{w}_{ii}$ or $\overline{m}_{ii}$, and is given, for any $i \in \{1, \ldots, n\}$, by

$$z_{ii} = \begin{cases} z_{1,2} & \Leftrightarrow i = 1, \\ \frac{1}{2}(z_{i-1,i} + z_{i+1,i}) & \Leftrightarrow 1 < i < n, \\ z_{n-1,n} & \Leftrightarrow i = n, \end{cases} \tag{11}$$

where z can be equal to $\overline{w}$ or $\overline{m}$. Notice that the LAAMs method always gives a number of candidate endmembers that is either equal or slightly less than the spectral dimensionality. In practice, contiguous columns are highly correlated being necessary to use some techniques to discard most of these potential endmembers. For example, minimum mutual information has been used to obtain a final set of endmembers (Graña et al, 2007); a matrix of linear correlation coefficients followed by a threshold process to get a subset of selected endmembers pairs with low correlation coefficients is introduced in (Ritter & Urcid, 2010). Here we use a simpler technique based on the fact that the LAAMs based method forms $\lfloor \sqrt{n+1} \rfloor$ subsets, each with $\lfloor \sqrt{n+1} \rfloor$ column vectors taken from W (respectively M); then, a representative from each group is selected as endmember. Although this technique provides a reasonable number of approximate true endmembers, in practical situations where a reduced number of materials comprises the hyperspectral scene, it is necessary to perform a final selection by considering those spectra that are spectrally different from the others. Therefore, in this application example, from the 20 endmembers candidates derived from $\overline{W} \cup \overline{M}$ a final selection of uncorrelated endmembers provided a reduced set containing 5 spectral vectors that forms the columns of S; thus, $S = \{\overline{w}^2, \overline{w}^{24}, \overline{w}^{43}, \overline{w}^{54}, \overline{w}^{79}\}$.

On the other hand, the VCA algorithm was applied to the same hyperspectral data set. According to the implementation, the algorithm requires as input parameter the number of endmembers to be determined; the corresponding output includes the endmembers spectra as well as the pixel positions in the image from they were extracted. Repeated iterations specifying the same number of endmembers produce almost the same output set, with differences in the order in which endmembers appear. After testing different input values, such as 5, 7, 9, and 10, we decided to use the number of endmembers determined with the LAAMs method as the input parameter to the VCA algorithm. Hence, the set S of endmembers identified with VCA is conformed by $S = \{x^{27876}, x^{90661}, x^{97850}, x^{84588}, x^{191634}\}$. Figure 4 displays three endmembers spectra determined from the columns selection from the set $\overline{W} \cup \overline{M}$, and whose spectral curves correspond to natural resources in the hyperspectral scene of the Gulf Coast. Similarly, Figure 5 shows three endmembers spectra obtained with application of the VCA algorithm. In both cases, normalization of reflectance data values in spectral distributions is linearly scaled from the range [0, 6000] to the unit interval [0,1]. Finally, observe that there is a similarity between spectral curves, which is indicated for curves drawn with the same colors.

Example 2. Belstville area.

The following example was performed using a hyperspectral cube registered over the Belstville area, located in northern Prince George's County in Maryland, USA. The area includes agriculture and vegetation samples. Similar to the previous image, the data cube used for this experiment is of size $600 \times 320 \times 90$, at approximately the same wavelength interval. The left part of Figure 6 displays a color composite image formed by combination of bands 54 (red, λ=693 nm), 34 (green, λ=583.9 nm), and 14 (blue, λ=469.5 nm) simulating

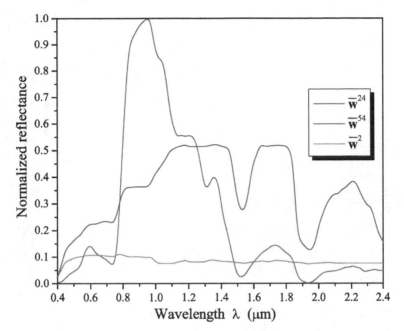

Fig. 4. Three endmembers spectra determined with application of the LAAMs method to the hyperspectral cube of the Gulf of Mexico. The associated column values selected from $\overline{W} \cup \overline{M}$ are: $\overline{w}^2, \overline{w}^{24}, \overline{w}^{54}$.

a true color image; the right part of the same Figure shows a combination of bands 81 (red, λ=851 nm), 54 (green, λ=693 nm), and 34 (blue, λ=583.9 nm) that allows to emphasize the vegetation areas in red tones. This way, the set of all spectral vectors of the image was formed by $X = \{x^1, \ldots, x^k\}$, where $k = 600 \times 320 = 192,000$ and $n = 90$. Following the same procedure described in the previous example for endmember determination, we compute the memories W_{XX} and $M_{XX} = -W_{XX'}^t$, as well as the vector bounds u and v used to obtain the matrices $\overline{W}$ and $\overline{M}$. Once the affine independence condition is checked, the resulting scaled memories are of size 90×90. According to the previous discussion, the spikes effects generated in the diagonal of both memories are removed using Eq. (11), and a selection of 20 endmembers candidates is made from the set $\overline{W} \cup \overline{M}$. Finally, the election of spectrally different column vectors is performed to form the final set of endmembers, whose column vectors are defined by $S = \{\overline{w}^{24}, \overline{w}^{37}, \overline{w}^{47}, \overline{w}^{64}, \overline{m}^{46}, \overline{m}^{57}, \overline{m}^{62}\}$. Therefore, the matrix S will be used in Eq. 3 to estimate the fractional abundance of each endmember.

The VCA algorithm was applied to the set X containing all the spectral vectors of the image. As the previous example, the number of column vectors selected from the LAAMs method was established as the input parameter of the algorithm. The resulting endmembers spectra determined by VCA were used to form the matrix $S = \{s^1, \ldots, s^7\}$, whose spectra are associated to the column vectors $\{x^{191845}, x^{191419}, x^{9630}, x^{111446}, x^{114301}, x^{191724}, x^{65969}\}$. Figure 7 displays four of the final endmembers set obtained with the selection of vectors from $\overline{W} \cup \overline{M}$; Figure 8 shows four of the endmembers spectra determined by the VCA algorithm that are similar to those computed with the LAAMs method. The similarity between spectral curves

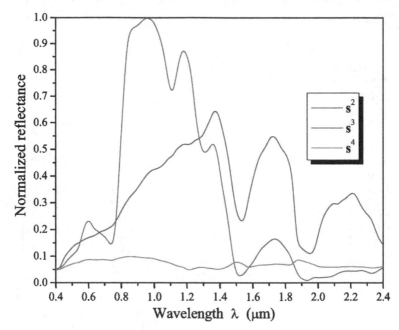

Fig. 5. Three endmembers spectra obtained with application of the VCA algorithm to the hyperspectral cube of the Gulf of Mexico. The column vectors s^j for $j = 1, \ldots, 5$ indicate the corresponding column of the S matrix.

can be identified for curves drawn in the same color. Although the spectral curves in both sets seem to be alike, a similarity measure must be applied in order to quantify these similarities. Here, we have computed the correlation coefficients between the sets obtained with the LAAMs method and the VCA algorithm, for each one of the application examples. Table 1 presents the correlation coefficients computed for the spectral curves displayed in Figures 4, 5 and 7, 8, respectively.

Gulf of Mexico		Beltsville	
VCA & LAAMs	Corr. Coef.	VCA & LAAMs	Corr. Coef.
s^2 and $\overline{w}^{24}$	0.980	s^1 and $\overline{w}^{37}$	0.965
s^3 and $\overline{w}^{54}$	0.974	s^2 and $\overline{w}^{24}$	0.944
s^4 and $\overline{w}^2$	0.641	s^3 and $\overline{w}^{64}$	0.912
—	—	s^4 and $\overline{w}^{47}$	0.939

Table 1. Correlation coefficients for similar spectra obtained with the LAAMs method and the VCA algorithm from the hyperspectral images of the Gulf of Mexico and Beltsville.

6.1 Constrained linear unmixing

The spectral unmixing process can be realized by means of the inversion expressed in Eq. (5), subject to the restrictions of full additivity and non-negativity of abundance coefficients. Notice that Eq. (3) is an *overdetermined* system of linear equations such that $n > p$. For the examples here discussed, both matrices $\overline{W}$ and $\overline{M}$ have full rank, thus their column vectors

Fig. 6. Color images of Beltsville hyperspectral scene used fot example 2. Left: image formed by combining bands 54 (red), 34 (green) and 14 (blue). Right: combination of bands 81 (red), 54 (green), and 34 (blue).

are linearly independent. In addition, the set of final endmembers, either determined with the LAAMs method or those identified by the VCA algorithm, is a linear independent set whose pseudoinverse matrix is unique. Although the unconstrained solution corresponding to Eq. (5), where $n > p$ ($n = 90$ and $p = 7$ or $p = 5$), has a single solution, some coefficients may be negative for many pixel spectra and do not sum up to unity. If full additivity is enforced, negative coefficients appear. Therefore, the best approach consists of imposing non-negativity for the abundance proportions, relaxing full additivity by considering the inequality $\sum_{i=1}^{p} a_p < 1$. For the examples here presented we use the *non-negative least squares* (NNLS) algorithm that solves the problem of minimizing the Euclidian norm $\|Sa - x\|_2$ subjected to the condition $a > 0$ (Lawson & Hanson, 1974).

Figure 9 displays the color abundance maps of the endmembers determined with each one of the methods here discussed. These maps were generated using the NNLS numerical method implemented in Matlab 7.6; in these images, brighter areas represent maximum distribution of the corresponding endmember. The left part of Figure 9 shows the distribution of four natural resources that were determined with implementation of the LAAMs method in the hyperspectral cube of the Gulf Coast. The right part of the same Figure displays the distribution of three natural resources that were determined using the VCA algorithm. In

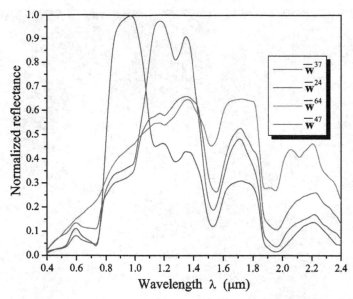

Fig. 7. Four endmembers spectra determined with application of the LAAMs method to the hyperspectral cube of Beltsville. The associated column vectors selected from $\overline{W} \cup \overline{M}$ are: $\overline{\mathbf{w}}^{24}, \overline{\mathbf{w}}^{37}, \overline{\mathbf{w}}^{47}, \overline{\mathbf{w}}^{64}$.

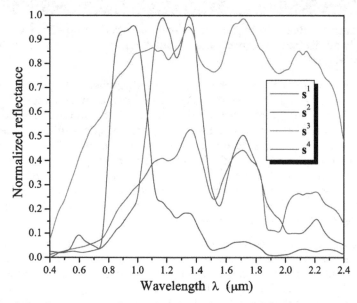

Fig. 8. Four endmembers spectra obtained with application of the VCA algorithm to the hyperspectral cube of Belstville. The column vectors $\mathbf{s}^j$ for $j = 1, \ldots, 7$ indicate the corresponding column of the S matrix.

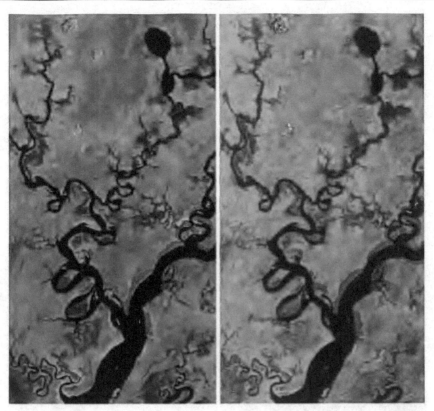

Fig. 9. Color abundance maps of natural resources determined with the autonomous identification of endmembers in the hyperspectral cube of the Gulf Coast. Left: abundances of four endmembers determined with the LAAMs method, whose distribution of colors corresponds to yellow = $\overline{w}^2$, magenta = $\overline{w}^{24}$, green = $\overline{w}^{43}$, blue = $\overline{w}^{54}$. Right: abundances of three endmembers determined with the VCA algorithm whose distribution of colors is yellow = s^4, magenta = s^2, green = s^3. Brighter areas mean higher distributions of the corresponding natural resource.

both cases we present only the abundance maps that provide meaningful information; thus, the maps presenting redundant information or predominant dark areas were not included. Although the region is characterized by the presence of wetlands, it has not been possible to use a set of reference spectra to identify the natural resources. Furthermore, Figure 10 displays the color abundance maps of the endmembers determined with the LAAMs method (left part), as well as the VCA algorithm (right part) from the hyperspectral data cube of Beltsville. Although the set S has conformed by seven endmembers spectra, we have included in both cases the abundance maps that best match the distribution of vegetation according to a visual inspection of Figure 6. In this example, it is evident that the identification of vegetation types, produced with each one of the methods, presents similar results for the segmentations colored in yellow and magenta. However, the results presents important differences particularly in the green and blue segmentations. These differences are mainly caused by the endmember search procedure used in each technique; in addition, the fact that spectral curves of vegetation types are alike, varying in certain absorption bands, contributes with the disagreements in these segmentation results.

Fig. 10. Color abundance maps of vegetation types obtained with the autonomous identification of endmembers in the Belstville hyperspectral image. Left: abundances of four endmembers determined with the LAAMs method whose distribution of colors corresponds to magenta = $\overline{w}^{24}$, yellow = $\overline{w}^{67}$, blue = $\overline{m}^{46}$, green = $\overline{m}^{62}$. Right: abundances of four endmembers determined with the VCA algorithm, whose distribution of colors is magenta = s^2, yellow = s^3, blue = s^5, green = s^7. Brighter areas correspond to higher distributions of the corresponding natural resource.

7. Conclusion

The use of high resolution image spectrometers for Earth observation purposes has given place to different applications oriented toward the identification, classification and monitoring of natural resources from remotely sensed data. In this chapter we have described the physical foundations behind the acquisition and calibration of hyperspectral imagery that constitute the basis of modern hyperspectral instruments, such as AVIRIS, HYDICE, and SpecTir's imaging spectrometers. Also, we have made a review of past, as well as recent methods for the autonomous endmember determination process based on the geometry of convex sets. The mathematical foundation behind these methods is to model the spectral mixtures acquired at a pixel basis as a linear combination of constituent materials. Hence, the aim of these techniques is to determine the constituent materials that are identified as the purest pixels in the scene. Among the methods discussed, we have emphasized a lattice algebra based method that uses two canonical associative memories, the min-W_{XX} and the max-M_{XX}, to determine a $2(n + 1)$-simplex enclosing the hyperspectral data set. Thus, any subset of vertices of the simplex can be used as the endmember set to perform the unmixing process. The application of the LAAMs method and the VCA algorithm for the autonomous segmentation of real hyperspectral scenes taken from the SpecTir's imaging spectrometer has shown the effectiveness of convex set approaches. Although there exist some differences in the results obtained with both methods, any of them can be used for unsupervised hyperspectral segmentation, in particular if there is no reference data of the area, as the example cases treated in this chapter.

8. Acknowledgments

The authors acknowledge to SpecTir for providing the data sets used in these experiments. Juan C. Valdiviezo-N thanks the National Council of Science and Technology (CONACYT) for doctoral scholarship # 175027. Gonzalo Urcid is grateful with the National Research System (SNI-CONACYT) for partial financial support through grant # 22036.

9. References

Boardman J.W. (1993). Automated spectral unmixing of AVIRIS data using convex geometry concepts, *Proceedings of AVIRIS Workshop*, Vol. 4, pp. 11–14, USA, JPL Publication.

Boardman J.W. (1995). Analysis, understanding and visualization of hyperspectral data as convex sets in n-space, *Proceedings of SPIE, Imaging Spectrometry*, Vol. 2480, pp. 14–22, Orlando, FL, USA, April, 1995, SPIE Press.

Chan, T-H. Chi, C-Y. Huang, Y-M. and Ma, W-K. (2009). A convex analysis based minimum volume enclosing simplex algorithm for hyperspectral unmixing. *IEEE Trans. on Signal Processing*, Vol. 57, Issue 11, June 2009, pp. 4418–4432.

Craig M.D. (1994). Minimum-volume transforms for remotely sensed data, *IEEE Trans. on Geoscience and Remote Sensing*, Vol. 32, No. 3, May 1994, pp. 542–552.

Esa (2010). European Space Agency, website: www.esa.int.

Farrand W.H. (2005). Hyperspectral remote sensing of land and the atmosphere, In: *Encyclopedia of Modern Optics*, Vol. 1, Robert D. Guenther editor, pp. 395–403, Elsevier, Academic Press.

Gallier J. (2001). Geometric Methods and Applications for Computer Science and Engineering, In: *Texts in Applied Mathematics*, Vol. 38, Springer, New York.

Graña, M. Jiménez, J.L. Hernàndez, C. (2007). Lattice independence, autoassociative morphological memories and unsupervised segmentation of hyperspectral images, *Proceedings of the 10th joint Conference on Information Sciences*, NC-III, pp. 1624-1631, World Scientific Publishing Co.

Graña, M. (2008). A brief review of lattice computing, *Proceedings of IEEE, World Congress on Computational Intelligence*, pp. 1777–1781, Hong Kong, June 2008, IEEE Press.

Graña, M. Villaverde, I. Maldonado, J.O. Hernández, C. (2009). Two lattice computing approaches for the unsupervised segmentation of hyperspectral images, *Neurocomputing*, Vol. 72, No. 10-12, June 2009, pp. 2111–2120.

Huebshman M.L., Schultz R.A., Garner H.R. (2005). Hyperspectral imaging, In: *Encyclopedia of Modern Optics*, Vol. 1, Robert D. Guenther editor, pp. 134–143, Elsevier, Academic Press.

Jensen, J.R. (2007). *Remote Sensing of the Environment: an Earth Resource Perspective*, 2nd edition, Pearson Prentice Hall.

Kaburlasos, V.G. Ritter G.X. (eds.) (2007), *Computational Intelligence based on Lattice Theory*, Vol. 67. Springer Verlag, Heidelberg, Germany.

Keshava N. (2003). A survey of spectral unmixing algorithms, *Lincoln Laboratory Journal*, Vol. 14, No. 1, pp. 55–78.

Keshava, N. and Mustard, J.F. (2002). Spectral unmixing, *IEEE Signal Processing Magazine*, Vol. 19, No. 1, January 2002, pp. 44–57.

Lawson, C.L. Hanson, R.J. (1974). *Solving least squares problems*, chap. 23, Prentice-Hall, Englewood Cliffs NJ.

Lay, S.R. (2007). *Convex Sets and Their Applications*, Dover Publications, New York, USA.

More, K.A. (2005). Spectrometers, In: *Encyclopedia of Modern Optics*, Vol. 1, Robert D. Guenther editor, pp. 324–336, Elsevier, Academic Press.

Nascimento, J.M.P. and Bioucas-Dias, J.M. (2005). Vertex component analysis: a fast algorithm to unmix hyperspectral data, *IEEE Trans. on Geoscience and Remote Sensing*, Vol. 43, No. 4, April 2005, pp. 898–910.

Ritter, G.X. Sussner, P. Díaz de León, J.L. (1998). Morphological associative memories, *IEEE Trans. Neural Networks*, Vol. 9, No. 2, March 1998, pp. 281–293.

Ritter, G.X. Urcid G., and Iancu, L. (2003). Reconstruction of noisy patterns using morphological associative memories, *Journal of Mathematical Imaging and Vision*, Vol. 19, No. 5, pp. 95–111.

Ritter, G.X. Gader, P. (2006). Fixed points of lattice transforms and lattice associative memories. In: *Advances in imaging and electron physics*, Vol. 144, P. Hawkes editor, 165–242. Elsevier, San Diego, CA.

Ritter, G.X. Urcid, G. Schmalz, M.S. (2009). Autonomous single-pass endmember approximation using lattice auto-associative memories, *Neurocomputing*, Vol. 72, Issues 10-12, June 2009, pp. 2101–2110.

Ritter, G.X., Urcid G. Lattice algebra approach to endmember determination in hyperspectral imagery. In: *Advances in Imaging and Electron Physics*, Vol. 160, Peter W. Hawkes editor, pp. 113-169, Elsevier Inc, Academic Press.

SpecTir (2009). SpecTir: end to end hyperspectral solutions, website: www.spectir.com.

Urcid, G. Valdiviezo-N., J.C. (2007). Generation of lattice independent vector sets for pattern recognition applications, *SPIE Proceedings, Mathematics of Data/Image Pattern Recognition, Compression, Coding, and Encryption X with Applications*, Vol. 6700, pp. 67000C :1–12, San Diego, CA, August 2007, SPIE Press.

Urcid, G. Valdiviezo, J.C. (2009). Color image segmentation based on lattice auto-associative memories, *Proceeding of IASTED, Artificial ingelligence and soft computing*, pp. 166-173, Palma de Mallorca, Spain, September 2009, Acta Press.

Valdiviezo, J.C. and Urcid, G. (2007). Hyperspectral endmember detection based on strong lattice independence, *Proc. SPIE, Applications of Digital Image Processing XXX*, Vol. 6696, pp. 669625 :1–12, San Diego, CA, USA, August 2007, SPIE Press.

Winter M.E. (1999). NFIND-R: an algorithm for fast autonomous spectral endmember determination in hyperspectral data, *Proceedings of SPIE, Imaging Spectrometry V*, Vol. 3753, pp. 266–275, Denver, CO, USA, July 1999, SPIE Press.

Winter M.E. (2000). Comparison of approaches for determining end-members in hyperspectral data, *Proceedings of IEEE: Aerospace Conference*, Vol. 3, pp. 305–313, Big Sky, MT, USA, March 2000, IEEE Press.

7

8-Band Image Data Processing of the Worldview-2 Satellite in a Wide Area of Applications

Cristina Tarantino, Maria Adamo, Guido Pasquariello,
Francesco Lovergine, Palma Blonda and Valeria Tomaselli
National Council of Researches (CNR),
Italy

1. Introduction

Recent years have seen advances in remote sensing in many fields with applications at a spatial scale which range from global to local. As a consequence, the need to observe the Earth with more specialized and sophisticated sensors and data analysis techniques to obtain more accurate information has increased. On the 8th October 2009 a new second next-generation Worldview-2 satellite was launched by DigitalGlobe: it represents the latest innovation among sensors for the acquisition of remote sensed imagery. It has an advanced agility due to control moment gyros (like Worldview-1) and combines an average revisiting time of 1.1 days around the globe with a large scale collection capacity. Moreover, it is also the first commercial satellite able to provide panchromatic imagery at 46 cm of spatial resolution and 8-band multispectral imagery at 1.84 m spatial resolution. In addition to the standard panchromatic and multispectral BLUE, GREEN, RED and NEAR INFRARED (NIR1) bands the Worldview-2 sensor has:

1. a shorter wavelength blue band, COASTAL, ranging from 400 to 450 nm, planned for bathymetric studies, for water color analyses and substantially influenced by atmospheric scattering;
2. a YELLOW band, ranging from 585 to 625 nm, significant for the "Yellowness" of vegetation both on land and water;
3. a RED EDGE band, ranging from 705 to 745 nm, strategically centered at the onset of the high reflectivity portion of vegetation response so potentially significant in the measurement of plant health;
4. a longer wavelength NEAR INFRARED band (NIR2), ranging from 860 to 1040 nm, partially overlapping the NIR1 band and sensitive to atmospheric water vapor absorption.

In literature, many studies deal with the use of the add on bands of the Worldview-2 sensor with respect to the traditional bands of the most common commercial satellites searching for new indexes in different application fields such as bathymetry [1], or vegetation and agricultural purposes ([2], [3]). In [4] the authors analyze the high correlation among some bands of the Worldview-2, like COASTAL and BLUE bands or

NIR1 and NIR2 bands which could mean redundant and useless information associated with some of the add on bands.

The aim of this work is the study of the performance of the whole spectral information offered by the Worldview-2 sensor for the characterization and the classification of some selected land cover targets. Three main land cover targets were recognized: "Water", "Bare lands" and "Vegetated lands". The Worldview-2 image was, firstly, used for a finer discrimination of different sub-classes on the ground belonging to the land cover targets with the application of an unsupervised approach and the help of a certified CORINE-like Land Use Map, at a 1:10.000 scale. A hyperspectral image acquired by the airborne MIVIS sensor was used to analyze the spectral profiles characterizing each distinct sub-class. Then a standard Maximum Likelihood classifier was applied to the Worldview-2 image with different input configurations as below:

1. the 4 bands (R,G,B,NIR1) common to the standard commercial multispectral sensors at very high spatial resolution;
2. the 4 bands R,G,B,NIR1 adding on, one at a time, the new bands;
3. the new complete configuration with 8 spectral bands.

The accuracy of the classification map was estimated using a set of test fields randomly selected on the ground truth map.

ITT ENVI© and GRASS software were used to analyze and process data.

2. The Worldview-2 data

The data set analyzed was a Worldview-2 image granted by DigitalGlobe over an area of 100 km² chosen by the authors among the available archive acquisitions. The scene, acquired on the 13th June 2010, includes the region known as the "Natural Oasis of Lago Salso", an area essentially wet and marshy, sited in the south-east of the Capitanata in the Apulia Region, Italy. The Natural Oasis of Lago Salso is characterized by the presence of a wetland of considerable importance (one of the most important in southern Italy) as a breeding and step birds station. The area falls in the Natura 2000 network, found within the boundaries of the Site of Community Interest (SCI) IT 9110005 "Zone umide della Capitanata" and of the Special Protection Area (SPA) IT9110038 "Paludi presso il Golfo di Manfredonia". The Natural Oasis of Lago Salso falls also within the Gargano National Park. The Natural Oasis has an extent of about 1040 ha and only 500 ha are wetland "sensu strictu", the remaining part is covered by cultivated or partially abandoned areas. Agricultural areas cover a wide surface formerly occupied by coastal lagoons (until the 1950s) and subsequently buried and used for agricultural purposes. SCI and SPA have an extent, respectively, of 14,109 ha and 14,437 ha. Water bodies are subject to fluctuations of water levels over the year, creating ecological gradients due to the variation of salt rates and moisture in soil. Soil salinity gradually increases with soil elevation, reaching a maximum just above mean high sea level (MHSL). Above the MHSL, the salinity tends to decrease due to progressively less frequent flooding. The zonation of the vegetation of salt marshes is typically associated with the tolerance to these ecological gradients.

Figure 1 shows an RGB composition in the visible spectrum of the Worldview-2 image.

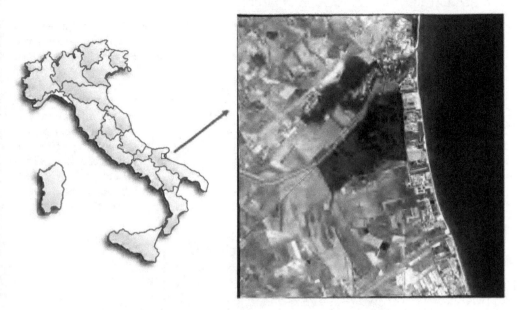

Fig. 1. RGB composition in the visible spectrum of Worldview–2 image.

2.1 Preprocessing

The image was calibrated in order to produce the reflectance image and to obtain the spectral profiles of some targets in the scene to compare with a previously acquired hyperspectral data set. The processing includes the following steps:

1. transformation of digital numbers into the spectral radiance values at TOA (Top Of Atmosphere). This first calibration step, known as absolute radiometric calibration, consists of multiplying radiometrically corrected image pixels by the appropriate absolute radiometric calibration factor to get band-integrated radiance (W/m^2·sr) and then dividing the result by the appropriate effective bandwidth to get spectral radiance (W/m^2·sr·µm). The absolute radiometric calibration factor and effective bandwidths are delivered with the image and available in the image metadata files (extension .IMD);
2. transformation of TOA spectral radiance into TOA reflectance. TOA reflectance is defined as the ratio of radiance reflected from a surface target to the solar irradiance incident on the surface. It is obtained using the formula:

$$\rho = \frac{d_{ES}^2 \pi L(\theta,\phi,\lambda)}{E_S(\lambda)\cos\theta_S} \qquad (1)$$

where θ_S is the Solar Zenith Angle, L is the spectral radiance for a defined pixel and wavelength and Es is the mean solar spectral irradiance. The term d_{ES} is the earth-sun distance in astronomical units as a function of the viewing day and time.

Reflectance values belong to the range [0, 1].

3. Selection and characterization of targets

An existing certified CORINE-like land use map at 1:10000 scale [5] was considered as ground truth. The map was produced in 2006. It originally showed a set of 40 land use thematic classes: after a first screening only land cover classes were selected. An unsupervised analysis was used to cluster the EO data into a certain number of spectrally different signatures. To accomplish this task the "K-Means" algorithm was considered and the 8 bands of the Worldview-2 image were used as input. After a few attempts, a number of 20 unlabelled classes (with a maximum number of 50 iterations until the convergence and 1% as the change threshold to end the iterative process) were selected. Comparing the clusters with the ground truth information resulted in the splitting/merging of certain classes, for a total number of 18 land cover classes. As shown in Figure 2 where a bathymetry map of the test site is represented, the 8 band segmented map reports 3 differentiated clusters in correspondence with the ground truth class labeled as "Sea". These three different signatures could be associated with different depth values of the sea.

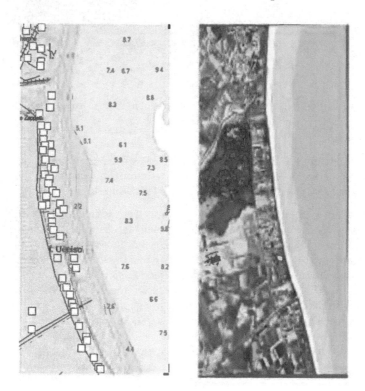

Fig. 2. Bathymetric map (a) and 8-band segmented map (b).

The 18 selected classes, grouped into 4 main land cover targets of interest were "Water", "Bare lands", "Vegetated lands" and "Artificial", as shown in Table 1. The target "Artificial" was eliminated due to its poor presence in the scene. A sample of each considered class is shown in Figure 3.

TARGET	N°	Label	TRAINING pixels (TOT.21948)	TEST pixels (TOT.26428)
WATER	1	SEA WATER 1 (deep)	1365	1270
	2	SEA WATER 2 (medium deep)	938	1845
	3	SEA WATER 3 (coastal)	2009	2067
	4	RIVER WATER	1322	2012
	5	MARSH WATER	1924	1991
BARE LAND	6	ARABLE LAND WITHOUT VEGETATION 1 (dark brown)	1224	1245
	7	ARABLE LAND WITHOUT VEGETATION 2 (light brown)	1104	1409
	8	ARABLE LAND WITHOUT VEGETATION 3 (orange)	1483	1646
	9	ARABLE LAND WITHOUT VEGETATION 4 (very light brown)	1316	1198
	10	SAND	857	730
VEGETATED LAND	11	ARABLE LAND WITH VEGETATION 1 (intense green)	1250	918
	12	VEGETATED MARSHY AREA 1 (dark green)	1272	537
	13	NOT ARBOREOUS VEGETATION	1031	3252
	14	FORESTED AREA	561	1693
	15	ARABLE LAND WITH VEGETATION 2 (light green)	1302	742
	16	VEGETATED MARSHY AREA 2 (less dark green)	1154	748
ARTIFICIAL	17	ARTIFICIAL STRUCTURES	984	1521
	18	ARABLE LAND WITH SCREENING COVERS	852	1265

Table 1. Different land cover classes selected for supervised classification.

3.1 Analysis of the targets' spectral profiles

The analysis of the targets' spectral profiles was carried out by means of a dataset composed by an MIVIS airborne system hyperspectral image, acquired on 25th May 2009 at 06:18 UTC. The selection of this image was possible because of the comparable period of acquisition with respect to the Worldview-2 image. MIVIS (Multispectral Infrared and Visible Imaging Spectrometer) is a hyperspectral sensor consisting of 4 spectrometers which acquire radiation coming from the surface in the VNIR (20 bands between 0.411 and 0.819 µm), in the NIR (8 bands between 1.145 and 1.54 µm), in the MIR (64 bands between 1.992 and 2.474 µm) and in the TIR (10 bands between 8.34 and 12.42 µm). The result of the MIVIS images pre-processing step is an image with pixels given in radiance values ($\mu W/cm^2 \cdot sr \cdot nm$). In order to compare Worldview-2 and MIVIS spectral profiles the analysis was focused on the 20 bands of the VNIR spectrometer which match with the Worldview-2 bands. Details of the VNIR MIVIS bands and comparison with the Worldview-2 bands are shown in Table 2.

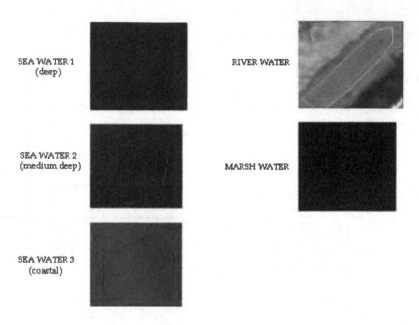

SEA WATER 1
(deep)

RIVER WATER

SEA WATER 2
(medium deep)

MARSH WATER

SEA WATER 3
(coastal)

Fig. 3a. The different classes grouped in the target "Water".

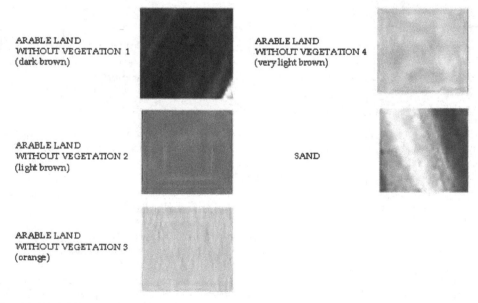

ARABLE LAND
WITHOUT VEGETATION 1
(dark brown)

ARABLE LAND
WITHOUT VEGETATION 4
(very light brown)

ARABLE LAND
WITHOUT VEGETATION 2
(light brown)

SAND

ARABLE LAND
WITHOUT VEGETATION 3
(orange)

Fig. 3b. The different classes grouped in the target "Bare Land".

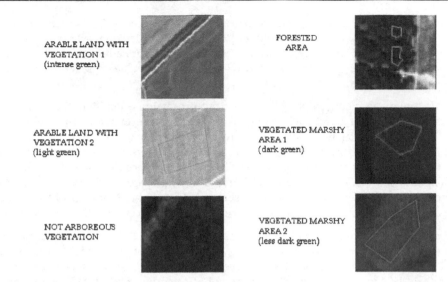

Fig. 3c. The different classes grouped in the target "Vegetated Land".

MIVIS			Worldview-2		
Bands	Centre (nm)	FWHM (nm)	Bands	Lower Edge (nm)	Upper Edge (nm)
1	441	20	COASTAL	400	450
2	460	20	BLUE	450	510
3	480	20			
4	500	20			
5	521	20	GREEN	510	580
6	541	20			
7	561	20			
8	581	20			
9	601	20	YELLOW	585	625
10	621	20			
11	641	20	RED	630	690
12	661	20			
13	681	20			
14	701	20	RED EDGE	705	745
15	721	20			
16	740	20			
17	760	20	NIR1	770	895
18	779	20			
19	798	20			
20	819	20			
			NIR2	860	1040

Table 2. MIVIS and Woldview-2 spectral details.

Because of its flexible airborne platform for remote sensing, the MIVIS system is able to acquire images with a good spatial resolution. The MIVIS acquisition used for this analysis was made at a height of 1.5 Km so the spatial resolution is 3 m at nadir. In order to compare MIVIS spectra with those produced by Worldview-2, the pixel values of MIVIS images were also converted into reflectance.

Due to the unknown quality of the MIVIS data pre-processing calibration, the comparison between the Worldview-2 and the MIVIS profiles has to be considered in terms of spectral profile trends. In addition, no atmospheric correction was made to both the images [6]. As a consequence, the consideration that the atmosphere contributes in a different way to the reflectance measured by sensors, due to the different day of acquisition and the different flight height of sensors, should be observed. In Figure 4, a subset of the MIVIS acquisition corresponding to the Worldview-2 image is shown. Close to the right edge of the frame a slight pattern of sunglint is visible. It is presumed that it influences the reflectance of the sea.

The spectral analysis was carried out by selecting regions which can be considered representative of the 18 classes identified by the unsupervised analysis and the ground truth map. A time interval of about one year between the MIVIS and Worldview-2 acquisitions restricts the selection of target areas to regions not affected by significant changes between the two dates. In fact there could be some arable lands where crops changed for agricultural practices or were covered with screening covers. Moreover, it should be considered that the two images were acquired in two different spring months corresponding to the different phenological status of the same crop.

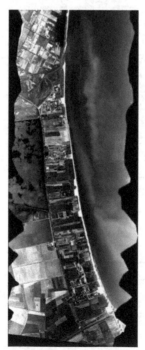

Fig. 4. MIVIS image acquired on 25th May 2009 at 06:18UTC.

In Figure 5, Worldview-2 (top) and MIVIS (bottom) water target spectral profiles are shown. As it can be noticed for all the target profiles, there is an atmospheric contribution to the Worldview-2 reflectance. In particular, in the range of shorter wavelengths of the COASTAL and the BLUE bands, the Rayleigh scattering is the prevailing contribution while for the longer wavelengths, like the NIR2 band, the water vapor absorption is dominant. The gaseous absorption results as visible also for the MIVIS profiles. It can be noted that:

- "Sea Water 1", "Sea Water 2" and "Sea Water 3" correspond to: deep water (far from the coast), intermediate deep water and coastal water, respectively. The Worldview-2 reflectance profiles, which tend to converge at shorter wavelengths due to atmospheric effects, show increasing reflectance values for sea regions closer to the coast due to the increasing contribution of suspended sediments [7];
- this behavior is not evident in MIVIS spectral profiles because of the presence of a slight sunglint contribution which implies an increase of reflectance values in the eastern part of the image. This portion of the image corresponds to a sea region with a high and intermediate depth. The presence of glint is also evident when considering the spectral profiles which show the typical solar irradiance trend;

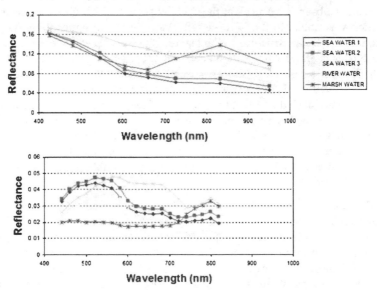

Fig. 5. Worldview-2 (top) and MIVIS (bottom) spectral profiles of water classes.

- "Sea Water 3" and "River Water" MIVIS spectral profiles are characterized by particulate material and/or a land-derived yellow substance which can influence the reflectance [8];
- with regard to the "Marsh Water" classified regions, they are relative to a series of waterbodies alternating with different kinds of vegetated areas belonging to the Natural Oasis of Lago Salso (Figure 6). Specifically, it encompasses three waterbodies with depths ranging from 50 to 170 cm, depending on the seasonal level and the regional operational necessities which are fed by a small river. This class shows a spectral profile which starts to be strongly affected by the bottom vegetation contribution.

Fig. 6. Natural Oasis of Lago Salso.

In Figure 7, Worldview-2 (top) and MIVIS (bottom) spectral profiles for vegetated land regions are shown. MIVIS spectral profiles of the six vegetated land classes show the typical vegetation trend. The absorption of chlorophyll in the BLUE and the RED regions of the spectrum can be observed. A peak at the GREEN region which gives rise to the green color of vegetation was noted. In the NIR the reflectance is much higher than that in the visible band due to the cellular structure in the leaves. The slope of the spectrum profile between RED and NIR is characteristic of the vegetation species and gives information about plant health [9]. Spectral profiles also show a reduction in band 20 (Table 1) due to atmospheric absorption. Analyzing MIVIS spectra some considerations about vegetated land classes can be made. The class labeled as "Arable Land with Vegetation 2" shows a reduced increase of reflectance in the wavelength range between RED and NIR and a peak of reflectance in correspondence with the GREEN range which is less evident with respect to the other profiles. Considering the particular color of the regions and the presence of an almost regular texture, it is possible that this class is related to arable fields covered by a thick net typical of local agricultural practices.

Classes labeled as "Vegetated Marshy Areas 1" and "Vegetated Marshy Areas 2" are relative to different kinds of vegetation characterizing the "Natural Oasis of Lago Salso" (Figure 6). The Worldview-2 profiles, due to atmospheric effects, do not show the typical trend of vegetation spectra in the visible range. The absorption peak in the BLUE range is suppressed by the Rayleigh scattering contribution which decreases with an increase in wavelength. On the contrary, the range of the spectrum from RED to NIR1 can be useful for vegetation characterization and, except for a few differences (which could be explained considering the time interval between the two acquisitions) MIVIS and Worldview-2 spectral profiles are sufficiently in agreement. The class labeled as "Forested Area "is mainly composed of the Siponto pine forest (Figure 8) sited in a coastal area on the Manfredonia Gulf. The spectral profile obtained by MIVIS and confirmed by Worldview-2 shows a low reflectance in the NIR spectral range which is correlated to lower vegetation LAI [10].

In Figure 9, Worldview-2 (top) and MIVIS (bottom) spectral profiles for bare land regions are shown. In this case the trend of MIVIS spectral profiles are in agreement with the Worldview-2 ones; although, the better spectral resolution of MIVIS is able to acquire finer spectral signatures for every class.

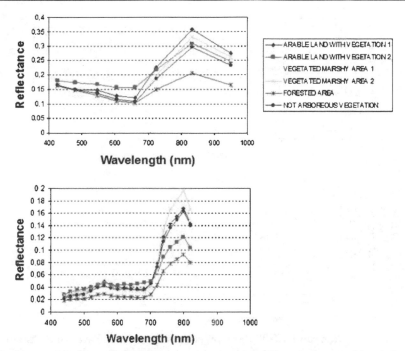

Fig. 7. Worldview-2 (top) and MIVIS (bottom) spectral profiles of vegetated land classes.

Fig. 8. Siponto Pine Forest.

It can be noted that "Arable Land without Vegetation 1", "Arable Land without Vegetation 2" and "Arable Land without Vegetation 3" are characterized by a spectral signature similar in shape but with an increasing average reflectance. This can be explained considering that the reflectance level decreases for soil with increasing moisture [11] and so the different targets could be associated with different moisture content. "Sand", instead, shows a spectral profile which differs from the other ones probably due to the extremely different composition of the soil.

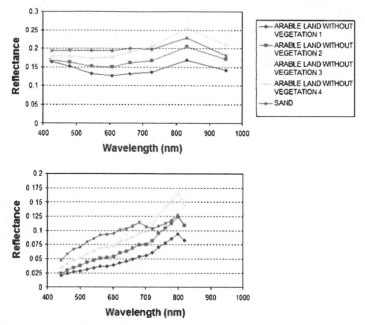

Fig. 9. Worldview-2 (top) and MIVIS (bottom) spectral profiles of Bare Land classes.

4. Processing, results and discussion

For the supervised analysis, the standard statistic "Maximum Likelihood" (ML) algorithm was considered. The 18 classes of table 1, recognized on the scene with the guide of the segmentation step and better characterized with the help of MIVIS data, were selected. Randomly selected training (TR) and test (TE) sets were used respectively to train the algorithm and to assess the accuracy of the produced maps. For the accuracy of all the classes, the Overall Accuracy percentage (OA%) (i.e. number of correctly classified pixels divided by the total of pixels) with the estimation of the relative confidence interval with a significance of 95% [12] as computed. For the accuracy of each class, the Mapping Accuracy percentage (MA%), [13], [14], was computed. It is defined as:

$$MA\% = \frac{pixels_{correctlyclassified}}{pixels_{correctlyclassified} + pixels_{omission} + pixels_{commission}} \cdot 100 \qquad (2)$$

where:

$pixels_{omission}$ is the number of pixels assigned to other classes along the row of the confusion matrix relevant to the class considered;

$pixels_{commission}$ is the number of pixels assigned to other classes along the column of the confusion matrix relevant to the class considered.

According to [15], many input configurations to the classifier were tested considering, firstly, the standard 4 spectral bands of the image and then adding a fifth band among the 4 add on bands of Worldview-2 in order to analyze the specific contribution of each band.

Finally all the 8 band contributions were considered. The results obtained for all the classes with different input bands to the supervised classifier are shown in table 3.

Input Configuration	OA_TR%	±δ_TR%	OA_TE%	±δ_TE%
4 BANDS	98.63	0.15	75.31	0.52
5 BANDS WITH COASTAL	99.60	0.08	78.09	0.50
5 BANDS WITH YELLOW	98.62	0.15	77.96	0.50
5 BANDS WITH RED EDGE	98.87	0.14	79.00	0.49
5 BANDS WITH NIR2	98.94	0.13	78.92	0.49
8 BANDS	99.71*	0.07	85.50*	0.42

Table 3. Results in the supervised classification.

In training and testing, with an increase in the number of the bands there is an increase in the OA% because more information was added as input to the classifier to improve discrimination among classes. Observing the generalization ability, in testing, an improvement of 10% was achieved with the use of 8 bands with respect to 4 bands. The asterisk indicates the best value.

The same analysis was carried out for each target ("Water"-"Bare land"-"Vegetated land") in order to evaluate the contribution that each of the add on 4 bands could give to characterize the specific target. A finer detailed discrimination among the classes is expected. For the target "Water" the MA% in testing in the different input configurations to the classifier are shown in Table 4.

Input Configuration	SEA WATER 1 (deep)	SEA WATER 2 (medium deep)	SEA WATER 3 (coastal)	RIVER WATER	MARSH WATER
4 BANDS	83.33	66.84	70.27	55.81	81.86
5 BANDS WITH COASTAL	80.41	66.39	72.48	65.53*	83.92
5 BANDS WITH YELLOW	86.68	84.97*	90.11*	59.01	81.81
5 BANDS WITH RED EDGE	96.28	77.83	76.87	62.80	84.13
5 BANDS WITH NIR2	99.52*	72.47	63.35	59.00	93.51*
8 BANDS	100**	91.70**	90.63**	82.88**	99.04**

Table 4. MA% in test for the classes belonged to the target Water.

The best MA% value is obtained with the use of 8 bands, as indicated with a double asterisk, with an average improvement of 20% with respect to the use of only 4 bands. Analyzing the contribution of each add on band to the single class, it emerged that (the best value due to the add on bands has been marked with a single asterisk):

- the discrimination of "Sea Water 1" (deep) and "Marsh Water" is improved by the NIR2 band. "Marsh Water" is water with the presence of vegetation under and over the surface and this could explain the role of NIR2;

- the discrimination of "Sea Water2 " (medium deep) and "Sea Water 3" (coastal) is improved by the YELLOW band that appears to be able to recognize water with hanging deposits;
- the discrimination of "River Water", substantially muddy water, is improved by the COASTAL band which appears able to recognize a mixture of water and mud.

For the target "Bare land", the MA% in testing in the different input configurations to the classifier are shown in Table 5.

Input Configuration	ARABLE LAND WITHOUT VEGETATION 1 (dark brown)	ARABLE LAND WITHOUT VEGETATION 2 (light brown)	ARABLE LAND WITHOUT VEGETATION 3 (orange)	ARABLE LAND WITHOUT VEGETATION 4 (very light brown)	SAND
4 BANDS	80.84	38.62	48.72	55.20	39.10
5 BANDS WITH COASTAL	75.05	38.08	49.48	58.74	41.40*
5 BANDS WITH YELLOW	81.31*	50.53*	48.51	56.31	39.07
5 BANDS WITH RED EDGE	78.71	49.78	55.96*	60.28	40.71
5 BANDS WITH NIR2	80.55	50.17	55.99*	65.01**	39.16
8 BANDS	86.15**	57.67**	61.17**	64.10	44.24**

Table 5. MA% in test for the classes belonged to the target Bare Land.

For all the different spectral signatures, the best MA% value is obtained with the use of 8 bands, as indicated with a double asterisk, with an average improvement of 10% with respect to the use of only 4 bands. The class "Arable Land without Vegetation 4" is an exception which can be justified by a high misclassification with the class "Arable Land without Vegetation 3". Analyzing the contribution of each add on band to the single class, it emerged that (the best value due to the add on bands has been marked with a single asterisk):

- the discrimination of "Arable Land without Vegetation 1" (dark brown) and "Arable Land without Vegetation 2" (light brown) is improved by the YELLOW band;
- the discrimination of "Arable Land without Vegetation 3" (orange) is improved by the NIR2 band and by the RED EDGE, whereas "Arable Land without Vegetation 4" (very light brown) is improved only by the NIR2 band;
- the discrimination of "Sand" is improved by the COASTAL band.

The different spectral profiles could be explained by the different pedological composition of soil or its different water content. For the "Vegetated land" target, the MA% test classification values obtained with different input bands are shown in Table 6.

The best MA% value is obtained with 8 bands, as evidenced by a double asterisk in the table, with an average improvement of about 3% for "Forested Area" and "Vegetated

Marshy Areas" and of about 30% for "Arable Land with Vegetation 1" and "Arable Land with Vegetation 2" with respect to the use of only 4 bands. For each class, the best result obtained by a specific band is evidenced by a single asterisk in the table.

Input Configuration	ARABLE LAND WITH VEGETATION 1 (intense green)	ARABLE LAND WITH VEGETATION 2 (light green)	VEGETATED MARSHY AREA 1 (dark green)	VEGETATED MARSHY AREA 2 (less dark green)	NOT ARBOREOUS VEGETATION	FORESTED AREA
4 BANDS	70.91	55.20	93.69	90.42	60.50	93.50
5 BANDS WITH COASTAL	75.89	86.48*	95.89*	92.90*	74.88*	95.80*
5 BANDS WITH YELLOW	74.07	53.30	95.20	92.49	60.97	94.69
5 BANDS WITH RED EDGE	85.25*	56.90	95.20	92.96	68.27	95.64*
5 BANDS WITH NIR2	75.02	60.96	92.86	90.79	69.68	94.88
8 BANDS	92.08**	89.83**	96.93**	95.38**	90.84**	97.02**

Table 6. MA% in test for the classes belonged to the target Vegetated Land.

5. Conclusions

This paper describes the experimental activity aimed at the exploitation of the new Worldview-2 sensor with respect to the effectiveness of the new add on COASTAL, YELLOW, RED EDGE and NIR2 bands. Firstly, an unsupervised analysis for data spectral clustering was applied to discriminate among the different spectral signatures, then a supervised image classification produced a land cover map. Standard/commercial tools were used. In the first step the clusters in the spectral domain were interpreted with the help of a detailed ground truth map and compared with a hyperspectral data set. This analysis showed that the 8-band sensor is extremely useful to better discriminate different spectral sub-signatures corresponding to the same land cover category. This means that the major capability of the new sensor resides in the capacity of investigating the "ground" diversity underlying the apparent homogeneity of conventional land cover/land use map categorization. From the supervised classification, it was possible to detect changes in the bathymetry for the "Sea Water" classes by using the COASTAL band; moreover, the lowest wavelength band appears to be significant for the recognition of mixed patterns of water and terrain. The YELLOW band appears significant to detect the presence of hanging deposits or to elicit terrain composition, as characterized by a certain degree of "yellowness". Finally, the RED EDGE and the NIR2 bands seem useful for a better discrimination of ground sites characterized by a mixing of water and vegetation. The increase in thematic accuracy was 10%, passing from the "traditional" 4-band to the new 8-band sensor.

6. Acknowledgements

This work was supported by the project "Flight Risks Mitigation and Nowcasting at Airports" (RIVONA) funded by the Apulia Region, POFESR 2007-2013.

The authors want to thank DigitalGlobe for having offered the opportunity to analyze images from the newest Worldview-2 sensor.

Special acknowledgements to Planetek Italia s.r.l. for supplying the MIVIS data set.

7. References

[1] Bramante, J. F.; Raju, D. K. & Tsai Min S., Derivation of bathymetry from multispectral imagery in the highly turbid waters of Singapore's south islands: A comparative study, *DigitalGlobe 8-Band Research Challenge 2010*, Available from: http://www.digitalglobe.com/downloads/8bc/8band_Challenge_TMSI.pdf.

[2] Ozdemir, I.; Karnieli, A. (2011), Predicting forest structural parameters using the image texture derived from WorldView-2 multispectral imagery in a dryland forest, Israel, *Int. Journal of Appl. Earth Observation and Geoinformation*, Volume 13, Issue 5, Pages 701-710.

[3] Borel, C. C., Vegetative canopy parameter retrieval using 8-band data, *DigitalGlobe 8-Band Research Challenge 2010*, Available from: http://www.digitalglobe.com/downloads/8bc/borel_8band_paper_12_14_10.pdf.

[4] Peroni, G.; Gachelin, J.P.; Saint-Pol, M., Legoff, V.; Fontanot, F. & Sannier C. (2010), New spectral data available for the controls in agriculture (CWRS) and for vegetation monitoring, *Proc. Of the 16th GeoCAP Annual Conference*.

[5] GIS Apulia Region, Italy, , Available from: http://www.sit.Apulia.it/

[6] Baraldi, A. (2009), Impact of radiometric calibration and specifications of spaceborne optical imaging sensors on the development of operational automatic remote sensing image understanding systems, *IEEE Journal of Selected Topics in Applied Earth Observations and Remote Sensing*, Vol. 2, No.2.

[7] Stramski, D.; Wozniak, S. B. & Flatau, P. J. (2004), Optical properties of Asian mineral dust suspended in seawater, *Limnol. Oceanogr.*

[8] Robinson, I.S. (2004), *Measuring the Oceans from Space - The principles and methods of satellite oceanography*, Springer.

[9] Govender, M.; Chetty, K. & Bulcock, H. (2007), A review of hyperspectral remote sensing and its application in vegetation and water resource studies, *Water SA*, 33(2), 1–8.

[10] Schlerf, M.; Atzberger, C. & Hill J. (2005), Remote sensing of forest biophysical variables using HyMap imaging spectrometer data, *Remote Sens. Environ.* 95, 177-194.

[11] Bowers, S.A. & Hanks, A.J. , Reflection of radiant energy from soil, *Soil Science*, 100: 130.

[12] Baraldi, A.; Puzzolo, V.; Blonda, P.; Bruzzone, L. & Tarantino C. (2006), Automatic Spectral Rule-based Preliminary Mapping of Calibrated Landsat TM and ETM+ Images, *IEEE Trans. On Geoscience and Remote Sensing*, Vol. 44, No. 9.

[13] Short, N.M., The Remote Sensing Tutorial, NASA, Available from http://rst.gsfc.nasa.gov.

[14] Congalton, R. & Green K. (1999), *Assessing the Accuracy of Remotely Sensed Data: Principles and Practices*, CRC/Lewis Press, Boca Raton.

[15] Puetz, A.M.; Lee, K. & Olsen R.C. (2009), Worldview-2 data simulation and analysis results, *Proc. of SPIE*, Vol. 7334.

Part 3

Remote Sensing and GIS
in Earth Observation Applications

Ocean Reference Stations

Meghan F. Cronin[1], Robert A. Weller[2], Richard S. Lampitt[3] and Uwe Send[4]

[1]*NOAA Pacific Marine Environmental Laboratory, Seattle WA*
[2]*Woods Hole Oceanographic Institution, Woods Hole, MA*
[3]*National Oceanography Centre, Southampton*
[4]*Scripps Institution of Oceanography, University of California, San Diego, La Jolla, CA*
[1,2,4]*USA*
[3]*UK*

1. Introduction

OceanSITES is an international network of deep ocean observatories that provide reference time-series for ocean and climate studies. While moorings form the backbone of the network, some stations comprise frequent shipboard observations. With dozens of advanced sensors on these platforms, the time-series are high quality, high resolution (hourly or better in many cases), and long (decades long in some cases). Most stations are interdisciplinary, measuring various aspects of the physical and biogeochemical environment from the sea floor to the atmosphere. All data are made publicly available, in a common format, many in near-real time. In this chapter we describe the motivation for, and the requirements and challenges of this network. Because the network includes more than 105 stations, for practical reasons, our overview of individual stations will focus on the subset of stations that make data available in near-real time. Our goal here is to provide an introduction to the network and provide information and links that will help the reader explore the network further.

2. Water world

We live on a water world. Over 70% of the Earth surface is covered by oceans. On the remaining 30%, human population is not distributed evenly, but instead is most dense in coastal regions. The oceans can affect climate and weather by absorbing, transporting, and emitting heat and gases such as carbon dioxide (Figure 1). Without the poleward heat transport by the ocean currents, the tropics would tend to steadily warm, while the poles would steadily cool. In the high latitudes, heat loss and ice formation generate very dense water at the surface that sinks to the interior and bottom of the ocean, driving the global thermohaline circulation. The oceans also absorb CO_2, thus reducing the effects of anthropogenic climate change.

Because of the high heat content of water, the ocean temperature has much less variability than the atmosphere, particularly the atmosphere over land. While at a given location over land surface air temperature can have a range of up to 90°C, air temperature at a given location over open ocean generally has a range of less than 20°C, and the overall ocean temperature range is roughly 30°C (Figure 2). As such, the oceans typically have a

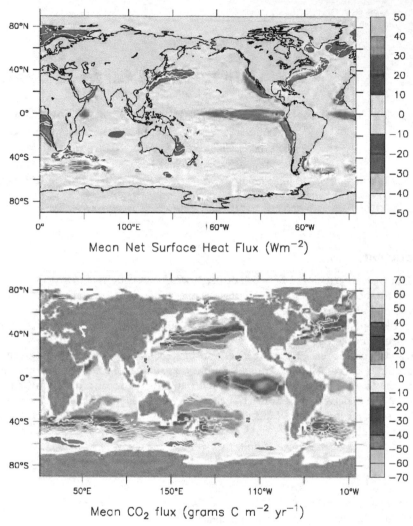

Fig. 1. Mean net surface heat flux in units watts per m2 based on ECMWF 40-year reanalyses (ERA40) (top) and net surface CO_2 flux in units grams of carbon per m2 per year, based on the Takahashi et al. (2009) air-sea CO_2 flux climatology (bottom). A positive flux indicates a flux from the ocean to the atmosphere. Note that the color scale is inverted for the CO_2 flux. At the equator, heat enters the ocean through the surface and CO_2 outgasses. White contours indicate mean dynamic sea level height (Rio & Hernandez, 2004).

moderating effect on weather and climate. Indeed, because 2.5 m of water has the same heat capacity per unit area as the whole height of the atmosphere, relatively small changes in the sea surface temperature distribution can have a significant influence on the atmosphere above, particularly in warm water regions such as the tropics, as shown in Figure 3 and discussed below. Approximately 41% of rainfall over land is of maritime origin (Oki & Kanae, 2006). Evaporation, which provides this precipitable water, is strongly dependent upon temperature.

Significantly more moisture is evaporated where the surface water is warm, fueling deep convection and precipitation (Figure 3). Small shifts in the location and temperature of very warm water can thus cause shifts in the atmospheric convection and weather patterns, both locally and global (Ding et al., 2011; Wallace & Gutzler, 1981).

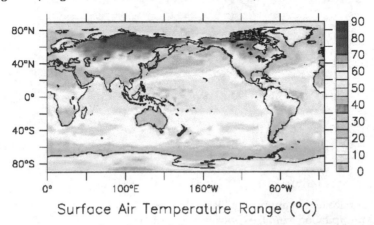

Surface Air Temperature Range (°C)

Fig. 2. Air temperature range based upon daily averaged ERA40 values in units degrees Celsius.

Because of the Earth's rotation, the direct ocean response to wind forcing is an upper ocean transport that is to the right of the wind in the Northern Hemisphere (NH) and to the left of the wind in the Southern Hemisphere (SH). The easterly trade wind and westerly jet stream, and the placement of the continents, thus tend to cause convergence and divergence patterns that result in higher sea level in the subtropics and lower sea level in the subpolar regions (Figure 1). To a certain extent, the sea level height anomalies can be considered as streamlines of the surface flow. Water, that would tend to flow downhill, is deviated to the right in the NH and to the left in the SH, so that the adjusted flow is along the anomalous sea level height isobars. Consequently, the trade winds and jet streams result in an anticyclonic subtropical gyre in each of the ocean basins. The NH westerly jet stream also supports a cyclonic subpolar gyre in the North Pacific and North Atlantic, while in the SH, the jet stream drives an eastward flowing Antarctica Circumpolar Current. Directly at the equator, the axis of rotation is perpendicular to the vertical axis (gravity), making vertical motion near the equator much more dynamic both in the atmosphere and ocean. This, together with the effects of the warm water on precipitable water, causes the ocean and atmosphere to be much more coupled in the tropics than elsewhere. Changes in the ocean surface temperature can result in changes in the atmospheric deep convection and winds, which can in turn affect the ocean temperature structure.

While the ocean drift in most parts of these gyres is slow (~25 cm/s), in some parts, such as the western boundary currents and the circumpolar current, speeds can be up to 2 m/s near the surface and 25 cm/s at depth, corresponding to a transport of order 100×10^6 m^3/s. While the analogy has its limits, these ocean currents can be considered as a conveyor belt, carrying heat, salt, and marine ecosystems. Warm currents carry heat poleward, and return currents and the deep thermohaline circulation carry cool water equatorward, resulting in a large-scale meridional overturning circulation.

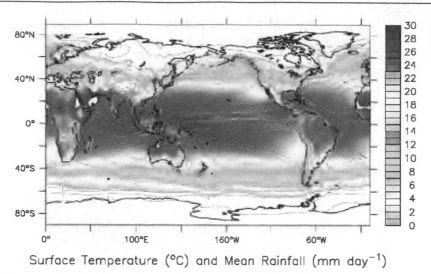

Surface Temperature (°C) and Mean Rainfall (mm day^{-1})

Fig. 3. Mean surface temperature from ERA40 in units °C (only values greater than 0 are shown), and precipitation from the Global Precipitation Climatology Project in units mm per day (contoured).

Marine life, which lives within this dynamic environment, can be quite sensitive to the ocean temperature. Many animals reproduce, feed, or migrate only within a limited temperature range. Temperature also affects the buoyancy of the water, which can trap nutrients and dissolved inorganic carbon within the euphotic zone where photosynthesis and primary production occur. During blooms, CO_2 is used in the production of both organic and inorganic biogenic particles, a portion of which sink into the deeper ocean and are regenerated into CO_2 through respiration and dissolution. This export of CO_2 is referred to as the "biological pump" of the carbon cycle. During respiration, oxygen is depleted. Anoxic water, devoid of dissolved O_2, is generally barren of macroscopic life. Temperature also affects the solubility of dissolved gasses and thus the concentrations of dissolved O_2 and CO_2: As the surface water cools, it can hold and absorb more CO_2. Thus as the surface water cools and sinks, atmospheric CO_2 is absorbed into the water and exported into the deep ocean, a process referred to as the "solubility pump" of the carbon cycle.

The distribution of CO_2 within the ocean is also critical to the pH of the water and the concentration of carbonate ions, which is a basic building block of skeletons and shells for a many marine organisms, including corals, shellfish, and marine plankton (Feely et al., 2004). As the ocean absorbs more anthropogenic CO_2, the CO_2 reacts with the seawater to form carbonic acid (H_2CO_3). This then dissociates to form a bicarbonate ion (HCO_3^-) and a hydrogen ion (H^+), which can react with carbonate ions (CO_3^{2-}) to form bicarbonate (HCO_3^-). The net effect of the increased CO_2 is thus a decrease in pH and a decrease in the carbonate ion concentration, a process referred to as ocean acidification (Feely et al., 2010). The reduction in carbonate ion affects the saturation state of calcium carbonate ($CaCO_3$) and is critically important as it directly affects the ability of some $CaCO_3$ secreting organisms to produce their shells or skeletons. When pteropods were exposed to undersaturated water, their $CaCO_3$ shells showed notable dissolution (Orr et al., 2005).

Like global weather maps of wind, barometric pressure, and atmospheric humidity and temperature properties, global maps of the ocean circulation, sea level, and temperature and salinity properties are needed to visualize, quantify, and understand the ocean physical variability (Cazenave et al., 2010; Schmitt et al., 2010; Talley et al., 2010; Wijffels et al., 2010). For understanding and quantifying the ocean and atmosphere interactions, maps of the air-sea fluxes of heat, moisture, momentum, and gasses are needed (Fairall et al., 2010; Gulev et al., 2010). Likewise, for understanding and quantifying changes to the carbon cycle, maps of the atmospheric and seawater pCO_2, dissolved O_2 concentration, pH, and nutrients are needed (Gruber et al., 2010). Monitoring and predicting O_2 concentration levels is critical for assessing the effects of the biological pump both on the carbon cycle and the ecosystem. Monitoring and mapping changes in the ocean acidification is likewise critical for understanding the biological impacts of increased of anthropogenic CO_2 (Feely et al., 2010).

To make these maps, satellites, ships, floats, drifters, and moored buoys gather data that are routinely ingested into numerical models (Eyre et al., 2010). However, across the broad ocean, as compared to the land, observations are sparse. To validate and assess these modeled fields, as well as to assess satellite remotely sensed fields, in situ observations are needed as reference data (Send et al., 2010). Reference data are also needed for evaluating the processes and mechanisms that affect the ocean environment and ecosystems, and for developing parameterizations of processes that cannot be fully resolved within the numerical models (Lampitt et al., 2010a).

3. Reference ocean data

3.1 Requirements

Reference data, by definition, must be high quality, with quantified uncertainty that is small relative to the signal that is being measured. Uncertainties are determined by the measurement resolution, by investigation of sensor performance in the field, and through calibrations that are traceable to a standard at national metrology institutes such as the U.S. National Institute of Standards and Technology. Measurement resolution is determined by the sensor's precision and sampling frequency. If the sampling frequency is lower than the Nyquist frequency of the signal, errors can arise due to aliasing. In particular, biases can result if the sample frequency is identical to the signal frequency. For example, in regions such as the tropics where the diurnal cycle is large, surface temperature measurements will be biased high if the samples are only during the daytime. Likewise, in regions where the annual cycle is large, measurements may be biased high if they are only sampled during the summer.

Figure 4 shows the spectral bandwidth of various oceanographic signals that have periodicities that range from order of seconds (surface waves), to minutes (internal waves in very high stratification), hours (diurnal cycle, tides, inertial oscillations, internal waves), days (storm forced variability, hydrodynamic instabilities, mesoscale eddies), months (planetary waves, seasonal cycle), years (El Niño, gyre-scale variability), decades (gyre-scale, meridional overturning), and longer (anthropogenic forcing). For some processes that depend upon variables in a nonlinear way, variability in these parameters at one scale may affect variability in the process at another scale. Turbulence generally causes a cascade of energy from low to high frequency. However, high-frequency variability can also, in some cases, rectify into the longer scales. For example, since the efficiency of surface forcing depends upon the stratification, large diurnal co-variations in the forcing and stratification in the tropics can rectify and impact intraseasonal and longer timescales, thus affecting the coupled ocean-atmosphere interactions (Bernie et al., 2007; Guemas et al., 2011; Shinoda, 2005).

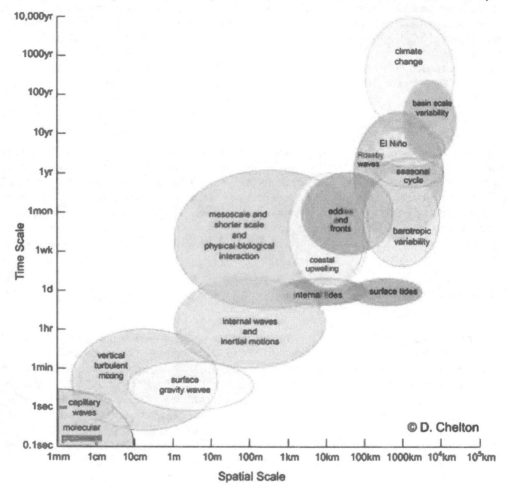

Fig. 4. Time and space scales of ocean variability (courtesy D. Chelton, Oregon State University, after Dickey (2001)).

3.2 A network of open ocean reference stations

For resolving high-frequency variability, moorings are the ideal platform, as the resolution of the moored sensors is generally limited only by constraints on the battery life and duration objectives. With the mooring refreshed at regular intervals (generally 6–12 months), these stations can provide long-term, high-resolution, accurate time-series. Moorings thus form the backbone of the global network of OceanSITES reference stations (Lampitt et al., 2010a; Send et al., 2010). The OceanSITES network, shown in Figure 5, is an element of the Global Ocean Observing System, which is a system within the Global Earth Observing Systems of Systems (GEOSS).

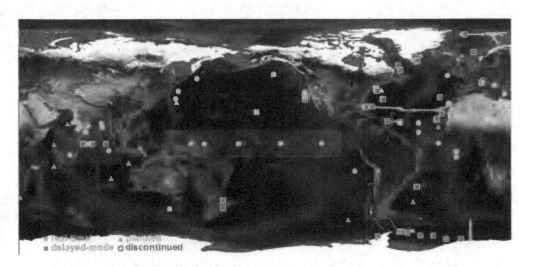

Fig. 5. OceanSITES network of reference stations, as of 2010 (figure courtesy http://www.oceansites.org/network/). Stations with near-real-time data are shown as green circles. Observatories without data telemetry are shown as blue squares. Transport stations are shown as small green squares and regular transport transects are shown as green lines. Planned stations are shown as orange diamonds; discontinued stations and transects are indicated in red.

At present, the OceanSITES network is a collection of stations operated by scientists throughout the world, supported through their national agencies, who agree to the basic requirements of data quality and open data with common formats. The vision is that the OceanSITES network would be interdisciplinary: "a worldwide system of deepwater reference stations: providing high resolution measurements, the full depth of the ocean, multi-year time scales, dozens of variables, real-time access." Indeed the OceanSITES acronym stands for OCEAN Sustained Interdisciplinary Timeseries Environment observation System. In practice, however, not every station monitors the full suite of physical and biogeochemical variables that characterize the local ocean environment. Within the array, moored buoys that carry meteorological sensors to characterize the exchanges or fluxes of heat, momentum, freshwater, and gases (e.g., carbon dioxide) across the air-sea interface are referred to as air-sea "flux" stations. These moorings also generally carry sensors on their anchor line to monitor the physical and sometimes biogeochemical environment in the upper ocean. Other moorings and frequently visited stations, referred to as "observatories," have as their primary objective monitoring the biogeochemical properties within much of the water column. Finally, the purpose of the "transport" stations is to monitor the ocean currents and transport.

3.3 Data latency

While some of the mooring stations have surface buoys that allow telemetry of near-real-time data; other mooring stations are entirely subsurface, and must be recovered to obtain the data. This can introduce a delay in the data availability of more than a year. With telemetered data, analyses can begin almost immediately, thus accelerating the research. Telemetry also acts as important insurance on the data. If the mooring is lost, the telemetered data may be the only source of the data. Having telemetry also allows the operators to identify and address problems in current and future deployments, thus minimizing data gaps. Finally, in some cases, the telemetered near-real-time data are used to assimilate into a short-term weather forecast, for which every hour of latency implies an hour of forecast.

Due to the sparse nature of oceanographic data, there is often a desire to assimilate all data possible, including reference data. Model operators often argue that an individual measurement is weighted in a way that it will not introduce a bullseye pattern in the fields and make the product appear falsely accurate when compared with the reference time-series. Reference data are, by definition, supposed to be independent of the products for which they are used to assess. A World Meteorological Organization (WMO) data identification number containing the digits "84" indicate that they are reference data. Protocols are being developed to identify when reference data are being assimilated.

The delayed mode data, available after internally recording instruments are recovered and processed, also have unique value. Because of limited bandwidth and technical challenges for telemetry of ocean data, the real-time data are only a subset of the data available on the moorings. Internally recorded data may have sampling rates of every 1 minute and faster, whereas hourly data may be what was telemetered. Further, the recovered instruments are post-calibrated; thus, the delayed mode data have less uncertainty associated with their accuracy. In general, the delayed mode data are the highest quality data at a reference stations.

4. The OceanSITES network of reference moorings

In the following section we provide an overview of individual stations within the OceanSITES network, focusing on stations that telemeter data to shore. These include all of the air-sea flux stations, many of which also serve as biogeochemical observatories or are coordinated with nearby observatory and transport stations. A few subsurface observatories also have a small surface buoy used exclusively for telemetry. Because of the complexity of the network it is not possible to describe the network in its entirety. Our purpose here is to provide an introduction and information for further exploration of the network. As a start, the reader is directed to the OceanSITES network website: http://www.oceansites.org.

At roughly $30,000-50,000 per day, shiptime is a significant component of the overall cost of the deep ocean mooring array. These costs and the limited number of global-class research vessels have necessitated efficient use of the fleet. For example, mooring maintenance cruises are often used for long-term coordinated observations. Likewise, while the stations themselves carry a suite of sensors for monitoring multiple variables, the stations also offer opportunities for other coordinated observations. Nearly all stations have been sites of process studies, involving multiple platforms (ships, extra moorings, drifters, floats, aircraft, etc.). In the following overview, we describe some of these activities, although a full list is not feasible.

4.1 Tropics

4.1.1 The Global Tropical Moored Buoy Array (GTMBA)

The network has its densest coverage in the tropics (Figure 6). The tropical moored buoy array began in the eastern equatorial Pacific in the early 1980s and expanded to cover 8°S–8°N across the Pacific with moorings and shiptime at present provided by the US National Oceanic Atmospheric Administration (NOAA) and the Japan Agency for Marine-Earth Science and Technology (JAMSTEC). The NOAA portion of the Pacific array is referred to as the Tropical Atmosphere and Ocean (TAO) array and the JAMSTEC portion is referred to as the Triangle Trans-Ocean Buoy Network (TRITON) array. As discussed below, the primary purpose of the TAO/TRITON array is to observe, better understand, and predict the El Niño-Southern Oscillation (ENSO). In 1997, the array expanded into the Atlantic with moorings from NOAA and shiptime provided by Brazil, France, and the US. The primary purpose of the Atlantic array, referred to as the Prediction and Research Moored Array in the Atlantic (PIRATA), is to observe, better understand, and predict seasonal, interannual, and longer variability, including both ENSO-like and meridional modes of variability. In 2000, the array expanded into the Indian Ocean, with moorings provided by the US, Japan, India, and China, and shiptime provided by India, Indonesia, France, Japan, and the Agulhas Somali Current Large Marine Ecosystems project. The primary purpose of the Research Moored Array for African-Australian Monsoon Analysis and Prediction (RAMA) is to advance monsoon research and forecasting.

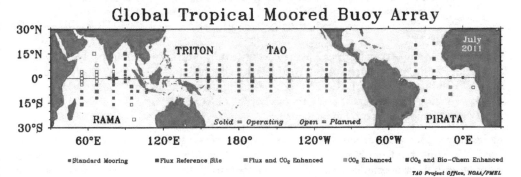

Fig. 6. Global Tropical Moored Buoy Array, as of July 2011. Flux reference stations are indicated by a blue square. Courtesy M. McPhaden, NOAA Pacific Marine Environmental Laboratory (PMEL).

Within the tropical Pacific, surface trade winds tend to blow from the cool waters off of South America to the warm waters off of Indonesia, where the wind converges and rises in deep convective clouds. As can be seen in Figure 7, as the warm water shifts eastward, the region of wind convergence and deep convection shifts eastward, resulting in the ENSO cycle, with teleconnections to global weather and climate patterns (Bouma et al., 1997; Diaz & Markgraf, 2000).

The standard suite of sensors on the tropical buoys includes wind speed and direction, air temperature and humidity, surface salinity, and surface and subsurface temperature. A number of the moorings, however, are enhanced with additional sensors to monitor the air-

Monthly Zonal Wind and Heat Content   2°S to 2°N Average

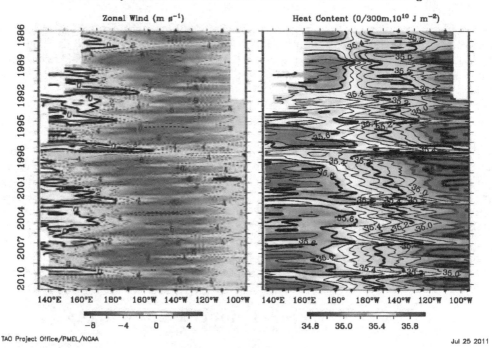

Fig. 7. Zonal wind (left) and upper 300 m heat content (right) time-series along the equatorial Pacific, as measured by the Pacific Tropical Atmosphere Ocean (TAO) /Triangle Trans-Ocean Buoy Network (TRITON) array. This figure was generated using the data display webpage, courtesy of the TAO project office of NOAA PMEL: http://www.pmel.noaa.gov/tao/jsdisplay/.

sea heat, moisture, and carbon dioxide fluxes; upper ocean salinity; and currents. The most heavily instrumented of these sites are designated as flux stations. The entire GTMBA, together with these specialized flux stations, contribute to the OceanSITES network of deep ocean reference stations (Figure 5).

Through the decades there have been several large international process studies built around the array, including the Coupled Ocean Atmosphere Response Experiment in the western tropical Pacific in 1992–1993 (Webster & Lukas, 1992), the Eastern Pacific Investigation of Climate in 2001 (Cronin et al., 2002; Raymond et al., 2004), and the GasEx 2001 study of physical, chemical, and biological factors controlling pCO_2 fluxes in the eastern equatorial Pacific (Sabine et al., 2004). In the Atlantic, African Monsoon Multidisciplinary Analyses (AMMA) occurred 2005–2007 (Lebel et al., 2011). The Cooperative INDian Ocean experiment on intraseasonal variability in the Year 2011 / Dynamics of the Madden-Julian Oscillation (CINDY/DYNAMO) experiment in the Indian Ocean is planned for 2011. Maintenance cruises for the array have also been opportunities for ship-based ancillary projects, including regular hydrographic and Acoustic Doppler Current Profiler (ADCP) sections (Johnson et al., 2002); water sample (Behrenfeld et al., 2006) and atmospheric boundary layer measurements (Fairall et al., 2008); underway surface pCO_2 (Feely et al., 2006) and chlorophyll fluorescence (Behrenfeld et al., 2006); and regular deployments of ARGO floats (Roemmich et al., 2009) and surface drifters (Lumpkin & Pazos, 2007), among other activities. For more information on the Pacific TAO/TRITON array, see McPhaden et al. (1998); for the Atlantic PIRATA array, see Bourlès et al. (2008); and for the Indian Ocean RAMA array, see McPhaden et al. (2009). Data and information can be accessed through the GTMBA project website: http://www.pmel.noaa.gov/tao/global/global.html.

4.1.2 Stratus reference station mooring west of Chile

The Stratus reference station mooring, located west of Chile at 20°S, 85°W in 4450 m depth water, was initiated in 2000. During the 1990s it became clear that nearly all coupled general circulation models had significant biases in the tropical Pacific that impeded their ability to properly reproduce the ENSO variability (Mechoso et al., 1995). In particular, nearly all models had too warm SST and too little stratus cloud in the eastern boundary region just west of Chile. As a result, these models tended to produce convective rainfall north and south of the equator, rather than just north of the equator as shown in Figure 3. The air-sea fluxes as well as the dynamics of the ocean and atmosphere in this data sparse region were poorly known, and any further progress required new data from the region. Thus, in 2000, with support from NOAA, a reference surface mooring, referred to as the "Stratus" mooring, was deployed at 20°S, 85°W. The mooring provides quality surface meteorology and air-sea fluxes of heat, freshwater and momentum, and CO_2. Annual cruises to maintain the buoy have provided opportunities for intensive ship-based measurements, particularly of the atmospheric boundary layer (Bretherton et al., 2004). In 2008, the international process study VAMOS Ocean-Cloud-Atmosphere-Land Study Regional Experiment (VOCALS-REx) (Wood et al., 2011) was anchored by the Stratus mooring. For more information on the Stratus reference mooring see Colbo & Weller (2007). The project website can be found at: http://uop.whoi.edu/projects/Stratus/stratus.html.

4.1.3 Northwest Tropical Atlantic Station (NTAS)

The NTAS surface mooring was established in 4700 m depth water near 15°N, 51°W to investigate surface forcing and oceanographic response in a region of the tropical Atlantic with strong sea surface temperature (SST) anomalies and the likelihood of energetic local air–sea interaction on interannual to decadal timescales. Two modes of coupled air-sea variability are found in the tropical Atlantic, a dynamic mode similar to the Pacific ENSO and a thermodynamic mode characterized by changes in the cross-equatorial SST gradient. Forcing for these modes may be by synoptic atmospheric variability, remote forcing from ENSO, and extratropical forcing from the North Atlantic Oscillation (NAO). Relationships between tropical SST variability, the NAO, and the meridional overturning circulation, as well as between the two tropical modes, are poorly understood.

The NTAS site is co-located with the easternmost subsurface mooring of the Meridional Overturning Variability Experiment (MOVE) "transport" array, which monitors the deep southward branch of the North Atlantic meridional overturning circulation west of the Mid-Atlantic Ridge. Annual cruises to NTAS are shared with MOVE. Funding for NTAS and MOVE is primarily from NOAA. For more information see Kanzow et al. (2008). The NTAS and MOVE project websites can be found at: http://uop.whoi.edu/projects/NTAS/ntas.html, and http://mooring.ucsd.edu/index.html?/projects/move/move_results.html.

4.1.4 Tropical Eastern North Atlantic Time-Series Observatory (TENATSO) and Cape Verde Atmospheric Observatory (CVAO)

CVAO meteorological and atmospheric chemistry measurements and TENATSO-moored physical and biogeochemical measurements in the tropical eastern North Atlantic were initiated in 2006. In 2008, routine ship visits to TENATSO were initiated to collect physical and biogeochemical measurements at TENATSO. CVAO is located on a small Cape Verde island (Sao Vicente) at 16.8°N, 24.9°W, while the TENATSO ocean station is located in 3600 m depth water ~93 km north of the island at 17.6°N, 24.2°W. Like other tropical stations, this is a region of intense air-sea interaction. Being downwind of the Mauritanian upwelling, the ocean and atmospheric data can be used to link biological productivity and atmospheric composition. The location is critical for climate and greenhouse gas studies and for investigating dust impacts on marine ecosystems. CVAO atmospheric reference data contribute to the Global Atmospheric Watch (GAW) program of the WMO, and TENATSO is part of the EuroSITES network (http://www.eurosites.info/), which contributes to the global OceanSITES network. CVAO and TENATSO are funded by Germany, UK, and the EU. For more information, see: Read et al. (2008). CVAO and TENATSO websites can be found at: http://ncasweb.leeds.ac.uk/capeverde/, http://tenatso.ifm-geomar.de/, and http://www.eurosites.info/tenatso.php.

4.2 North Pacific

4.2.1 Kuroshio Extension observatories and JAMSTEC biogeochemical observatories K2 and S1

The NOAA Kuroshio Extension Observatory (KEO) surface mooring is located south of the Kuroshio Extension jet at 32.3°N, 144.5°E in 5700 m depth water, and the JAMSTEC KEO (JKEO) surface mooring is located north of the jet at 38°N, 146.5°E in 5400 m depth water. KEO was initiated in 2004 during the Kuroshio Extension System Study (KESS) (Donohue et al., 2008) and JKEO was initiated in 2007. Both KEO and JKEO monitor the air-sea fluxes of heat, moisture, momentum, and carbon dioxide, as well as the upper ocean temperature, salinity, and near-surface currents in the region of very large ocean heat loss in the western North Pacific (Figure 1). The large heat fluxes occur during winter, when cold, dry continental air blows over the warm ocean current. As can be seen in Figure 1, similar regions of high ocean surface heat loss are seen in all basins (Cronin et al., 2010). This strong oceanic warming of the atmosphere can affect the surface winds, clouds, storm development, and, potentially, the storm track. The large air-sea heat fluxes also can affect the formation of water masses, or mode water. The KEO site is located in the subtropical mode water formation region and the JKEO site is located in the central mode water formation region (Oka et al., 2011a, 2011b). Mode waters are formed and modified at the surface, and, after they subduct beneath the surface layer, they generally preserve these characteristics as they circulate through the ocean (Hanawa & Talley, 2001; Oka & Qiu, 2011).

Beginning in 2011, KEO will be enhanced with additional sensors to monitor ocean acidification and the net biological production of oxygen in the surface waters. The carbon cycle and its biological pump are also being monitored at the JAMSTEC biogeochemical observatories, which include K2 in the western subarctic Pacific at 47°N, 160°E in 5200 m depth water and S1 in the western subtropical gyre at 30°N, 145°E in 5900 m water depth.

K2 was initiated in 2001 and S1 was initiated in 2010. Both the western and eastern regions of the subarctic Pacific are expected to experience significant effects of ocean acidification during the next century from the absorption of anthropogenic CO_2 (Orr et al., 2005).

Routine measurements from the JAMSTEC mooring maintenance cruises have included hydrographic, atmospheric profile sounding sections, and underway meteorological and oceanographic measurements, among other activities (Tokinaga et al., 2009). All four stations are visited regularly during JAMSTEC biogeochemical cruises. For more information on the KEO array, see Cronin et al. (2008) and Konda et al. (2010). For Station K2, see Kawakami et al. (2007). Project websites can be found at: http://www.pmel.noaa.gov/keo/, http://www.jamstec.go.jp/iorgc/ocorp/ktsfg/data/jkeo/ and http://www.jamstec.go.jp/res/ress/hondam/index_e.html.

4.2.2 Hawaii Ocean Time-series (HOT)

One of the most iconic long time-series is the famous "Keeling" curve showing the increase in atmospheric CO_2 observed at Mauna Loa since 1958 (Figure 8). While the atmospheric CO_2 has seasonal peak-to-peak variations of ~7 parts per million (ppm), over the past five decades, the CO_2 concentration has steadily increased by more than 10 times that amount due to anthropogenic sources.

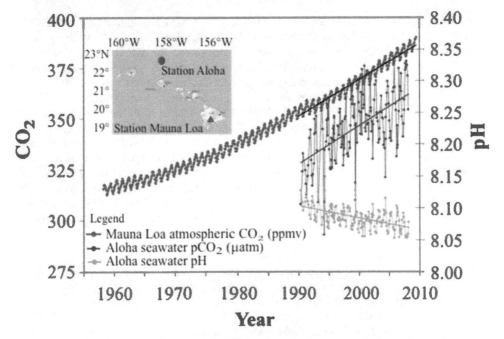

Fig. 8. Time-series of atmospheric CO_2 at Mauna Loa, in parts per million volume (ppmv; red), surface ocean pCO_2 (µatm; blue) and surface ocean pH (green) from the Hawaii Ocean Time-series Station ALOHA. Note that the increase in oceanic CO_2 over the past 17 years is consistent with the atmospheric increase within the statistical limits of the measurements. From Doney et al. (2009), after Feely et al. (2008).

A corresponding oceanic time-series at an observatory station in 4780 m depth water, 100 km north of the island of Oahu, was initiated in 1988 through the Hawaii Ocean Time-series (HOT) program, funded primarily by the US National Science Foundation (NSF). As shown in Figure 8 (Doney et al., 2009; Feely et al., 2008) the rise in pCO_2 is observed in the surface waters, and because it interacts with seawater to form carbonic acid as discussed in Section 2, this rise is also associated with a decrease in the water's pH. Essentially, the absorption of anthropogenic CO_2 is causing the waters to become more corrosive.

As shown in Figure 8, the sea surface CO_2 has rapid natural variability due to variations in the ocean temperature, mixing, upwelling, and biological processes. While much of this variability is captured in the monthly cruises to the HOT ALOHA (A Longterm Oligotrophic Habitat Assessment) observatory, there is significant variability at higher frequencies. Thus, with funding from NOAA and additional funding from NSF, in 2004, a surface reference station flux mooring was deployed at the observatory site. The mooring measures surface oceanic and atmospheric pCO_2 at three hourly intervals, and meteorological and other physical measurements even more frequently. For more information on the HOT program, see Karl et al. (2003). The project websites can be found at: http://hahana.soest.hawaii.edu/hot/hot_jgofs.html and http://uop.whoi.edu/projects/WHOTS/whots.html.

4.2.3 Station Papa in the eastern subarctic Pacific

One of the oldest ocean time-series is Ocean Station Papa, which began in December 1949 as part of the ocean weathership program. Station Papa is located at 50°N, 145°W in the eastern subarctic Pacific in 4260 m depth water. During its first year the site was occupied by a US Coast Guard ship. For the next three decades it was occupied continuously by Canadian ships on 6-week rotations. Taking meteorological and oceanic measurements, information was radioed to shore and contributed to the weather forecasts during this period. With the advent of the satellite era in the early 1980s, the Canadian Weathership Program was terminated. The Line-P program funded by Canadian Fisheries and Oceans, however, continued to make shipboard measurements on transects from Victoria, Canada, to Station Papa 3–6 times per year. Standard Line-P measurements include hydrography (Crawford et al., 2007), O_2 (Whitney et al., 2007), phytoplankton biomass and nutrient samples (Peña & Varela, 2007), zooplankton net tows (Mackas et al., 2007), chlorophyll, transmissivity, as well as dissolved inorganic carbon and total alkalinity (Wong et al., 2002), among other measurements. The present program samples three times per year, in February, May-June, and August-September.

Through the decades, Station Papa has been the location of numerous process studies, including, among others: the Mixed Layer Experiment in 1977 (Davis et al., 1981), Subarctic Pacific Ecosystem Research in 1984 (Miller, 1993), Storm Transfer and Response Experiment in 1980 and 1981 (Large et al., 1986), Ocean Storms in 1987 (Paduan & Niiler, 1993), and the SOLAS/SERIES iron enrichment experiment in 2003 (Boyd et al., 2004; de Baar et al., 2005). From 2007 to 2009, an NSF funded Carbon Cycle process study included support for a flux reference station mooring at Station Papa to monitor the carbon cycle and ocean acidification, in addition to the physical and meteorological environment (Emerson et al., 2011). In order to continue the mooring station on an ongoing basis, in 2009, support for the reference station mooring was transferred to NOAA. Shiptime for annual mooring maintenance has been provided by the Canadian Line-P program. The US NSF Ocean Observatory Initiative (OOI) plans to enhance this station in the coming years with additional moorings and sensors to make station Papa one of its four global nodes. For more information on Station Papa and Line-P, see Freeland (2007) and Peña & Bograd (2007). The project websites can be found at: http://www.pmel.noaa.gov/stnP/ and http://www.pac.dfo-mpo.gc.ca/science/oceans-eng.htm.

4.2.4 Monterey Bay Aquarium Research Institute (MBARI) moorings, California Current Ecosystem (CCE) moorings, and the California Oceanic Cooperative Fisheries Investigation (CalCOFI)

The MBARI moorings were in the California Current system at 36.7°N, 122°W in 1600 m and 36.7°N, 122.4°W in 1800 m water depth were first deployed in 1989. The moorings carry physical, meteorological (air-sea flux), and biogeochemical sensors. Ecosystem productivity and the biogeochemical cycling of elements in the California upwelling regions is regulated by physical processes that vary on daily to multidecadal time scales. As with other observatories described here, through these concurrent measurements of physics, chemistry, and biology, changes in biological and chemical fluxes associated with the physical variability can be estimated and used to develop predictive models. These moorings are funded primarily through support from the David and Lucile Packard Foundation, with support for bio-optical measurements from NASA. For more information, see Chavez et al. (1997). The project website can be found at: http://www.mbari.org/oasis/.

With funding from NOAA, two multi-disciplinary moorings, CCE1 and CCE2, are being sustained off Point Conception at 33.5°N, 122.5°W and 34.3°N, 120.8°W in 4000 m and 800 m of water, respectively. CCE1 was initiated in 2008 and CCE2 was initiated in 2010 and carry physical, meteorological, biogeochemical, and ecosystem sensors. The moorings contrast the productive upwelling regime near the coast and the open-ocean regime in the center of the southward flowing low-salinity Californian Current, and are co-located with repeat stations of the CalCOFI shipboard sampling grid, and a glider repeat transect. CCE1 and CCE2 provide real-time data and connectivity to sensors along the mooring wire down to several hundred meters depth, and have spare capacity for adding and telemetering additional community-provided sensors. Ground-truthing for chemical and optical/acoustic ecosystem observations is provided by CalCOFI cruises. The CalCOFI program began in 1949 for the purpose of studying the ecological aspects of the sardine population collapse off California. Initially monthly cruises, the present sampling is quarterly cruises to 75 stations in a 1.9 x 10^5 km^2 grid located off the coast of Southern California and provides unique long-term time-series at select locations in the southern California Current. For more information on CalCOFI, see: Ohman and Venrick (2003). The CCE and CalCOFI project websites can be found at http://mooring.ucsd.edu/cce/.

4.3 North Atlantic

4.3.1 Bermuda Atlantic Time-series Study (BATS)

Biweekly ship-based observations at "Hydrostation S", in 3300 m depth water 25 km SE of Bermuda, began in 1954, making this one of the few ocean time-series that exceeds 50 years. In October 1988, monthly cruises were extended to the BATS station located in 4500 m depth water approximately 80 km SE of Bermuda. These monthly (and biweekly during spring bloom periods) BATS cruises had a broader focus on the biogeochemistry and hydrography of the Sargasso Sea ecosystem. The site is located within the North Atlantic subtropical gyre, similar to the HOT location in the center of the North Pacific subtropical gyre. From 1994 through 2007, a surface mooring at this site, referred to as the "Bermuda Testbed Mooring" (BTM), carried a suite of meteorological, physical, and biogeochemical sensors. At present the BATS observations are supported primarily through NSF research grants. Funding cuts to the BTM, however, caused this long, high-resolution time-series to be discontinued in 2007. As discussed later, one of the main challenges of the reference station network is securing sustained funding. For more information on hydrostation S, see Phillips & Joyce (2007); for BTM and BATS, see Dickey et al. (2001). Project websites can be found at: http://www.bios.edu/research/bats.html and http://www.opl.ucsb.edu/btm.html.

4.3.2 Central Irmingir Sea (CIS)

The CIS observatory, established in 2002, is located about 200 km east of the southern tip of Greenland, at 59.4 °N, 39.4 °W in a water depth of 2800 m. The instrumentation is optimized for resolving physical and biogeochemical processes in the mixed layer, with sensors that monitor temperature, salinity, currents, nitrate, pCO_2, O_2, and fluorescence, among other variables. Wintertime surface cooling can be intense and very deep mixed layer depths have been observed, indicating deep water formation. Because weather conditions have been a perpetual challenge, the mooring has a small surface element for real time data transmission, but does not carry meteorological sensors.

The NSF OOI plans to enhance this station in the coming years with additional moorings and sensors to make the CIS station one of its four global nodes. Currently, CIS is funded by Germany and the EU. For more information, see: http://www.eurosites.info/cis.php.

4.3.3 Porcupine Abyssal Plain (PAP)

The PAP observatory is located in 4850 m depth water south of the North Atlantic Current, at 49°N, 16.5°W, in a region with a relatively flat seafloor. The mooring, equipped with sediment traps at three depths, was first deployed in 1989 to study and monitor the open ocean water column biogeochemistry, physics, and benthic biology. Capability has steadily increased to include upper ocean biogeochemical variables such as CO_2, chlorophyll and nutrients in 2002. In 2009, the station was enhanced to monitor surface meteorology and thus the observatory became an air-sea flux station as well. PAP is located in a region with large ocean absorption of atmospheric CO_2. Surface mixed layers are deep during winter, and during springtime the mixed layer becomes shallow, supporting a widespread phytoplankton bloom. PAP observations thus allow monitoring of the carbon cycle from the atmosphere to the abyss and its physical and biological pumps.

PAP is funded primarily by the UK Natural Environment Research Council (NERC) and the EU, and is part of the EuroSITES network. For more information on PAP see Lampitt et al. (2010b). The project webpage can be found at: http://www.noc.soton.ac.uk/pap.

4.3.4 European Station for Time-series in the Ocean Canary Islands (ESTOC)

The ESTOC observatory, located about 100 km north of the Canary Islands at 29.2°N, 15.5°W in 3610 m depth water, was initiated in 1994 with monthly ship visits to the station that included a sediment trap mooring and nearby subsurface current meter mooring. Since 2002, the station has been occupied by a surface mooring that measures upper-ocean physical and biogeochemical variables and surface meteorology. In 2007, the sediment trap mooring was terminated and in 2008, the surface mooring was upgraded to monitor air-sea fluxes. As it is windward of the Canary Islands, the station avoids wake effects of the Canary Current and northeast trade winds. It is also far enough from coasts and islands to serve as a reference for satellite images and altimetry.

Funding for ESTOC has come from the EU, the German Research Foundation (DFG), and national and regional projects from Spain and the Canary Islands. At present, funding from the governments of Spain and the Canary Islands comes primarily through the Canary Oceanic Platform (PLOCAN; http://plocan.eu). For more information on ESTOC, see Neuer et al. (2007) and González-Dávila et al. (2010). ESTOCS is part of the EuroSITES network and its websites can be found at: http://www.eurosites.info/estoc.php and http://www.estoc.es/.

4.4 Mediterranean Sea

4.4.1 Mediterranean Moored Multi-sensor Array (M3A) network

The M3A network includes three reference stations which contribute to the EuroSITES and OceanSITES networks: POSEIDON/E1-M3A in the south Aegean Sea at 35.8°N, 24.93°E (initiated in 2000), E2-M3A in the Adriatic Sea at 41.84°N, 17.76°E (initiated in 2004), and the W1-M3A in the Ligurian Sea at 43.81°N, 9.12°E (also initiated in 2004). All three moorings carry suites of sensors to monitor the surface meteorology and air-sea fluxes, directional wave parameters, upper ocean temperature, salinity, currents, and biochemical parameters in the euphotic zone. Biogeochemistry within the Ligurian Sea is also monitored by the DYFAMED station described below. All three M3A stations are in water depth greater than 1200 m. The M3A network is funded by Italy, Greece, and the EU. For more information, see: http://www.eurosites.info/.

4.4.2 Dynamics of the Atmospheric Fluxes in the MEDiterranean (DYFAMED) station in the Ligurian Sea

The DYFAMED station in the Ligurian Sea at 43.42°N, 7.87°E was initiated in 1988 with the deployment of a mooring with sediment traps at 200 m and 1000 m, in water depth of 2350 m. Since 1991, monthly cruises have been performed as well to observe the physical and biogeochemical variability throughout the water column. In 1999, a nearby surface mooring was deployed by Météo-France to monitor the surface meteorology and wave parameters. Ocean physical parameters are also measured at present by sensors mounted on the sediment trap mooring. DYFAMED is currently funded by France and the EU. For more information, see Marty (2002) and the project websites: http://www.obs-vlfr.fr/dyfBase, and http://www.eurosites.info/dyfamed.php.

4.5 Southern Ocean

4.5.1 Southern Ocean Time-Series (SOTS)

SOTS (Trull et al., 2010) commenced in 1998 with a sediment trap mooring program (SAZ; Trull et al., 2001) located in the Sub-Antarctic Zone 650 km south of Tasmania at 46.75°S,

142°E, in 4600 m of water. The site was expanded in 2003 with the addition of the Pulse mooring, to understand biogeochemical processes in the surface ocean, and again in 2010 with the addition of the Southern Ocean Flux Station (SOFS; Schulz et al., 2011) climate mooring, autonomous drifting profilers, and gliders. The Southern Ocean "Roaring Forties" is notorious for its storms, waves, and strong currents. Its Circumpolar Current is a route by which water can be carried from the South Atlantic Ocean to the South Indian Ocean and the South Pacific. As waters, formed at the surface in the Subantarctic Zone, sink and flow under warmer subtropical and tropical waters, they carry CO_2 into the deep ocean, out of contact with the atmosphere. Through this subduction process, oxygen and nutrients are also supplied to deep ocean ecosystems throughout much of the global ocean. It should be noted that this is the only OceanSITES surface mooring south of the Tropic of Cancer. SOTS is funded through the Australian Integrated Marine Observing System (IMOS; Hill, 2010; Meyers, 2008). For more information, see: http://imos.org.au/sofs.html.

5. Challenges facing the network

5.1 Long term commitment

Obtaining long time-series requires commitment: organizational, institutional, and scientific. Funding organizations that can support a long-term project do not always exist. In many cases, these long time-series are funded through 3–5 year research grants and the time-series is vulnerable to the funding cycle. If the research proposal with a 3-year time horizon is rejected, the long time-series is discontinued, as was the case of the 13-year surface mooring time-series at the BATS observatory discussed above. Likewise, for the very long time-series, the scientists who initiated the time-series may no longer be involved. During the transition in leadership, the institution's interest in the station can play a critical role in the ultimate success of the transition. Ultimately, the value of the station is determined by how the data are used, which depends upon the scientific importance of the station, the suite of measurements and their quality, the data latency and availability, and the ease with which the data can be used (Karl, 2010).

5.2 Public data and common data formats

OceanSITES has an active data management group that developed a self-documented netCDF (network common data form) format that all station operators agree to use. (For more information, see http://www.oceansites.org/data/). All station operators also agree to submit their data in this common format to Data Assembly Centers (DACS) that, in turn, forward data to two Global Data Assembly Centers (GDACS) that mirror each other: one at the NOAA National Data Buoy Center (NDBC) in the US and one at the Institut Français de Recherche pour l'exploitation de la MER (IFREMER) in France. Both GDACS can be accessed through the OceanSITES website provided above.

5.3 Governance

OceanSITES began as a volunteer group. Recently it has become an Action Group of the Data Buoy Cooperation Panel (DBCP) of the Joint WMO and International Oceanographic Commission's (IOC) Technical Commission for Oceanography and Marine Meteorology (JCOMM)   (http://www.jcomm.info/index.php?option=com_content&task=view&id=76

&Itemid=76). With support of a technical staffer at JCOMM and staff from the NOAA NDBC in the US and IFREMER in France, OceanSITES has made significant progress on developing governance. The executive committee includes representatives for each ocean, for the physical and biogeochemical communities, and from the data management panel. The OceanSITES data management team includes scientists and technical staff from the various DACS and GDACS. An emphasis has been placed on making all data openly and easily available at no cost to the user.

The OceanSITES also has a scientific steering team that includes the principal investigators (station operators) from all OceanSITES reference stations. The scientific steering team is charged with developing and reviewing the network and its data requirements and data management, coordinating the implementation of the network, identifying gaps in the network and synergies with other programs, and ensuring the integration of the network into the overall global ocean observing system. While many of the stations were initiated prior to or independently from the OceanSITES network, by becoming part of the network, the stations can significantly increase their user base and thereby increase the value of the station. Admittance into the network, however, carries responsibility, particularly in terms of providing open and easy access to the data.

5.4 High latitudes

As can be seen in Figures 5 and 6, most open ocean surface moorings are located in the tropics. While this is in part because the tropical environment is much more benign (it is much easier to maintain a mooring in tropical conditions than in the "Roaring Forties"), the primary reason is that, as discussed earlier, the ocean and atmosphere are highly coupled in the tropics. The tropical oceans can thus have a strong influence on the tropical and global atmosphere. However, higher latitudes are important to monitor as these source regions form various different water masses, are living environments for important fisheries, and are where the CO_2 solubility pump occurs and is the driver for the downwelling limb of the thermohaline circulation. Furthermore, model studies indicate that ocean acidification will lead to the high latitude surface waters becoming undersaturated with respect to calcium carbonate biominerals (e.g. aragonite, calcite) within a matter of decades (Orr et al., 2005). This would have a detrimental effect on the high latitude ecosystems and reference stations are needed to quantify these changes.

As we seek to use ocean and coupled ocean-atmosphere models to investigate the ocean's role in climate variability and change, there is great interest in knowing the fluxes across the ocean's surface integrated over its entire surface and in assessing whether the models accurately represent those surface integrals. Yet, the high latitudes have few reference stations and our knowledge of the regional surface meteorology, air-sea exchanges, and physical and biogeochemical dynamics, is poor. It is thus a high priority to expand ocean reference stations to the high latitudes.

6. The future

OceanSITES seeks to encourage the sustained support of ocean reference stations. As discussed above, many of the stations are supported through partnerships that involve multiple scientists, institutions, agencies, and nations. Hope for expansion into the high latitudes is at hand, as is shown by the Australian site south of Tasmania, SOTS, mentioned above. The US NSF OOI is also initiating four deep-ocean, full water column, interdisciplinary reference stations. These stations, referred to as "global nodes," would be located at strategic sites within the three-dimensional circulation of the global oceans, including two currently existing OceanSITES reference stations: Station Papa in the eastern subarctic Pacific, and CIS, in the Irminger Sea southeast of Greenland. The two other global nodes are in the Argentine Basin at 42°S, 42°W, and in the Southern Ocean, southwest of Chile at 55°S, 90°W. Further contributions to the high latitude sites and continued efforts to develop common, multidisciplinary instrumentation to be deployed at each site would complete the global array of ocean reference stations.

7. Conclusion

We live on a water world. Weather and climate over land cannot be isolated from that over and within the ocean. In order to understand the global heat balance, hydrological cycle, and carbon cycle, it is necessary to observe, understand, and map the physical, chemical, and ecosystem environment with sufficient temporal, horizontal, and vertical resolution. This is the purpose of the Global Ocean Observing System (GOOS), which is a system within the GEOSS. The network of OceanSITES reference stations is an integral part of the GOOS. Data from these reference stations detect rapid changes and episodic events as well as long-term changes. These reference data are made available to the public to further our understanding of our changing world. The data are used to validate and assess satellite products and improve our ability to monitor the globe remotely. Scientific researchers are using these data to study mechanisms controlling the climate and ecosystems and to test and improve numerical models used for predicting future changes. Our ability to plan, adapt, and cope with future changes in weather, climate, and ecosystem depends crucially upon our ability to monitor and predict these changes. Through dedication and commitment, the OceanSITES network provides the baseline data for these efforts.

8. Acknowledgement

Much of the information in section 4 was provided to the OceanSITES project office by the operators of the station. The authors thank Hester Viola for compiling this information into white papers that are available on the OceanSITES website. The authors also thank M. McPhaden, L. Carpenter, M. Honda, Y. Kawai, M. Church, B. Crawford, F. Chavez, M. Lomas, J. Karstensen, A. Cianca, V. Cardin, L. Coppola, R. Bozzano, E. Schulz, S. Bigley, K. Ronnholm, and Z. Yu for their helpful feedback on this manuscript.

9. References

Behrenfeld, M.J.; Worthington, K.; Sherrell, R.M.; Chavez, F.P.; Strutton, P.; McPhaden, M.J. & Shea, D.M. (2006). Controls on tropical Pacific ocean productivity revealed through nutrient stress diagnostics. *Nature*, Vol. 442, pp. 1025-1028

Bernie,D. J.; Guilyardi, E.; Madec, G.; Slingo, J.M. & Woolnough, S.J. (2007). Impact of resolving the diurnal cycle in an ocean-atmosphere GCM. Part 1: A diurnally forced OGCM. *Climate Dynamics*, Vol. 29, pp. 575-590

Bouma, M.J.; Kovats, R.S.; Goubet, S.A.; Cox, J.S.H. & Haines, A. (1997). Global assessment of El Niño's disaster burden. *Lancet*, Vol. 350, No. 9089, pp. 1435-1438

Bourlès, B.; Lumpkin, R.; McPhaden, M.J.; Hernandez, F.; Nobre, P.; Campos, E.; Yu, L.; Planton, S.; Busalacchi, A.J.; Moura, A.D.; Servain, J. & Trotte, J. (2008). The PIRATA Program: History, accomplishments, and future directions. *Bulletin of the American Meteorological Society*, Vol. 89, No. 8, pp. 1111-1125, doi: 10.1175/2008BAMS2462.1

Boyd, P.W. & Co-Authors (2004). Evolution, decline and fate of an iron-induced phytoplankton bloom in the subarctic Pacific. *Nature*, Vol. 428, pp. 549-553, doi: 10.1038/nature02437

Bretherton, C.S.; Uttal, T.; Fariall, C.W.; Yuter, S.E.; Weller, R.A.; Baumgardner, D.; Comstock, K.; Wood, R. & Raga, G.B. (2004). The EPIC 2001 stratocumulus study. *Bulletin of the American Meteorological Society*, Vol. 85, pp. 967-977

Cazenave, A. & Co-Authors (2010). Sea level rise – Regional and global trends. *Proceedings of the "OceanObs'09: Sustained Ocean Observations and Information for Society" Conference (Vol. 1)*, Venice, Italy, September 2009, Hall, J.; Harrison, D.E. & Stammer, D. (Eds.), ESA Publication WPP-306, doi:10.5270/OceanObs09.pp.11

Chavez, F.P.; Pennington, J.T.; Herlien, R.; Jannasch, H.; Thurmond, G.; & Friederich, G.E. (1997). Moorings and drifters for real-time interdisciplinary oceanography. *Journal of Atmospheric and Oceanic Technology*, Vol. 14, pp. 1199-1211

Colbo, K. & Weller, R. (2007). The variability and heat budget of the upper ocean under the Chile-Peru stratus. *Journal of Marine Research*, Vol. 65, No. 5, pp. 607-637

Crawford, W.; Galbraith, J. & Bolingbroke, N. (2007). Line P ocean temperature and salinity, 1956-2005. *Progress in Oceanography*, Vol. 75, pp. 161-178

Cronin, M.F.; Bond, N.; Fairall, C.; Hare, J.; McPhaden, M.J. & Weller, R.A. (2002). Enhanced oceanic and atmospheric monitoring underway in the eastern Pacific. *EOS Transactions AGU*, Vol. 83, No. 19, pp. 205, 210-211

Cronin, M.F.; Meinig, C.; Sabine, C.L.; Ichikawa, H. & Tomita, H. (2008). Surface mooring network in the Kuroshio Extension. *IEEE Systems Special Issue on GEOSS*, Vol. 2, No. 3, pp. 424-430

Cronin, M.F. & Co-Authors (2010). Monitoring ocean-atmosphere interactions in western boundary current extensions. *Proceedings of the "OceanObs'09: Sustained Ocean Observations and Information for Society" Conference (Vol. 2)*, Venice, Italy, September 2009, Hall, J.; Harrison, D.E. & Stammer, D. (Eds.), ESA Publication WPP-306, doi:10.5270/OceanObs09.cwp.20

Davis, R.E.; deSzoeke, R.; Halpern, D. & Niiler, P. (1981). Variability and dynamics of the upper ocean during MILE; Part I: the heat and momentum balances. *Deep-Sea Research*, Vol. 28A, pp. 1427-1451

de Baar, H.J.W. & Co-Authors (2005). Synthesis of iron fertilization experiments: From the Iron Age in the Age of Enlightenment. *Journal of Geophysical Research*, Vol. 110, C09S16, doi:10.1029/2004JC002601

Diaz, H.F. & Markgraf, V. (Eds.) (2000). *El Niño and the Southern Oscillation: Multiscale Variability and Global and Regional Impacts*, Cambridge University Press, ISBN: 0-521-62138-0, Cambridge, UK

Dickey, T. (2001). New technologies and their roles in advancing recent biogeochemical studies. *Oceanography*, Vol. 14, No. 4, pp. 108-120

Dickey, T.; Zedler, S.; Frye, D.; Jannasch, H.; Manov, D.; Sigurdson, D.; McNeil, J.D.; Dobeck, L.; Yu, X.; Gilboy, T. ; Bravo, C.; Doney, S.C.; Siegel, D.A. & Nelson, N. (2001). Physical and biogeochemical variability from hours to years at the Bermuda Testbed Mooring site: June 1994– March 1998, *Deep-Sea Research Part II: Topical Studies in Oceanography*, Vol. 48, pp. 2105-2140

Ding, Q.; Wang, B.; Wallace, J.M. & Branstator, G. (2011). Tropical-extratropical teleconnections in boreal summer: Observed interannual variability. *Journal of Climate*, Vol. 24, pp. 1874-1896, doi: 10.1175/2011JCLI3621.1

Doney, S.C.; Balch, W.M.; Fabry, V.J. & Feely, R.A. (2009). Ocean acidification: A critical emerging problem for the ocean sciences. *Oceanography*, Vol. 22, No. 4, pp. 16-25

Donohue, K.A. & Co-Authors (2008). Program studies the Kuroshio Extension. *EOS Transactions AGU*, Vol. 89, No. 17, pp. 161-162

Emerson, S.; Sabine, C.; Cronin, M.F.; Feely, R.; Cullison, S. & DeGrandpre, M. (2011). Quantifying the flux of $CaCO_2$ and organic carbon from the surface ocean using in situ measurements of O_2, N_2, pCO_2 and pH. *Global Biogeochemical Cycles*, Vol. 25, GB3008, 12 pp., doi:10.1029/2010GB003924

Eyre, J.; Andersson, E.; Charpentier, E.; Ferranti, L.; Lafeuille, J.; Ondrá, M.; Pailleux, J.; Rabier, F. & Riishojgaard, L. (2010). Requirements of numerical weather prediction for observations of the oceans. *Proceedings of OceanObs'09: Sustained Ocean Observations and Information for Society (Vol. 2)*, Venice, Italy, September 2009, Hall, J.; Harrison, D.E. & Stammer, D. (Eds.), ESA Publication WPP-306, doi:10.5270/OceanObs09.cwp.26

Fairall, C.W.; Uttal, T.; Hazen, D.; Hare, J.; Cronin, M.F.; Bond, N.; Veron, D.E. (2008). Observations of cloud, radiation, and surface forcing in the equatorial eastern Pacific. *J. Clim.*, Vol. 21, No. 4, pp. 655-673, doi: 10.1175/2007JCLI1757.1

Fairall, C.W. & Co-Authors (2010).Observations to quantify air-sea fluxes and their role in climate variability and predictability. *Proceedings of the "OceanObs'09: Sustained Ocean Observations and Information for Society" Conference (Vol. 2)*, Venice, Italy, September 2009, Hall, J.; Harrison, D.E. & Stammer, D. (Eds.), ESA Publication WPP-306, doi:10.5270/OceanObs09.cwp.27

Feely, R.A.; Sabine, C.L.; Lee, K.; Berelson, W.; Kleypas, J.; Fabry, V.J.; & Millero, F.J. (2004). Impact of anthropogenic CO2 on the CaCO3 system in the oceans. *Science*, Vol. 305, No. 5682, pp. 362-366

Feely, R.A.; Takahashi, T.; Wanninkhof, R.; McPhaden, M.J.; Cosca, C.E.; Sutherland, S.C. & Carr, M.-E. (2006). Decadal variability of the air-sea CO_2 fluxes in the equatorial Pacific Ocean. *Journal of Geophysical Research*, Vol. 111, C08S90, doi:10.1029/2005JC003129

Feely, R.A.; Fabry, V.J.; & Guinotte, J.M. (2008). Ocean acidification of the North Pacific Ocean. *PICES Press*, Vol. 16, No. 1, pp. 22-26.

Feely, R.; Fabry, V.; Dickson, A.; Gattuso, J.; Bijma, J.; Riebesell, U.; Doney, S.; Turley, C.; Saino, T.; Lee, K.; Anthony, K.; & Kleypas, J. (2010). An international observational network for ocean acidification. *Proceedings of the "OceanObs'09: Sustained Ocean Observations and Information for Society" Conference (Vol. 2)*, Venice, Italy, September 2009, Hall, J.; Harrison, D.E. & Stammer, D. (Eds.), ESA Publication WPP-306, doi:10.5270/OceanObs09.cwp.29

Freeland, H. (2007). A short history of Ocean Station Papa and Line P. *Progress in Oceanography*, Vol. 75, pp. 120-125

González-Dávila, M.; Santana-Casiano, J.M.; Rueda, M.J. & Llinás, O. (2010). The water column distribution of carbonate system variables at the ESTOC site from 1995 to 2004. *Biogeosciences*, Vol. 7, pp. 3067-3081

Gruber, N. & Co-Authors (2010). Towards an integrated observing system for ocean carbon and biogeochemistry at a time of change. *Proceedings of the "OceanObs'09: Sustained Ocean Observations and Information for Society" Conference (Vol. 1)*, Venice, Italy, September 2009, Hall, J.; Harrison, D.E. & Stammer, D. (Eds.), ESA Publication WPP-306, doi:10.5270/OceanObs09.pp.18

Gulev, S. & Co-Authors (2010). Surface energy, CO_2 fluxes and sea ice. *Proceedings of the "OceanObs'09: Sustained Ocean Observations and Information for Society" Conference (Vol. 1)*, Venice, Italy, September 2009, Hall, J.; Harrison, D.E. & Stammer, D. (Eds.), ESA Publication WPP-306, doi:10.5270/OceanObs09.pp.19

Guemas, V.; Salas-Mélia, D.; Kageyama, M.; Giordani, H. & Voldoire, A. (2011). Impact of the ocean mixed layer diurnal variations on the intraseasonal variability of sea surface temperatures in the Atlantic Ocean. *Journal of Climate*, Vol. 24, pp. 2889-2914, doi: 10.1175/2010JCLI3660.1

Hanawa, K. & Talley, L.D. (2001). Mode waters, In: *Ocean circulation and climate*, Seidler G. & Church, J. (Eds.), pp. 373-386, Academic, New York

Hill, K., (2010). The Australian Integrated Marine Observing System (IMOS). *Meteorological Technology International*, Vol. 1, pp. 114-118

Johnson, G.C.; Sloyan, B.M.; Kessler, W.S. & McTaggart, K.E. (2002). Direct measurements of upper ocean currents and water properties across the tropical Pacific during the 1990s. *Progress in Oceanography*, Vol. 52, pp. 31-61

Kanzow, T.; Send, U. & McCartney, M. (2008). On the variability of the deep meridional transports in the tropical North Atlantic. *Deep-Sea Research Part I: Oceanographic Research Papers*, Vol. 55, pp. 1601-1623.

Karl, D.M. (2010). Oceanic ecosystem time-series programs: Ten lessons learned. *Oceanography*, Vol. 23, No. 3, pp. 104-125

Karl, D.M., Bates, N.; Emerson, S.; Harrison, P.J.; Jeandel, C.; Llinas, O.; Liu, K.; Marty, J.-C., Michaels, A.F., Miquel, J.C., Neuer, S., Noriji, Y. & Wong, C.S. (2003). Temporal studies of biogeochemical processes determined from ocean time-series observations during the JGOFS era, In: *Ocean Biogeochemistry: The Role of the Ocean Carbon Cycle in Global Change*, Fasham, M.J.F. (Ed.), pp. 239-267, Springer, ISBN: 978-3-540-42398-0, New York

Kawakami, H., Honda, M. C., Wakita, M. & Watanabe, S. (2007). Time-series observation of dissolved inorganic carbon and nutrients in the northwestern North Pacific. *Journal of Oceanography*, Vol. 63, pp. 967-982

Konda, M.; Ichikawa, H.; Tomita, H. & Cronin, M.F. (2010). Surface heat flux variations across the Kuroshio Extension as observed by surface flux buoys. *Journal of Climate*, Vol. 23, pp. 5206-5221

Lampitt, R.S. & Co-Authors (2010a). In situ sustained Eulerian observatories. *Proceedings of the "OceanObs'09: Sustained Ocean Observations and Information for Society" Conference (Vol. 1)*, Venice, Italy, September 2009, Hall, J.; Harrison, D.E. & Stammer, D. (Eds.), ESA Publication WPP-306, doi:10.5270/OceanObs09.pp.27

Lampitt, R.S.; Billett, D.S.M. & Martin, A.P. (2010b). The sustained observatory over the Porcupine Abyssal Plain (PAP): Insights from time series observations and process studies (preface) [In special issue: Water Column and Seabed Studies at the PAP Sustained Observatory in the Northeast Atlantic]. *Deep Sea Research Part II: Topical Studies in Oceanography*, Vol. 57, No. 15, pp. 1267-1271, doi:10.1016/j.dsr2.2010.01.003

Large, W.G.; McWilliams, J.C. & Niiler, P.P. (1986). Upper ocean thermal response to strong autumnal forcing of the northeast Pacific. *Journal of Physical Oceanography*, Vol. 16, pp. 1524-1550

Lebel, T.; & Co-Authors (2011) The AMMA field campaigns: accomplishments and lessons learned. *Atmospheric Science Letters*, Vol. 12, pp. 123-128, doi: 10.1002/asl.323

Lumpkin, R. & Pazos, M. (2007). Measuring surface currents with Surface Velocity Program drifters: the instrument, its data, and some recent results. (Chapter 2) In: *Lagrangian Analysis and Prediction of Coastal and Ocean Dynamics*, Griffa, A.; Kirwan, A.D.; Mariano, A.; Özgökmen, T. & Rossby, T. (Eds.), pp. 39-67, Cambridge University Press, ISBN: 978-0-521-87018-4, Cambridge, UK

Mackas, D.L.; Batten, S. & Trudel, M. (2007). Effects on zooplankton of a warmer ocean: Recent evidence from the northeast Pacific. *Progress in Oceanography*, Vol. 75, pp. 223-252

Marty, J.C., (2002). The DYFAMED time-series program (French-JGOFS). *Deep-Sea Research II*, Vol. 49, No. 11, pp. 1963-1964

McPhaden, M.J.; Busalacchi, A.J.; Cheney, R.; Donguy, J.-R.; Gage, K.S.; Halpern, D.; Ji, M.; Julian, P.; Meyers, G.; Mitchum, G.T.; Niiler, P.P.; Picaut, J.; Reynolds, R.W.; Smith, N. & Takeuchi, K. (1998). The Tropical Ocean Global Atmosphere observing system: A decade of progress. *Journal of Geophysical Research*, Vol. 103, No. C7, pp. 14,169–14,240, doi: 10.1029/97JC02906

McPhaden, M.J.; Meyers, G.; Ando, K.; Masumoto, Y.; Murty, V.S.N.; Ravichandran, M.; Syamsudin, F.; Vialard, J.; Yu, L. & Yu, W. (2009). RAMA: The Research Moored Array for African-Asian-Australian Monsoon Analysis and Prediction. *Bulletin of the American Meteorological Society*, Vol. 90, pp. 459-480

Mechoso, C.R. & Co-Authors (1995). The seasonal cycle over the tropical Pacific in coupled ocean-atmosphere general circulation models. *Monthly Weather Review*, Vol. 123, pp. 2825-2838

Meyers, G. (2008). The Australian Integrated Marine Observing System. *Journal of Ocean Technology*, Vol. 3, pp. 80-81

Miller, C.B. (1993). Pelagic production processes in the Subarctic Pacific. *Progress in Oceanography*, Vol. 32, pp. 1-15

Neuer, S. & Co-Authors (2007). Biogeochemistry and hydrography in the eastern subtropical North Atlantic gyre. Results from the European time-series station ESTOC. *Progress in Oceanography*, Vol. 72, No. 1, pp. 1-29

Ohman, M.D. & Venrick, E.L. (2003). CalCOFI in a changing ocean. *Oceanography*, Vol. 16, No. 3, pp. 76-85

Oka, E. & Qiu, B. (2011). Progress of North Pacific mode water research in the past decade. *Journal of Oceanography*, doi:10.1007/s10872-011-0032-5

Oka, E.; Suga, T.; Sukigara, C.; Toyama, K.; Shimada, K. & Yoshida, J. (2011a). "Eddy-resolving" observation of the North Pacific subtropical mode water. *Journal of Physical Oceanography*, Vol. 41, pp. 666–681

Oka, E.; Kouketsu, S.; Toyama, K.; Uehara, K.; Kobayashi, T.; Hosoda, S. & Suga, T. (2011b). Formation and subduction of central mode water based on profiling float data, 2003–08. *Journal of Physical Oceanography*, Vol. 41, pp. 113–129, doi: 10.1175/2010JPO4419.1

Oki, T., & Kanae, S. (2006). Global hydrological cycles and world water resources. *Science*, Vol. 313, pp. 1068-1072, doi:10.1126/science.1128845

Orr, J.C. & Co-Authors (2005). Anthropogenic ocean acidification over the twenty-first century and its impact on calcifying organisms. *Nature*, Vol. 437, No. 7059, pp. 681–686, doi:10.1038/nature04095

Paduan, J.D. & Niiler, P.P. (1993). The structure of velocity and temperature in the northeast Pacific as measured with Lagrangian drifters in fall 1987. *Journal of Physical Oceanography*, Vol. 23, pp. 585-600

Peña, M.A. & Bograd, S.J. (2007). Time series of the northeast Pacific. *Progress in Oceanography*, Vol. 75, pp. 115-119

Peña, M.A. & Varela, D.E. (2007). Seasonal and interannual variability in phytoplankton and nutrient dynamics along Line P in the NE subarctic Pacific. *Progress in Oceanography*, Vol. 75, pp. 200-222

Phillips, H.E. & Joyce, T.M. (2007). Bermuda's tale of two time series: Hydrostation S and BATS. *Journal of Physical Oceanography*, Vol. 37, pp. 554-571, doi:10.1175/JPO2997.1

Raymond, D.J.; Esbensen, S.K.; Paulson, C.; Gregg, M.; Bretherton, C.S.; Petersen, W.A.; Cifelli, R.; Shay, L.K.; Ohlmann, C. & Zuidema, P. (2004). EPIC2001 and the Coupled Ocean-Atmosphere System of the Tropical East Pacific. *Bulletin of the American Meteorological Society*, Vol. 85, pp. 1341-1354, doi: 10.1175/BAMS-85-9-1341

Read, K.A.; Mahajan, A.S.; Carpenter, L.J.; Evans, M.J.; Faria, B.V.E.; Heard, D.E.; Hopkins, J.R.; Lee, J.D.; Moller, S.; Lewis, A.C.; Mendes, L.; McQuaid, J.B.; Oetjen, H.; Saiz-Lopez, A.; Pilling, M.J. & Plane, J.M.C. (2008). Extensive halogen-mediated ozone destruction over the tropical Atlantic Ocean. *Nature*, Vol. 453, pp. 1232-1235

Rio, M.-H. & Hernandez, F. (2004). A mean dynamic topography computed over the world ocean from altimetry, in situ measurements, and a geoid model. *Journal of Geophysical Research*, Vol. 109, C12032, doi:10.1029/2003JC002226

Roemmich, D.; Johnson, G.C.; Riser, S.; Davis, R.; Gilson, J.; Owens, W.B.; Garzoli, S.L.; Schmid, C. & Ignaszewski, M. (2009). The Argo Program: Observing the global oceans with profiling floats. *Oceanography*, Vol. 22, No. 2, pp. 34-43

Sabine, C.L.; Feely, R.A.; Johnson, G.C.; Strutton, P.G.; Lamb, M.F. & McTaggart, K.E. (2004). A mixed layer carbon budget for the GasEx-2001 experiment. *Journal of Geophysical Research*, Vol. 109, No. C8, C08S05, doi: 10.1029/2002JC001747

Schulz, E.; Grosenbaugh, M.A.; Pender, L.; Greenslade, D.J.M. & Trull, T.W. (2011). Mooring design using wave-state estimate from the Southern Ocean. *Journal of Atmospheric and Oceanic Technology*, in press

Send, U.; Weller, R.A.; Wallace, D.; Chavez, F.; Lampitt, R.; Dickey, T.; Honda, M.; Nittis, K.; Lukas, R.; McPhaden, M. & Feely, R. (2010). OceanSITES. *Proceedings of the "OceanObs'09: Sustained Ocean Observations and Information for Society" Conference (Vol. 2)*, Venice, Italy, September 2009, Hall, J.; Harrison, D.E. & Stammer, D. (Eds.), ESA Publication WPP-306, doi:10.5270/OceanObs09.cwp.79

Shinoda, T. (2005). Impact of the diurnal cycle of solar radiation on intraseasonal SST variability in the western equatorial Pacific. *Journal of Climate*, Vol. 18, pp. 2628-2636

Schmitt, R. & Co-Authors (2010). Salinity and global water cycle. *Proceedings of the "OceanObs'09: Sustained Ocean Observations and Information for Society" Conference (Vol. 1)*, Venice, Italy, September 2009, Hall, J.; Harrison, D.E. & Stammer, D. (Eds.), ESA Publication WPP-306, doi:10.5270/OceanObs09.pp.34

Takahashi, T. & Co-Authors. (2009). Climatological mean and decadal change in surface ocean pCO_2 and net sea-air CO_2 flux over the global oceans. *Deep-Sea Research Part II: Topical Studies in Oceanography*, Vol. 56, pp. 554-577, doi:10.1016/j.dsr2.2008.12.009

Talley, L., Fine, R., Lumpkin, R., Maximenko, N. & Morrow, R. (2010). Surface ventilation and circulation. *Proceedings of the "OceanObs'09: Sustained Ocean Observations and Information for Society" Conference (Vol. 1)*, Venice, Italy, September 2009, Hall, J.; Harrison, D.E. & Stammer, D. (Eds.), ESA Publication WPP-306, doi:10.5270/OceanObs09.pp.38

Tokinaga, H.; Tanimoto, Y.; Xie, S.-P.; Sampe, T.; Tomita, H. & Ichikawa, H. (2009). Ocean frontal effects on the vertical development of clouds over the western North Pacific: In situ and satellite observations. *Journal of Climate*, Vol. 22, pp. 4241-4260

Trull, T.W.; Sedwick, P.N.; Griffiths, F.B.; & Rintoul, S.R. (2001) Introduction to special section: SAZ Project. *Journal of Geophysical Research*, Vol. 106, pp. 31,425–31,430

Trull, T.W.; Schulz, E.W.; Bray, S.G; Pender, L; McLaughlan, D. & Tilbrook, B. (2010). The Australian Integrated Marine Observing System Southern Ocean Time Series facility. OCEANS '10 IEEE Sydney Conference Volume, May 2010, 7 pp., doi:10.1109/OCEANSSYD.2010.5603514

Wallace, J.M. & Gutzler, D.S. (1981). Teleconnections in the geopotential height field during the northern hemisphere winter. *Monthly Weather Review*, Vol. 109, pp. 784-812

Webster, P.J. & Lukas, R. (1992). TOGA COARE: The Coupled Ocean-Atmosphere Response Experiment. *Bulletin of the American Meteorological Society*, Vol. 73, pp. 1377-1416

Whitney, F.A.; Freeland, H.J. & Robert, M. (2007). Persistently declining oxygen levels in the interior waters of the eastern subarctic Pacific. *Progress in Oceanography*, Vol. 75, pp. 179-199

Wijffels, S. & Co-Authors (2010). Progress and challenges in monitoring ocean temperature and heat content. *Proceedings of the "OceanObs'09: Sustained Ocean Observations and Information for Society" Conference (Vol. 1)*, Venice, Italy, September 2009, Hall, J; Harrison, D.E. & Stammer, D. (Eds.), ESA Publication WPP-306, doi:10.5270/OceanObs09.pp.41

Wong, C.S.; Waser, N.A.D.; Whitney, F.A.; Johnson, W.K. & Page, J.S. (2002). Time-series study of biogeochemistry of the North East subarctic Pacific: reconciliation of the Corg/N remineralization and update ratios with the Redfield ratios. *Deep-Sea Research Part II: Topical Studies in Oceanography*, Vol. 49, pp. 5717-5738

Wood, R. & Co-Authors (2011). The VAMOS Ocean-Cloud-Atmosphere-Land Study Regional Experiment (VOCALS-REx): goals, platforms, and field operations, *Atmospheric Chemistry and Physics*, Vol. 11, pp. 627-654, doi:10.5194/acp-11-627-2011

9

Forest Fires and Remote Sensing

Abel Calle and José Luis Casanova
University of Valladolid,
Spain

1. Introduction

The use of remote sensing techniques for the study of forest fires is a subject that started already several years ago and whose possibilities have been increasing as new sensors were incorporated into earth observation international programmes and new goals were reached based on the improved techniques that have been introduced. Three main topics can be distinguished, in which remote sensing provides results that can be applied directly to the subject of forest fires: risk of fire spreading, detection of hot-spots and establishment of fire thermal parameters and, finally, cartography of affected areas. In the last years, other two important topics are getting increasing interest; the first one is the estimation of severity, related to the post-fire phase, and the other one is the atmospheric impact of fire emissions.

With respect to the risk of fires, remote sensing has provided very valuable results in real time, which was the required aim. However, in order to be able to predict the existence of fires, it is necessary to incorporate indicators of very heterogeneous types which sometimes fall out of the field of earth observation studies; indicators related to economy, social and human activities or historical statistics among others, should, for example, be taken into account. That's why remote sensing must be restricted to a very limited aspect which makes it only suitable for the estimation of the spreading risk related to the vegetation dryness and surface temperature values. The main magnitude used as an indicator is the vegetation index, above all, the NDVI (Normalized Difference Vegetation Index). The first results in the estimation of the fire risk, although not in real time, were obtained through analyses by the satellites belonging to NOAA (National Oceanographic and Atmospheric Administration) series, by means of AVHRR (Advanced Very High Resolution Radiometer) sensor. Later on, further indicators coming from the same sensors were incorporated so as to improve the algorithms and include the information relative to meteorological conditions like the surface temperature obtained through satellites. The combination of the NDVI with the surface temperature has given place to a mixed index in which the lineal regression slope in both magnitudes established cells of terrain, presents a good correlation with the vegetation evapotranspiration and water stress (Nemani & Running, 1989). The use of the slope in this relation has been incorporated through different algorithms by different authors in order to establish another risk indicator (Illera *et al.*, 1996); thus, Casanova *et al.*, (1998) introduced it to work in real time within the operation in Mediterranean countries. The possibility of using the spectral information in the middle infrared, in the 1.6 µm region, has given place to the introduction of other indicators related to the fuel's moisture since the vegetation's reflectivity in this wavelength interval is strongly influenced by the water contained in it.

Hunt & Rock (1989) suggested a new vegetation index similar in the equation to the NDVI but including the reflectance in the near infrared and the reflectance in the 1.6 μm region, an index indicating the fuel's moisture. At first, this index could only be applied to the Landsat-TM (Thematic Mapper) sensor for the creation of fuel maps (Chuvieco *et al.*, 2002). Today, it can be used in real time on the AVHRR and MODIS (Moderate Resolution Imaging Spectroradiometer) sensors to be incorporated to the risk maps as a new indicator.

The detection of hot spots and, together with it, the establishment of fire parameters, is the most complex task of the ones presented here, due to the orbital configuration of the current spacecrafts. The methodologies are very clear from the point of view of physics, but the restrictions of the current sensors introduce difficulties in order to get quality results. By detection, it is understood the task of determining the location of a hot spot independently of its size. By monitoring, it is understood the establishment of the most important fire parameters with a view to obtain relevant information on this phenomenon. Among these parameters are the fire's temperature, the area taken by the fire, the energy intensity and, when the sensor's capacity allows it, the establishment of the advancing fire line. In order to place this subject of study in its appropriate context, it must be pointed out that fire detection with an aim to create alarms that facilitate a rapid extinction is a necessity that hasn't been fully resolved yet. Despite its limitations the NOAA-AVHRR sensor has been the most important for fire detection and has provided a benchmark for subsequent sensors. An excellent revision of the algorithms used on AVHRR can be found in Li *et al* (2001). The case of the European sensor (A)ATSR (Advanced Along Track Scanning Radiometer) and the World Fire Atlas from 1997 published by the ESA with the ERS-1 and ERS-2 (European Remote Sensing Satellite) satellites data (Arino & Rosaz, 1999) has been used to demonstrate its suitability to fire detection and assessment of vegetation fire emissions. The appearance of the MODIS sensor heralded a significant step forward in the observation of forest fires (Giglio *et al.*, 2003) and, at this moment, the MODIS fire product is a consolidated product and a reference for global Earth observation. Fire product has been identified as an important input for global change analysis; however, although the radiometric availability is satisfactory, the main problem is the time resolution to operate in real time. Detection of high temperature events through geostationary satellites has been taken into account with the different perspective. The improvements introduced in the sensors have allowed us to use geostationary satellites beyond their meteorological capabilities, adapting them to Earth observation; this is the response to the need for series of stable fire activity observations for the analysis of global change, changes in land use and risk monitoring. The GOES (Geostationary Operational Environmental Satellite) has been the worldwide reference for fire monitoring through geostationary platforms. Since 2000, the Geostationary Wildfire Automated Biomass Burning Algorithm (WF_ABBA) has been generating products for the western hemisphere in real-time with a time resolution of 30 minutes and this detection system has been operational within the NOAA NESDIS programme since 2002. The GOES-East and GOES-West spacecrafts are located in the Equator, providing diurnal coverage of North, Central and South America and data based on fire and smoke detection. The results provided by the GOES programme have been the starting point of a global geostationary system for fire monitoring, initially comprising four geostationary satellites that were already operational: two GOES platforms, from the USA, the European MSG (Meteosat Second Generation) and the Japanese MTSAT (Multifunctional Transport Satellite, covering Southeast Asia and several parts of India as observation regions. The minimum fire detection sizes of GOES, MSG and MTSAT, with time resolution less than 30 minutes has allowed the international community to think in a global observation network in real time. The implementation of this network is the aim of the Global Observations of Forest Cover and Land Cover Dynamics (GOFC/GOLD) FIRE Mapping and Monitoring program, internationally focussing on decision-taking concerning research into Global Change. The GOFC/GOLD FIRE program and the Committee on Earth Observation Satellites (CEOS) Land Product validation held a workshop dedicated to the applications of the geostationary satellites for forest fire monitoring (Prins *et al.*, 2004).

The cartography of areas affected by fires is a subject that has been dealt with in depth. Remote sensing has proved to be very useful in the study of forest fires cartography and severity since the time resolution does not prevent the subsequent evaluation of the consequences. Different radiometric procedures have been used based on the application of fixed thresholds to the NDVI in the case of low spatial resolution; the results are satisfactory but the difficulty of these procedures lies in the search of a fixed threshold value, because what seems quite probable is the dependence of the threshold values according to the area analysed and the time of the year in which the study is carried out. Methodologies based on neural networks (Al-Rawi *et al.*, 2001) have also been applied although they have the difficulty of training the neural net so that the final results will depend on the variability of the statistical sample used in the preparation of the neural net. The multi-temporal use of the NDVI for the radiometric analysis and the establishment of thresholds on the NDVI through a spatial contextual analysis is a procedure has been used for low spatial resolution. In the case of high spatial resolution, several procedures have been suggested using many different methodologies to be applied to the TM sensor, on board of Landsat satellites being one of the most frequent the spectral classification. However, an important problem is that it requires the distribution of data probability; another disadvantage is that in order to obtain higher quality results a supervised classification of the zones must be carried out, requiring interaction by user. Within the automatic methodologies, the lineal transformations have shown a great capacity for the obtaining of results. Thus, the application of the Principal Components to the reflectance bands obtains almost immediate cartographic results, since they can be analysed visually through a RGB (Red Green Blue) composite of the output components. An issue linked to the fire cartography is the estimation of severity. Each summer large fires affect to the Mediterranean Europe due to changes in traditional land use patterns which have led to an unusual accumulation of forest fuels, notably increasing fire risk and fire severity. According with Roldán-Zamarrón *et al.* (2006), there is interest in finding a quick and affordable methodology for obtaining fire severity maps that can be made available only a few days after the fire, as this information could prove very valuable in the early stages of rehabilitation planning for large fires. These maps should be based on independent data sources, such as remote sensing, employ automatic or semiautomatic methods, and produce results of an acceptable reliability. Remote sensing techniques are a useful tool in order to generate maps showing different degrees of damage affecting vegetation after a large wildfire in an effective manner. Objective of these severity maps is to locate priority intervention areas and plan forest restoration works.

2. Fires and climate

Following the GCOS (Global Climate Observing System) document "Systematic Observation Requirements for Satellite-based products for Climate", and ESA Climate

Initiative, the emissions of greenhouse gases (GHGs) and aerosols from fires are important climate forcing factors, contributing on average between 25-35% of total CO_2 emissions to the atmosphere, as well as CO, methane and aerosols. Hence, estimates of GHG emissions due to fire are essential for realistic modelling of climate and its critical component, the global carbon cycle. Fires caused deliberately for land clearance (agriculture and ranching) or accidentally (lightning strikes, human error) are a major factor in land-cover changes, and hence affect fluxes of energy and water to the atmosphere. Burnt area, as derived from satellites, is considered as the primary variable that requires climate-standard continuity. It can be combined with information on burn efficiency and available fuel load to estimate emissions of trace gases and aerosols. Measurements of burnt area can be used as a direct input to climate and carbon cycle models, or, when long time series of data are available, to parameterize climate-driven models for burnt area. Burnt area, combined with other information (burn efficiency and available fuel load) provides estimates of emissions of trace gases and aerosols. Measurements of burnt area can be used as a direct input to climate and carbon-cycle models, or, when long time series of data are available, to parameterise climate-driven models for burnt area (fire is dealt with in many climate and biosphere models using the latter approach). Fire-induced emissions are a significant terrestrial source of GHGs, with large spatial and interannual variability. Detection of active fires serves as part of the validation process for burnt area (i.e., is the burnt area associated with previous observations of active fire). Detection of active fires provides an indicator of seasonal, regional and interannual variability in fire frequency and shifts in geographic location and timing of fire events. Strong empirical relations exist between the FRP (Fire Radiative Power) and rates of combustion; so, the use of multiple FRP observations to integrate over the lifetime of the fire provides an estimate of the total CO_2 emitted. FRP provides a means to derive a CO_2 emissions estimate from remotely-sensed observations without relying on difficult-to-acquire ancillary data on fuel load and combustion completeness factors.

3. Fire detection

3.1 Physic principle of fire detection

As is to be expected, the process of the detection of hot spots is based on the use of bands in the middle and thermal infrared spectrum. There are three laws of Physics that govern the detection process: law of Plank, Wien's displacement law and Stefan-Boltzmann's law. The radiance emission corresponding to a body with a temperature of 300 K, as can be the Earth's mean temperature, will have a maximum value close to 10 μm, and a spectral band situated in this one would receive a very strong signal. For a temperature of 800 K, the maximum value will have displaced to wavelengths close to 3.6 μm and whereas the signal here would be very intense, it wouldn't be nearly as intense in higher wavelengths. The fire detection is based precisely on this inversion, which is possible to detect with thermal bands situated in the spectral regions of 11-12 μm and 3-4 μm. Figure 1, shows this physic principle graphically. The location of two generic bands, in the middle and thermal infrared, are also shown. It's clear that at a temperature of 300K, the radiance received in MIR (Middle InfraRed) is lower than the one received in TIR (Thermal InfraRed). However, at 500 K, this behaviour has inverted and now the radiance is higher in MIR.

The basic principle followed for the location of the spectral bands in a sensor, is described through two questions; first: what I want to see? And second: where the atmosphere allows

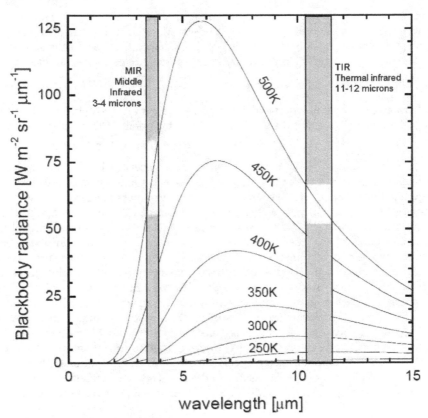

Fig. 1. Law of Planck, showing blackbody emission for different temperatures values of source, and location of MIR and TIR spectral bands (adapted from Li *et al*, 2001).

me to do? Fortunately, atmospheric absorption is selective in several spectral bands; the water vapour absorption is very strong below 3.4 μm but there is a atmospheric window in the interval [3.5-4.2 μm]; so, the MIR bands must be located preventing the absorption of water vapour in the 3-4 μm region. In the case of TIR region, there is a strong absorption band centred at 9.6μm, due to ozone, but there is an atmospheric window in the interval of [10-12 μm], with a weak effect of water vapour; its easy to remove this effect by means of two spectral bands located in this window (split-window technique, Price 1984).

A technical problem to take into account is that the radiance obtained by the sensor, coming from a concrete pixel in which there is a fire, does not only depend on the fire's temperature and, as it is logical, on the temperature of the surrounding surface, but also on the location of the fire inside the pixel since, at the end, the sensor's PSF (Point Spread Function) will determine the filtering that is carried out on the original image. In any case, to obtain a positive detection from an active fire is easy. The main problem to detect fires is to obtain a positive detection when the fire does not exist; that is: a false alarm. The false detections, in the 4 μm region, are due to radiance not only is coming from the emission, but there also exists a component due to the effect of reflection. That's why we can find ourselves in situations in which a high radiance signal does not necessarily correspond to a high-temperature pixel, except in the case of night observation, when the reflection component, evidently, does not exist. As is to be expected, the radiance that gets to the sensor in that part of the spectrum where the reflection and the emission effects superimpose, as is the 4 µm case. The value of the reflected component $L_{reflection}$ is $\dfrac{\alpha E_0 \cos(\theta_{sun})}{\pi}$, with θ_{sun} being the sun's zenith angle, E_0 the extraterrestrial sun's irradiance in that spectral band and α the spectral reflectance (the atmospheric effect are not included). The emission component, $L_{emission}$ is $(1-\alpha) \cdot B(\lambda, T_{surface})$, in which $B(\lambda,T)$ is the function of Planck, λ the wavelength in that spectral band and $T_{surface}$ the surface temperature observed by the sensor. Note that $\varepsilon=(1-\alpha)$ and $\alpha=(1-\varepsilon)$, with ε the emissivity. If a surface has a high value of reflectance in the MIR region, radiance coming from source is high and brightness temperature will be higher than surface temperature in several Kelvin. As an example, it can be observed that for a reflectance value of 20% and a sun's zenith angle of 30° the brightness temperature can increase more than 20 K higher due to reflectance effect. When the contribution of the reflection component is very marked and as a consequence the radiance increases in the middle infrared band, a pixel appears with an apparent high temperature that can be mistaken with a hot spot, producing a false alarm. These situations are more frequent in the highest spatial resolution sensors and when high reflectance surfaces coincide with sun-satellite geometrical situations close to specular reflection conditions. This is the case of small water surfaces, for example, and it is called *sun glint*. It must be pointed out that the problem of the appearance of false alarms is more difficult to solve than the detection itself due to the difficulty in separating both effects. Finally, it must be mentioned that clouds are also an important source of false alarms due to the sun's reflection. Their reflectance is high and they cause a strong signal in the MIR spectral band in situations of very high sun zenith angles.

3.2 Fire detection using heliosynchronous platforms

Not only was the NOAA-AVHRR sensor the first one to provide results, but it has also been a research platform in the development of hot-spot detection algorithms. This has been possible thanks to its high time resolution (among the polar heliosynchronous sensors) and to which the physic principles of detection mentioned above can be applied. It must be pointed out, however, that the AVHRR sensor has important limitations. The most important of these is the low saturation level, 320-331 K (Robinson, 1991), of the main band involved in the detection, the 3.7µm band. This limit is so low that a fire with a temperature of 1000 K on a non-reflective surface of 300K only needs a 13x 13 m^2 surface to reach the pixel's saturation. This important drawback makes the sensor suitable for the detection of hot-spots but in most cases, it makes it unsuitable for the analysis at a sub-pixel level. In spite of its limitations it is unavoidable to use this sensor as a comparative reference for subsequent, more operative sensors such as MODIS (Ichoku *et al.*, 2003). The detection has been developed through different algorithms that can be schematically classified into algorithms based on fixed thresholds and contextual algorithms, whose parameters have been adapted to the different zones of study. Both types of algorithms have advantages and disadvantages and their application will depend on the type of sensors to which they are going to be applied. The detection algorithms based on fixed thresholds, also called multi-channel, are based on the establishment of minimum temperature values in different spectral bands from which the detection is established. The most common scheme is to consider that a pixel is affected by a fire when the following conditions are fulfilled simultaneously:

$$T_{MIR} > V_{MIR} \; ; T_{MIR} - T_{TIR} > V_{DIF} \; ; T_{TIR} > V_{TIR} \; ; R_{NIR} < V_{NIR} \tag{1}$$

where T_{MIR} and T_{TIR} refer to the brightness temperature in the spectral bands of the 3.7 μm and 11 μm regions respectively, and V is the adopted threshold. In the former test, the first two conditions are the ones that carry out the detection of hot-spots strictly speaking according to the physic principles previously stated. The T_{MIR} test is for fire detection and the T_{MIR}-T_{TIR} test is to carry out the differentiation between the fire, which has high values in the MIR, and the hot surfaces which have high values both in the MIR and TIR. The T_{TIR} test is a cloud filter to apply the test to images in which the cloud cover has not been removed through other procedures. The R_{NIR} test is to filter the reflectance in sun-glint situations that are responsible for the appearance of false alarms. The threshold values established are varied. They depend on the algorithm and, above all, on the geographic area due to the influence of the background temperature. Thus, normally low surface temperature values use lower MIR threshold values without the appearance of false alarms. Two examples of this type of algorithms, operating on NOAA-AVHRR, are the used by the CCRS (Canadian Centre of Remote Sensing) (Li et al., 2000) and the ESA (European Space Agency) (Arino and Mellinote, 1998). The disadvantage of the algorithms based on fixed thresholds is that the values established depend on the zone of study and their environmental temperatures. In order to avoid this dependence, contextual algorithms can be used. They are based on the obtaining of threshold values carrying out a statistical analysis of the environment. The basic scheme is summarised in the following test:

$$T_{MIR} > \mu_{MIR} + f \cdot \sigma_{MIR} \; ; T_{MIR} - T_{TIR} > \mu_{DIF} + f \cdot \sigma_{DIF} \; ; R_{NIR} < \mu_{NIR} - f \cdot \sigma_{NIR} \tag{2}$$

where μ and σ are the mean values and the standard deviation in the environment of the pixel analysed and f is a factor that has to be established. The environment is analysed in a matrix with a size of NxN pixels, being N an odd value depending on the sensor to which it is applied. Two examples are the IGBP (International Geosphere and Biosphere Programme) algorithm (Justice & Malingreau, 1993), and an adaptation of the current algorithm on MODIS (Kaufman et al., 1998). Contextual algorithms have the advantage of making the detection process independent from the season and the zone analysed, since the thresholds are obtained by means of a statistical analysis of the environment. However, they have a serious drawback when they are applied to images in which the clouds have not been filtered since cloud edges cause false alarms. A variant to the basic contextual algorithm exposed is the one suggested by Lasaponara et al. (1998), in which the mean statistical parameters and the standard deviation are determined by using not just the spatial environment but also the temporal one, extending the matrix of analysis to the images of previous days, looking for the changes in the brightness temperature not only in a spatial scale but in a temporal interval too.

The launch of the MODIS sensor in 1999 on the Terra platform and in 2002 on Aqua with 36 different-spatial-resolution spectral bands has provided much more reliable results in detection. This sensors includes two spectral bands in the 4 μm spectral zone with saturation values very high to the MIR AVHRR band and the applied algorithm uses a large number of bands to consolidate the results. Another important characteristic of MODIS is an excellent radiometric resolution of 12 bits (instead AVHRR sensor with 10 bits) very interesting to fire monitoring. The original algorithm has been improved (Giglio *et al.*, 2003) and it carries out three test phases: cloud cover filtering, detection and consolidation test. The cloud-and-water-filtering phase uses three spectral bands: the reflectances in bands 1 and 2 with a spatial resolution of 250 meters, centred in 0.65 μm and 0.86 μm, respectively, and the temperature in the band of 12 μm, T_{12}. Thus, the pixels fulfilling any of the three following conditions will be rejected: having a T_{12} value lower than 265 K or a sum of reflectances higher than 0.9 or T_{12} lower than 285 K and sum of reflectances higher than 0.7 simultaneously. First, the detection phase establishes the potential pixels that must be analysed according to the criteria used by AVHRR with fixed thresholds, analysing the temperature in band 21, around 4 μm, establishing a threshold of 310 K and the difference of this band with band 31, around 11 μm, with a threshold difference of 10 K. Later, the identification of fire pixels, among the potential ones, is carried out through two procedures: first, an absolute test for the ones that have a $T_{4μm}$ value higher than 360 K during the day and 320 K at night. Secondly, an alternative test which carries out a characterisation of the environment's temperature through a contextual analysis on the pixels that were not considered potential and with a variable window until a significant number of points is obtained. This contextual analysis is similar to the one used by the AVHRR algorithms, but it follows additional steps to eradicate false alarms and it differentiates between day and night pixels. The methodology considers three different sources of false alarms: the first one is the possibility of sun-glint, which is solved through a geometrical analysis with the sun-pixel-satellite directions in order to reject situations of specular reflection; the second one are hot desert pixels and the third one the coast lines. The two latter are solved through the establishment of temperature and reflectance thresholds simultaneously. Finally, the consolidation phase establishes a statistical analysis to obtain well-confirmed pixels affected by a fire. This is due to the fact that the spatial resolution of MODIS in the thermal is 2 km, with step of 1km. This may cause that the same fire, located in a zone where two pixels are superimposed can be revealed by both of them. The consolidation test is carried out on the pixels adjacent to the one which is being analysed. More details about the algorithm can be seen in (Giglio *et al.*, 2003). It must be pointed out that MODIS has two bands in the MIR region used in detection: bands 21 and 22, both centred in the 3.9 μm. The difference is that the saturation level for the first one is 500K whereas for the second one it is 331. However, band 22 has less noise and a smaller error in the calibration. That's why, if the pixel is not saturated, the algorithm uses band 22. Otherwise, band 21 is used.

In spite of the tools we have shown for fire detection, it must be said that their results have not been brought into operation due to the lack of continuity in the monitoring of heliosynchronic satellites. Several monitoring programmes have been designed based on the co-ordination of several satellites in different orbital planes in order to increase the number of daily observations on a concrete place. Projects such as FUEGO originally and FUEGOSAT nowadays, are funded by ESA in order to obtain a product that can be put into operation for the monitoring of forest fires. On the other hand, the effectiveness in detection, of the sensors mentioned, could be improved through the design of sensors specially dedicated to fire detection. There was a prototype satellite fulfilling these characteristics and that has provided results to be analysed. It is the BIRD (Bi-spectral Infrared Detection), designed by the DLR German laboratory, as a sensor prototype, and its detection capacities have been very satisfactory thanks to its design (Briess *et al.*, 2003); the HSRS (Hot Spot Recognition Sensor System), with a visual field of 19° (190 km), a spatial resolution of 370 m and a radiometric resolution of 14 bits. Apart from the new spatial resolution in the thermal and its excellent radiometric resolution, this sensor is able to establish a dynamic rank of calibration that is completed with two successive expositions of the scene with a short time of integration; this makes it possible to establish a saturation limit close to 1000K, with a temperature resolution in the interval [0.1-0.2 K]. The algorithm includes 5 consecutive tests through which different threshold values of analysis are established: an adaptive test in the MIR to detect potential hot-spots, a threshold in the NIR to reject the sun reflection, which is a source of false alarms during day observations, a threshold adaptive to the MIR/NIR fraction of radiances to reject clouds and other high-reflective objects, a threshold adaptive to the MIR/TIR fraction of radiances to reject hot surfaces and finally, the gathering of pixels that are adjacent to the fire to obtain the fire's temperature and area parameters. It is important to mention that all the adaptive thresholds mentioned are obtained through the contextual spatial analysis. BIRD satellite must be considered as a very low-cost prototype to operate with several units in orbital co-ordination. The fire parameters provided by BIRD have been able to locate the flaming front very accurately (Wooster *et al.*, 2003).

3.3 Fire detection using geostationary platforms

As we have mentioned in the section "Introduction", the geostationary sensors can improve the fire detection results, due to its very short revisit time, even when spatial resolution is very limited due to location in space of geostationary platforms. Currently, the users international community feels that a real-time global observation network may become a reality by means of geostationary sensors such as GOES, MSG and MTSAT. This is one of the objectives of the Global Observations of Forest Cover and Land Cover Dynamics (GOFC/GOLD) FIRE Mapping and Monitoring program, focussing internationally on decision-taking concerning research into Global Change and its ecological and environmental implications. Major efforts are also being made by ESA-EUMETSAT to increase the use of MSG in environmental observation tasks. SEVIRI (Spinning Enhanced Visible and Infrared Imager) on board MSG platforms is a very interesting example of suitable sensor to perform forest fire monitoring in real time (Calle *et al.*, 2006). Some analyses are shown in the particular case of the geographical latitude of the Mediterranean Europe where, during the last years, detection campaigns and dissemination of results in real time have been carried out. The theoretical analysis of the minimum detectable size, including atmospheric effects and saturation conditions, are especially important to delimit the operational range of this sensor in Mediterranean latitudes, where the effects of forest fires are increasingly devastating each year, both in terms of financial as well as human losses. MSG-SEVIRI is geostationary sensor with a time resolution of 15 minutes; so, the comparison between successive scenes provides reliable results once the difference temperature threshold is established for such an interval. Thus, if a Time Thermal Gradient, TTG, higher to the one considered as normal, is detected, we will have a high temperature event. In order to estimate this gradient let's consider a day's thermal evolution as a sinusoidal curve responding to the form: $MIR\_Temp = A \cdot \sin(wt - \delta) + B$; $w = \dfrac{2\pi}{T}$; where T is the day's period in units of 15 minutes (T=96), A is semi-daily thermal oscillation and B is a not relevant coefficient. According to this model, the maximum difference in the MIR standard temperature between two consecutive SEVIRI scenes is ±1.5K for a diurnal-cycle thermal oscillation of around 30K, which is typical of summer days in middle latitudes. This estimation agrees with the experimental values found in the analysis of the series of MIR temperature evolution curves selected for different test sites in the Mediterranean Europe, during summer. Like this, the maximum temperature difference found, in absolute values, in the 98.2% of cases was lower than 2K. The averaged of differences found, only considering the intervals with thermal variability [05:00-11:00 GMT] and [14:00-20:00 GMT], was 1.2K, with a standard deviation of 0.5 K. So, we have considered appropriate to establish a threshold of 4K as the temperature increase value to detect the beginning of a fire without providing false alarms. In any case, it must be pointed out that there are two daily periods very well defined: from sunrise to midday, in which the temperature is increasing and where the estimation of 4K is appropriate, and the second one between midday and sunset, for which a value of 2-4K would be enough, being a negative gradient. During night periods detection is easier. In order to estimate minimum fire size detectable by the SEVIRI sensor, simulations have been done by means of MODTRAN radiative transfer code (Berk et al., 1996) by introducing different surface and fire temperatures according to different time thermal gradient values. Radiance observed by sensor was simulated as: $L_{sensor} = p \cdot L_{fire} + (1-p) \cdot L_{surface}$

where p is the surface fraction affected by fire and where two homogeneous phases have been considered: fire and surface; L_{fire} and $L_{surface}$ are the radiances incoming from fire and surface. Spectral radiance was integrated with 20 cm^{-1} resolution by means of spectral response function and considering different atmospheric attenuation conditions. Results are shown in figure 2, for a standard atmosphere of middle latitude summer and aerosol depth according to visibility 23 km. Abscissa axis shows the potential fire temperature and ordinate axis shows the minimum detectable area expressed in ha. Different magnitudes of influence must be analysed separately, being the most important the threshold of TTG considered, $\partial T/\partial t$, but geographic latitude of observation too. The figure contains the results for three different values of the gradient: 4, 6 and 2K/15_minutes and for two locations-type, at 20° and 50° latitude. With respect to the latitude, it must be taken into account that although the pixels's area in the nadir point is 9 km^2, at latitude of 20° it is 10km^2 and at 50° it has increased up to 18 km^2. Thus, for a required gradient of 4K/15_min. and a fire of 600K, the detectable area at 20° latitude is 0.5 ha, whereas at 50° latitude it would be 1ha. The geographic longitude has not been analysed since it has a very low distortion in the pixels' area. With respect to the thermal gradient, 4K/15_min is the reference for the analysis carried out in previous paragraphs. The figure shows results for a value 2K/15_min that can be applied in the descendant period of daily thermal evolution [14.00-20:00], because during this period $\partial T/\partial t < 0$ is expected and the value 2K/15_min could be enough. This means that during the evening, fires are more easily detected through this methodology and the fire starting can be established at 600K with 0.24 ha at 20° latitude and 0.48 ha at 50° latitude. As can be seen, the detectable sizes during the day at 20° latitude are similar to the ones in the evening at 50° latitude. Latitude has influence in the variability of the pixel's area and in the atmospheric transmittance, with the cenital angle, which has also been taken into account to obtain results. Results obtained for different atmospheric profiles do not differ too much. Another very important magnitude to be considered is the surface temperature since the considered methodology is presented with continuity throughout the day and night, a period for which different values are presented. It must be pointed out that lower surface temperatures make the detection considerably easier. Thus, if we go down from 300K to 290K, there is a decrease in the minimum detectable area of around 20-23%. This value is constant for different fire temperatures and also independent from the latitude considered.

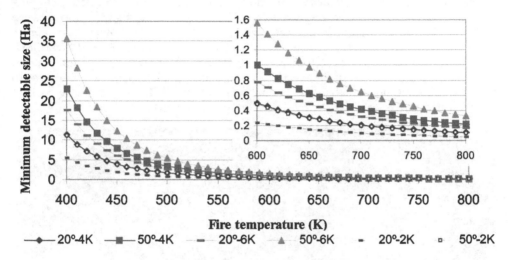

Fig. 2. Minimum size of fire (ha) to be detected by SEVIRI, for different fire temperature and latitude, applying TTG values of 2, 6 and 4K/15_minutes, taking into account atmospheric attenuation (taken from Calle *et al.*, 2006).

Establishing the outbreak of a fire, as accurately as possible, is crucial to alerting fire-fighting teams as quickly as possible. If the detection process takes into account the comparison with the previous image the delay can be up to 30 minutes in the worst cases. To show some representative results we have analyzed the day on which Spain's worst fire in the previous decades in terms of human losses occurred. This fire, which started between 12:30 and 12:45 on 16th July 2005, spread for over five consecutive days and devastated around 13,000ha. Figure 3 shows the image of the 3.9 μm spectral band corresponding to a few hours after the fire. The visual analysis of the image shows the existence of many fires in Spain and Portugal. Given their importance, two have been highlighted and shown. Number 1 is the fire in Guadalajara (Spain) and number 2, one of the fires that affected the natural park of Lago de Sanabria (Zamora, Spain) during the summer of 2005, whose initial characteristics, as will be seen, differ from the first. In the figure, we have indicated the wind direction in fire #1 from the smoke plume, which is perfectly visible and which will be useful later to analyze the spread of the fire. Below in the same figure are the two thermal evolution diagrams corresponding to these fires. The diagram shows the temperature evolution of band 3.9 μm, in °C, in the primary axis of the ordinate according to the time of the day, between 06:00 and 16:00 GMT. The secondary axis of the ordinate shows the evolution of the time thermal gradient of the same band, in °C/15_minutes. If we compare both temperature evolution curves, we can see that they are practically identical on the primary axis up to the moment at which the fire starts, at 12:30 in #1 and at 13:45 in #2 despite being different vegetation covers with different fuel moisture content since they occur in different climate zones. The analysis of the curve of the time thermal gradient is much more conclusive. The change in the temperature value is 1.5°C/15_minutes in both curves prior to the outbreak of the fire reaching a maximum of 2.3 in #1 and 1.8 in #2, which are exceptional considering the rest of the values. Case #2 was a fire that started with a time thermal gradient of 4.2°C/15_minutes in the first scene at 14:00 GMT, immediately jumping to 15°C/15_minutes in the following scene at 14:15 GMT. It is clear that it began between 13:45 and 14:00 as the figure shows. The case of fire #1 presents a much more abrupt beginning, with a time thermal gradient of 8°C/15_minutes in the first scene at 12.45 GMT. In this case, the fire broke out between 12:30 and 12:45 GMT. Apart from its initial causes, the characteristics of a fire at its onset depend on the combustible material and moisture. In this comparison, it is not surprising that the outbreak was slower in case #2, whose gradient was below #1, as this was a climate zone with higher moisture content.

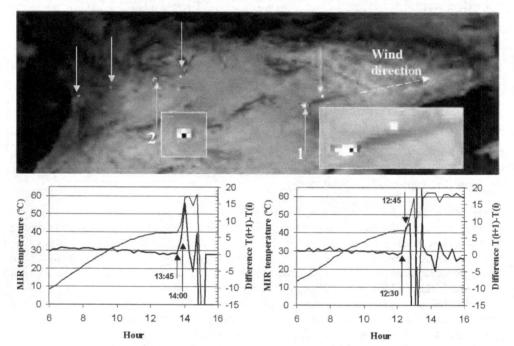

Fig. 3. This figure shows the methodology to detect the start of a fire for two different cases. The upper part of the figure shows the 3.9 μm band, highlighting several fires validated by MODIS) as well as wind direction. The second part shows the thermal evolution, in the left scale, and the time thermal gradient, in the right scale, in °C/15_minutes, for the two selected cases. (Calle *et al.*, 2006).

The methodology proposed to detect the beginning of the fire is no longer valid as the fire keeps developing since the temperature differences between the different scenes experiment strong variations. Even the frequent appearance of saturated pixels causes sharp changes that cannot be analysed. Further, for the subsequent monitoring of the fire, a methodology for detecting hot spots (after the starting) is required. Detection methods on other sensors used as a reference are sometimes based on physical models. However, experimental statistical models have shown better results and are easier to apply as the contextual models operating on AVHRR and MODIS, as we have seen in the paragraph before.

3.4 Spatial characterization of fire detection

The pixel dimension is the main parameter that characterises sensors concerning their spatial resolution. However, the radiance quantification and image interpretation need an appropriate analysis to obtain several physical parameters. The review of Cracknell (1998), describes spatial and radiometric considerations regarding the pixel precisely. In order to answer the question "what's in a pixel?", title of the mentioned paper, it is firstly necessary to carry out an accurate analysis of the target area that emits the radiance reaching the sensor which, in fact, never coincides exactly with the spatial resolution assigned to it nor with the square shape that it is imagined for the matrix elements of an image. The simplified concept of image as a mosaic of elements is quite far from the reality, something that becomes evident when trying to observe image detail or compare images from different sensors with similar spatial resolution. Moreover, the concept of spatial resolution is often identified with Ground Sampling Distance (GSD), defined as the distance between centres of neighbour pixels, or the use of Instantaneous Geometric Field of View (IGFOV), the geometric size of the image projected by the detector on the ground through the optical system introducing confusion in the sensor's spatial characterization. Since there are sensors with similar IGFOV but different Modulation Transfer Function (MTF), it is more realistic to define a quantity in the topic of MTF. The concept of Effective Instantaneous Field Of View (EIFOV) introduced by NASA, 1973, is defined as the resolution corresponding to a spatial frequency for which the MTF system is 0.5. The MTF shape in the frequency domain and, consequently, the Point Spread Function (PSF) in the spatial domain has not a special relevance when the surface observed shows a homogeneous distribution of radiance; nevertheless, when there are heterogeneous distribution of radiance inside the pixel, as is frequently the case of forest fires, PSF and deconvolution processes must be considered. In this paragraph, results by using real MTF functions of the SEVIRI sensor, are shown.

On the other hand, many thermal parameters in remote sensing are estimated by solving multi-spectral processes, such as the estimation of the temperature using split-window procedures or the estimation of thermal parameters in hot-spots through Dozier's method (Dozier, 1981; Matson and Dozier, 1981). In these estimations, it is assumed that the pixels of the bands involved correspond to the same spatial target and contribute with the same sensitivity to the radiance measurement. However, even in the case of a perfect co-registration between bands, this assumption would not be true since each band has a different PSF. This is one of the problems mentioned by Wooster et al. 2005, in order to propose a single-channel method to estimate the fire temperature instead of applying a bi-spectral method. In addition, the influence of the PSF has been highlighted as responsible for the differences in the Fire Radiative Power (FRP) when different sensors are compared. Concerning geostationary satellites, MSG is providing operational results in fire detection and biomass burning in Africa (Wooster et al., 2005) and Mediterranean countries (Calle et al., 2006) and Geostationary Operational Environmental Satellites (GOES) are used operationally in South-Central-North-America (Prins and Menzel, 1994). The issue of fire detection is understood in the framework of global geostationary fire monitoring applications and requires evaluating the impact of the MTF's shape in the estimation of thermal parameters.

In order to estimate the impact of PSF shape on detection suitability, it's interesting to analyze a sensor with low spatial resolution, as SEVIRI sensor onboard of MSG satellite (Calle *et al.*, 2009). Pixel affected by fire appears a a typical cross shape when fire is detected, due to PSF effects and overlapping between pixels. Figure 4 shows a three-dimensional graph where the brightness temperature in the 3.9 μm band (vertical axis) is shown versus the fire temperature (left part of figure) and background temperature (right part of figure) and the distance from the pixel centre (PSF impact), where the background temperature is 300 K (left), the fire temperature is 500 K (rigth) and the one-dimensional burning area is 50m (both cases). Saturation plane is shown in the figures. Note that for low fire temperatures (below 450 K, taking into account that we are talking of flaming and smouldering mixed phases, the PSF impact is not noticeable. However, large differences in brightness temperature are found in hotter fires. In order to explain the importance of a 10 K-difference in the 3.9 μm band, note that if a contextual detection algorithm is applied the detection will be lost when the standard brightness temperature deviation around the pixel is higher than 3 K.

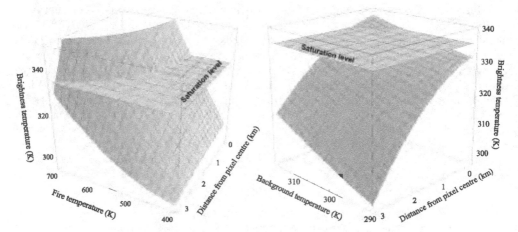

Fig. 4. Left part: Brightness temperature in the 3.9 μm band (vertical axis) *versus* fire temperature and distance from pixel centre (PSF impact); background temperature 300 K is considered. Rigth part: Brightness temperature in the 3.9 μm spectral band (vertical axis) versus background temperature and distance from pixel centre (PSF impact). fire temperature of 500 K is considered. One-dimensional burning size of 50 m. (Calle *et al.*, 2009)

4. Fire monitoring

The concept of detection is very clear, but it is not so clear the concept of monitoring. It could be said that monitoring comprises all the aspects related to the knowledge of a fire while it is taking place. Thus, we can talk of the fire temperature, the active area, the fire's energy intensity and the fire's front. However, all these parameters are subject to the technical possibilities of the spatial sensor used, especially the spatial resolution in the thermal spectrum. The main problem with monitoring tasks lies in the necessity of having available the time resolution typical of geostationary satellites in order to be able to know not just the instant value of the above mentioned parameters, but also their evolution throughout the fire's development. However, these sensors are currently very far from providing detailed results. Next, we will see the type of information we can get according to the capacities of different sensors.

For the knowledge of fire parameters, we need first an analysis at a sub-pixel level through the application of Dozier's methodology (1981). This methodology allows us to establish both the fire temperature and the fraction of the area that is burning simultaneously. This procedure can be applied to any sensor and it is based on the solution of the following system of equations: given a pixel affected by a fire at a temperature T_f that occupies the fraction of the pixel p, and it is surrounded by a surface at a temperature T_{surf}, then the radiances detected in the MIR and TIR bands will be given by the expressions:

$$\begin{cases} L_{MIR}=p\,B\left(\lambda_{MIR},T_f\right)+(1-p)B\left(\lambda_{MIR},T_{surf}\right) \\ L_{TIR}=p\,B\left(\lambda_{TIR},T_f\right)+(1-p)B\left(\lambda_{TIR},T_{surf}\right) \qquad 0<p<1 \end{cases} \tag{3}$$

where L_{MIR} and L_{TIR} are the radiances observed by the sensor in the spectral regions of 3.7μm and 11μm respectively and $B(\lambda,T)$ is the function of Planck. This system of equations provides the fire temperature value and the fraction of the pixel that is burning.

Before analysing some approximations taken in this methodology, we must point out two very important restrictions concerning its operating capacity. In the first place, it must be said that the equations are based on the establishment of the radiance emitted by the thermal spectrum. The 11 μm region has no other nature, but the radiance observed in the MIR region has a reflection component that has been analysed in the false alarms section. That's why, the application of these equations to diurnal images should include an additional solar term. Otherwise, they would only be valid for night images. On the other hand, in order to obtain reliable results, it is necessary to avoid saturation as much as possible.

Dozier's system of equations is very simple to understand although many of the approximations it takes are not realistic and should be analysed. In the first place, the pixel observed is divided into two parts, fire and surface, those are considered homogenous, but this is not the case, especially because of the surface's heterogeneity. On the other hand, the atmospheric effects have been neglected in this scheme. The most serious approximation with respect to the error magnitude is probably found in the establishment of a surface temperature value. Dozier suggested for this value the mean value of the pixels surrounding the fire but not affected by it. It must be highlighted that the results obtained depend to a great extent on this parameter. Simulations carried out on a real fire changing the surface temperature value (Calle *et al.*, 2005) show that the error in the surface temperature affects the fire temperature with a value multiplied by 10. Finally, another approximation taken is not to include in the equations the emissivity of the radiance received by the sensor. Although it is true that the fire performance is very similar to that of a blackbody, the same does not happen with the non-affected surface, which seems to have variable values. The deduction of the emissivity is justified in the fact that the zones observed for fire purposes are always forest zones and the emissivity values in this kind of environment are comprised in the interval [0.983-0.995] for the TIR band. A more realistic scheme derived from Dozier's methodology is the one suggested by Giglio & Kendall (2001). This scheme modifies the former one by including terms of emissivity, atmospheric effects and sun reflection in the radiance equation of the MIR band. The following are the modified equations of Dozier:

$$\begin{cases} L_{MIR} = \tau_{MIR} p\, B\left(\lambda_{MIR}, T_f\right) + (1-p)L_{surf,MIR} + pL_{atm,MIR} \\ L_{TIR} = \tau_{TIR} p\, B\left(\lambda_{TIR}, T_f\right) + (1-p)L_{surf,TIR} + pL_{atm,TIR} \qquad 0 < p < 1 \end{cases} \tag{4}$$

where $L_{atm,MIR}$ and $L_{atm,TIR}$ are the radiances emitted by the atmosphere to the sensor in the MIR and TIR bands respectively. These terms are worthless with respect to the radiances emitted by the surface, $L_{surf,MIR}$ and $L_{surf,TIR}$, and can be disregarded. τ is the atmosphere's spectral transmittance. The difference in these equations with respect to the original ones lies in the intervention of the radiances of the surrounding pixels instead of the temperature and finally, although they are taken into account, the surface's emissivity and temperature are not usually known explicitly. The techniques mentioned for the obtaining of fire parameters imply some difficulties related to the errors that are made. In the first place, they are not analytic equations so that their solution must be found by means of numerical calculation techniques. However, it must be said that their solution comes, in the end, from a convergent system. Other important sources of errors have their origin in different magnitudes that have been analysed by Giglio & Kendall (2001) and that will be mentioned here next.

First, a source of error in the results is the error in the calculation of the surface's radiance introduced in the equations. The values for the fire temperature and size are more sensitive to errors in the radiance of 11.0μm than in the 3.7μm. At low temperatures, this is not a big error, but, with a high fire temperature, the error increases noticeably both in the fraction of the pixel affected and in the fire temperature itself. Another source of error to consider is the one corresponding to the atmospheric transmittance. However, in this case, the errors made in the temperature and fraction of the pixel affected, are compensated in the MIR and TIR bands as long as such errors are caused by either an underestimation or an overestimation in both cases. Otherwise, the errors in the results will add up. Thus, an overestimation in the MIR transmittance overestimates the temperature calculated whereas an overestimation in the TIR transmittance produces the opposite effect. A third source of error in the calculations is due to the instrument's noise, although in this case it introduces an accidental systematic error. Finally, the omission of the atmospheric radiance that reaches the sensor is less important than the causes considered formerly, so that in no case does the temperature go over 1.5K or the area over 2%. A very interesting aspect in the theory developed is the one that refers to the fire's emissivity. A fire has always been considered as a blackbody. In fact, and strictly speaking, this is only true when the length of the flame seen from the sensor is larger than 6 metres (Langaas, 1995). This would make us reconsider this aspect in the case of smaller fires so that in these cases we should consider the fire as a grey body. In these cases that separate from the characteristics of a blackbody, the errors made for considering that the fire has an emissivity one, result in an underestimation of both the fire temperature and the fire area, and they are independent from the fraction of the pixel that is affected and the fire temperature. In spite of all the methodology developed, it is important to point out that a forest fire is, in reality, a very complex phenomenon in which different series of phenomena overlap. In this situation, we could ask ourselves what the parameter we call "fire temperature" is exactly and what the "burning area" is. The model presented is a simplification from the real phenomenon since, up until now, only two phases have been differentiated: the fire flame and the surface. In reality, it should at least be considered the middle phase corresponding to the smouldering. However, it must be taken into account that the introduction of further terms in the model would imply having more spectral bands available in order to obtain more equations and to be able to find all the unknown quantities. We are going to consider this aspect so as to reach some conclusions in relation with the appropriate spectral information. Kaufman *et al.*, (1998) introduced a modification in Dozier's methodology in order to include the flame phase, which is hotter, and the smouldering phase, which is in the middle between the surface and the flame. Thus, if we call p_f and p_s to the fractions of the pixel corresponding to the flame and the smouldering respectively, the bi-spectral equations will be as follows:

$$\begin{cases} L_{MIR} = \tau_{MIR} \left[p_f B\left(\lambda_{MIR}, T_f\right) + p_s B\left(\lambda_{MIR}, T_s\right) \right] + (1-p)L_{surf,MIR} + pL_{atm,MIR} \\ L_{TIR} = \tau_{TIR} \left[p_f B\left(\lambda_{TIR}, T_f\right) + p_s B\left(\lambda_{TIR}, T_s\right) \right] + (1-p)L_{surf,TIR} + pL_{atm,TIR} \qquad 0 < p < 1 \end{cases} \qquad (5)$$

so that $p_f + p_s = p$ is fulfilled. The analysis and discussion will be done through the flaming ratio function, f, defined as $f = \dfrac{p_f}{\left(p_f + p_s\right)}$. This relation is related to the importance that the flame phase has in the fire observed. Since in order to obtain more detailed information, more observation wavelengths are needed, Giglio & Justice (2003) established the errors found according to the pair of wavelengths used to solve the bi-spectral equations so as to establish the most appropriate pair for this purpose, always considering the atmospheric windows for the observation. These authors carried out simulations with combinations of wavelengths in the interval [1.6, 3.8 μm] for the MIR region and in the interval [2.4, 11 μm] for the TIR region so that λ_{TIR} was always higher than λ_{MIR}. The most relevant conclusions of this analysis were that the shortest pairs of wavelengths provided higher fire temperatures and smaller areas since the decrease in λ implies a major importance in the flame phase whereas an increase in λ gives more importance to the smouldering phase. It is also interesting that the results of the pair [3.8, 11.0 μm] and of the pair [3.8, 8.5 μm] are practically identical, with differences inferior to 5K for the temperature and 5% for the area. This means that the spectral difference in the AVHRR and MODIS sensors, which correspond to the first pair, and in BIRD, which corresponds to the second, are not significant.

MODIS has several bands situated in the spectral region of 4μm. One of them has a saturation value of 500K (band 21), which makes this sensor especially suitable for the establishment of fire parameters since it is difficult to find saturated pixels. It must be taken into account that it is very rare when this monitoring phase can be applied to the AVHRR sensor since, although the detection is possible, band 3 is very frequently saturated. Likewise, BIRD prototype is especially suitable for the obtaining of parameters and the establishment of the FRP (Free Radiative Power) (Kaufman and Justice, 1998). By definition, the FRE (Fire Radiative Energy) is basically the portion of chemical energy released during the burning of the vegetation and emitted as radiation during the combustion process. These parameters are comprised within the goal of fire analysis and the FRP is precisely the most important one because it contains information both on the emissions produced in the atmosphere by these events (Kaufman *et al.*, 1996) and on the fire's destructive power. These authors have suggested that the quantification of the radiated energy during the combustion process in the fire could supply a measurement related to the quantity of vegetation consumed per unit of time. Consequently, it would provide a measurement of the emissions produced during fires and, therefore, it would provide valuable support information to the processes of climate change obtained through remote-sensing. In spite of being a qualitative measurement of great value, we must take into account that the combustion phase is a mixture of physic processes through which the fire's energy is distributed, apart from the radiation phase, as is the case of the air mass convection above the fire and the conduction towards the interior of the earth.

For the MODIS sensor case, Kaufman & Justice (1998), have suggested an empiric expression in order to fix the intensity, in MWatts, from the brightness temperature of the pixel affected by the fire. This expression corresponds to:

$$FRP = 4.34 \cdot 10^{-19} \left(T^8_{MIR} - T^8_{MIR,b} \right) \qquad (6)$$

where T_{MIR} is the brightness temperature of the band of the 4 μm of the pixel affected by the fire and $T_{MIR,b}$ is the same temperature in the adjacent pixels. In order to carry out a validation of the results obtained at a sub-pixel level, that is, the fire's temperature and the fire's area, a comparison with the intensity values calculated through Stefan-Boltzmann's law and the previous formula has been carried out. This has been exclusively done for the MODIS sensor and on the large fires that affected Spain and Portugal during the summer of 2003. Besides, the comparison has been done for the Terra and Aqua spacecraft and at two processing levels: level of individual burning pixels and level of averaged clusters. The results of this comparison can be found in Calle *et al.* (2005), in which the two processing levels are represented separately. The almost exact coincidence of the values, which are even better in the case of the analysis at a cluster level, proves the reliability of the magnitudes temperature and area of fire. It is very important to highlight the fact that the coincidence between the empiric expression and Stefan-Boltzmann's law, after applying Dozier methodology) are coming from the analysis of clusters. However, when results are compared at the level of individual pixels, the differences are much more noticeable; so, the use of empiric expression is recommended. When sensor has a high spatial resolution in the thermal bands, the sub pixel analysis is a useful tool in order to discriminate the increasing direction of fire: that is, the flaming front. The figure 5 shows the results of the application of the sub-pixel analysis on one of the active fires that have been described. It corresponds to the superposition of the fire's temperatures on the BIRD sensor over NIR image, showing the affected fire area.

The real usefulness of remote-sensing in the early detection of fires will take place when the time resolution of the sensors implied is around 15 minutes or less. At present, this characteristic is only available in the geostationary satellites, but they have the problem of their low spatial resolution. The advantage of geostationary sensors is that it's possible to obtaining, not only the FRP but the FRE too. The fire radiative energy will be: $FRE = \int FRP\, dt$. In any case, the comparison of FRP results among sensors is only valid for qualitative purposes since in certain fires the lowest spatial resolution implies an important underestimation of this magnitude. This happens for example when comparing MSG and MODIS or MODIS and BIRD. With respect to the latter ones, Wooster *et al* (2003) found differences of up to 46%.

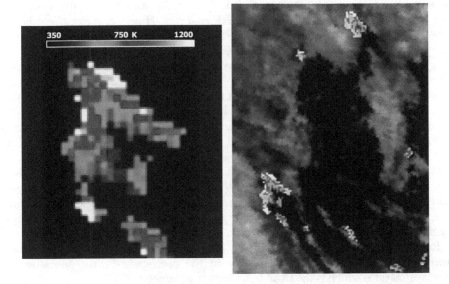

Fig. 5. This figure shows the fire temperatures, obtained by means of Dozier methodology. At this spatial resolution is very clear to recognize the flaming front and the spreading direction of fire (Calle *et al.*, 2005).

5. Atmospheric impact of fire emissions

The gases belonging to carbon cycle, CO and CO_2, are trace gases located in the atmosphere, mostly as the result of anthropogenic activities. Despite not being a greenhouse gas, the carbon monoxide plays a significant role in the carbon cycle; it is not a direct precursor of CO_2, but it essentially affects the budgets of OH radicals and O_3 present in the atmosphere (see Bergamaschi *et al.*, 2000, for an extended explanation about the modelling of the global CO cycle). The anthropogenic activities related to release carbon into the atmosphere can be divided in two well-defined groups: on the one hand, the urban pollutant emissions from vehicles and other industrial processes; on the other, from fires and global biomass burning emissions. The estimation of CO profiles and CO total column has been identified as a very important objective in order to improve our understanding of climate global system. The EOS (Earth Observing System) Science Steering Committee has proposed: "The fate of carbon monoxide, remotely detected from space, in conjunction with a few other critical meteorological and chemical parameters, is crucial to our understanding of the chemical reaction sequences that occur in the entire troposphere and govern most of the biogeochemical trace gases" (EOS, 1987). In the same line, the WMO (World Meteorological Organization) has proposed: "Definition of trends and distributions for troposphere CO is essential. A satellite-borne CO sensor operating for extended periods could help enormously" (WMO, 1985). The global estimation of CO based on satellite imagery involves a series of technical difficulties; the most important one is the associated error of the measurements.

The combustion by fire is a chemical reaction with heat release where the main products generated are, if combustion is completed, H_2O, CO_2, and N_2. In the case of high combustion temperatures, NO_2 and NO are released too. However it must be pointed out that the main cause of CO fire-related emissions is the incomplete or inefficient burning of wood, biomass and fossil fuels. Concerning wildfires, two phases are considered: the flaming phase (in which CO_2 and nitrogen are released), and smouldering phase (in which CO and hydrocarbons are released). Two procedures provide estimations of CO emissions, a direct procedure and an indirect one. The indirect method estimates CO mass from the knowledge of the previous burned biomass. This value can be obtained from satellite cartography of fire-affected areas and the vegetation index, which is the main indicator of biomass quantity on a global scale. The adjustment of the measurements is carried out by introducing the combustion efficiency coefficients of this particular gas. This procedure was first proposed by Seiler and Crutzen (1980), who estimated CO emissions according to the following indirect parameters: i) burned land cover area (m^2), ii) above-ground biomass density of burned area (kg-dry-matter/m^2), iii) burning efficiency of the above-ground biomass (that is, the fraction of biomass burned) dimensionless, and iv) the emission factor (g of CO [kg dry matter]$^{-1}$), which varies according to the type of vegetation and ecosystem. Note that many errors arise, from this indirect procedure, due to the uncertainty in the coefficients and, especially, in the biomass estimation, which is the main quantitative parameter.

The second procedure is the direct estimate of carbon content in the atmosphere by means of remote sensing. The SCIAMACHY (SCanning Imaging Absorption SpectroMeter for Atmospheric CHartographY) onboard the European satellite ENVISAT (Bovensmann *et al.*, 1999) has provided more measurements, of the most important trace gases, than any other sensor up to the present. The CO total column is retrieved from a small spectral fitting window located in SCIAMACHY channel 8 (2.324-2.335 μm); finally, the results of its measurements are adjusted according to the parameters of trace gas. Dils *et al.*, (2006) have carried out a series of comparisons between SCIAMACHY measurements and ground-station data. In the case of CO and CH_4, with similar algorithms, they have shown that the measurements provide good description of seasonal and latitudinal variability. However, they show important discrepancies in concrete cases. Besides, they show long periods in which the algorithm does not provide any data. The MOPITT (Measurements of Pollution in the Troposphere) instrument, onboard the Terra spacecraft, has proved to be the most operative sensor for the continuous estimation of CO. On the other hand, scientists from the NCAR (National Centre for Atmospheric Research), funded by NASA, have spread data and results concerning the global distribution of CO based on MOPITT measurements (http://www.acd.ucar.edu/), which have revealed the seasonal dynamics of CO throughout the planet and direct correlations between the increase in the CO total column measured by MOPITT and large fires. The validation of the results reveals the suitability of the MOPITT's spatial scale for monitoring continuously (at regional and global scale) observations of the spatial oscillations related to the atmospheric CO. Hereby, large horizontal gradients in the distribution of CO at the synoptic scale have been observed. These variations in CO can be as large as 50–100% and occur over spatial scales of around 100 km. These events, usually during several days, can span horizontal distances of 600-1000 km, and can appear over a range of pressure levels from 850 to 150 hPa (Liu *et al.*, 2006).

The biomass burning is a very important source of ozone and methane precursors and the main factor of CO emissions. High levels of carbon monoxide pollution are found around the world, and they result from different types of biomass burning in different locations. High levels of CO are linked to widespread fire activity, such as agricultural burning in central Africa in January through March, or in Central America in April through June. Carbon monoxide molecules can last from a few weeks to several months in the atmosphere, and they travel long distances, without regard for national or international boundaries. Emissions from biomass burning accounts for about one quarter of the CO released to the atmosphere, with an average of around 600 Mt CO per year (Khalil *et al.*, 1999). The occurrence of biomass burning, the size of fire, the different phases of fire considered (e.g. smouldering and flaming) and fire parameters (e.g. fire radiative power and temperature) vary greatly with time and space. Andreae and Merlet (2001) estimated that mean CO emission from vegetation fires in savanna and tropical forests is 342 Mt CO per year, while the total CO emission for all non-tropical forest fires is 68 Mt CO per year.

The pattern of fire occurrence in Africa and Amazonia is quite different to others regions in the planet with higher population density. The fire occurrence, in Africa and Amazonia, is dominated by the displacement of ITCZ (Inter Tropical Convergence Zone). During the winter of North hemisphere the ITCZ, and therefore the tropical rain, is located in the South of equator and Amazonia; so, the fire occurrence is stronger in the North of equator and vice versa. The figure 6 shows the results of seasonal study of CO in the North equatorial Africa ([4.5N-15N] and [17W-37E]), South equatorial Africa ([22S-3S] and [10E-40E]) and Amazonia ([20S-7.5S] and [65W-50W]). The bar diagram shows fire occurrence from MODIS (Giglio *et al.*, 2003; Davies *et al.*, 2009) in the period 2003-2008. In the background, in grey colour, the original data of CO total column, from MOPITT, are shown. In black colour, the inverse Fast Fourier Transform calculated by means of main harmonics with higher spectral energy. Finally, CO values from SCIAMACHY, averaged for each month, are displayed for the period 2003-2005, in order to compare results between MOPITT and SCIAMACHY sensors. Comparison between CO from MOPITT and SCIAMACHY have been carried out by Buchwitz *et al.* (2007) showing results over cities; so, this comparison over large fires, is a complementary result in order to know the spatial capabilities of these source of data.

Concerning analysis of results over Africa, northward equator, two main harmonics with maximum spectral energy, for each year, can be observed. First maximum is located in the period of January and February, showing a very good correlation with fire occurrence. The second maximum, weaker, is located in August, exactly when fire occurrence in the South of equatorial Africa is stronger. Concerning comparison between MOPITT (daily data) and SCIAMACHY CO total column is similar between them (having MOPITT data more amplitude). This is an expected difference, once SCIAMACHY data are averaged values. As we have underlined in the paragraph before, during the summer of North hemisphere the displacement of ITCZ is the responsible of a stronger fire occurrence in the South equator. Three main maximums, for each year, can be observed. The first maximum, with the shape of a peak, is located at the end of September, showing a very good correlation with the main fire occurrence in the year. The second maximum, weaker, is located at the end of January when fire occurrence in the North of equatorial Africa is stronger (see discussion before). The main difference with North equatorial Africa is the presence of a third harmonic providing an increasing tendency, of CO values, during June-September. As it's possible to

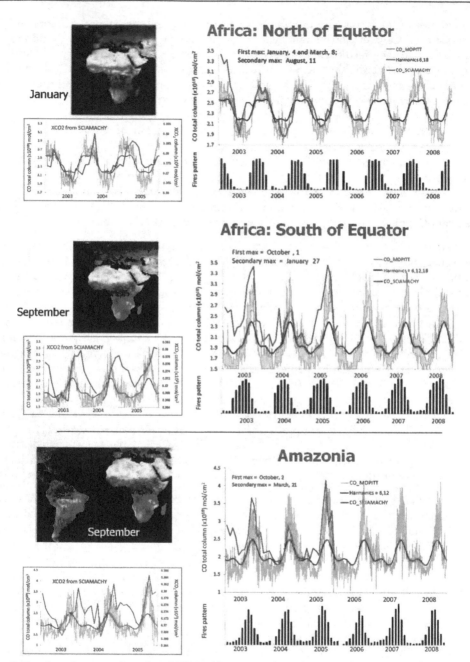

Fig. 6. Results of seasonal study of CO in Africa (North and South of equator) and Amazonia. A comparison between CO emissions and fire occurrences is shown. CO total column original values and Inverse FFT transform is underlined. Left part of each graph contains XCO₂ evolution for 2003-05. (Calle *et al.*, 2011).

observe in the figure 6, both geographical bands present a correlation between CO values and fire occurrence. But CO maximum values have a delay of 15-20 days with respect to maximum fire occurrence; additionally the local maximum of North band presents a coincidence with main maximum of South band; that is: influence between them due to CO transport processes in the atmosphere. In any case, the influence of North over the South is stronger. Concerning comparison between MOPITT (daily data) and SCIAMACHY CO total column is very similar between them. The pattern of fire occurrence in Amazonia is the same of the South of equatorial Africa, due to the ITCZ behaviour.

6. Conclusion

In the light of the results, the geostationary sensors prove to be a highly efficient tool in real-time forest fire management and monitoring. Despite not being originally designed as an Earth observation tool, but as a meteorological satellite, its excellent time resolution has proved useful for the detection of events which vary due to radiometric rather than spatial characteristics, as is the case of forest fires. On-going parameterization of fires has a strong influence on the subsequent treatment of forest regeneration. Major efforts are currently being made in the establishment of fire severity, where the main magnitude involved is the FRE in large fires for subsequently establishing intensity and include this magnitude in atmospheric emission models. This correlation between FRE and severity was not possible with polar sensors due to their lack of continuous observation. Another important magnitude that can be established from the FRE is the height of the flame, including some characteristics of the fuel, which could help the analysis of the fire front and other magnitudes linked to its advance. This is an essential magnitude since it is used by fire fighting services to determine the infrastructure necessary to combat fires. Both, the EOS Science Steering Committee and the WMO, have pointed out, as a main objective, the measurement and control of carbon monoxide as part of the control framework of trace gases involved in the carbon cycle. Forest fires are an important source of CO and CO_2 worldwide. However, the global estimates carried out have been based on indirect methods which require the previous determination of the burned areas and the introduction of burning efficiency coefficients, which are difficult to determine. In order to apply direct methods for emissions estimating, atmospheric sensors as MOPITT and SCIAMACHY have proven their ability to extract important conclusions about carbon cycle gases at global scale.

7. References

Al-Rawi, K-L-, Casanova, J.L. & Calle, A. (2001). Burned area mapping system and fire detection system, based on neural networks and NOAA-AVHRR imagery. *International Journal of Remote Sensing*, 22, 2015-2032. ISSN: 0143-1161

Andreae, M. O. and Merlet, P. (2001). Emission of trace gases and aerosols from biomass burning. *Global Biogeochemical Cycles*, 15:955–966. ISSN: 0886-6236

Arino, O. and Rosaz, J.M. (1999), 1997 and 1998 World ATSR FIRE Atlas using ERS-2 ATSR-2 Data, *Proceedings of the Joint Fire Science Conference*, Boise, 15-17, June 1999.

Arino, O., and Mellinotte, J.M. (1998). The 1993 Africa fire map, *International Journal of Remote Sensing*, 19:2019-2023. ISSN: 0143-1161

Bergamaschi, P., Hein, R., Heimann, M. and Crutzen, P. J. (2000): Inverse modelling of the global CO cycle, 1. Inversion of CO mixing ratios. *Journal of Geophysical Research*, 105:1909–1927. ISSN: 0148-0227

Berk, A., Bernstein, L.W. and Robertson, D.C. (1996), MODTRAN: A moderate resolution model for LOWTRAN 7, Philips Laboratory, Report AFGL-TR-83-0187, Hanscom ARB, MA.

Bovensmann, H., Burrows, J. P., Buchwitz, M., Frerick, J., Nöel, S., Rozanov, V. V., Chance, K. V. and Goede, A. (1999). SCIAMACHY- Mission Objectives and Measurement Modes. *Journal of Atmospheric Sciences*, 56:127–150. ISSN 0022-4928

Briess, K., Jahn, H., Lorenz, E., Oertel, D., Skrbek, W. & Zhukov, B. (2003). Fire recognition potential of the bi-spectral detection (BIRD) satellite. *International Journal of Remote Sensing*, 24, 865-872. ISSN: 0143-1161

Buchwitz, M., Khlystova, I., Bovensmann, H., and Burrows, J.P. (2007). Three years of global carbon monoxide from SCIAMACHY: comparison with MOPITT and first results related to the detection of enhanced CO over cities. *Atmospheric Chemistry and Physics*, 7:2399–2411. ISSN: 1680-7316

Calle, A., Romo, A., Sanz, J. & Casanova, J.L. (2005). Analysis of forest fire parametres using BIRD, MODIS and MSG-SEVIRI sensors. *New Strategies for European Remote Sensing*, Millpress, Rotterdam, ISBN 90 5966 003.

Calle, A., Casanova, J.L. and Romo, A. (2006). Fire detection and monitoring using MSG Spinning Enhanced Visible and Infrared Imager (SEVIRI) data. *Journal of Geophysical Research*, 111, G04S06, doi:10.1029/2005JG000116.

Calle, A., Casanova, J.L. and Romo, A. (2009). Impact of point spread function of MSG-SEVIRI on active fire detection. *International Journal of Remote Sensing*, 30(17), 4567–4579. ISSN: 0143-1161

Calle, A., Salvador, P. and González, F. (2011). Study of the impact of wildfires emissions, through MOPITT CO total column, at different spatial scales. *International Journal of Remote Sensing* (in press). ISSN: 0143-1161

Casanova, J.L., Calle, A. and González-Alonso F. (1998). A Forest Fire Risk Assessment obtained in real time by means of NOAA satellite images. *Forest Fire Research. III. International Conference on Forest Fire Research and 14th Conference on Fire and Forest Meteorology*. Vol I: 1169-1179. ISBN: 972-97973-0-7

Chuvieco, E., Riaño, D., Aguado, I. and Cocero, D. (2002). Estimation of fuel moisture content from multitemporal analysis of Landsat Thematic Mapper reflectance data: applications in fire danger assessment. *International Journal of Remote Sensing*, 23 (11):2145-2162. ISSN: 0143-1161

Cracknell, A.P. (1998). Review article synergy in remote sensing-what's in a pixel? *International Journal of Remote Sensing*, 19, 2025-2047. ISSN: 0143-1161

Davies, D.K., Ilavajhala, S., Wong, M.M., and Justice, C.O. (2009). Fire Information for Resource Management System: Archiving and Distributing MODIS Active Fire Data. *IEEE Trans. on Geoscience and Remote Sensing*, 47 (1):72-79. ISSN: 0196-2892

Dils, B., et al. (2006). Comparisons between SCIAMACHY and ground-based FTIR data for total columns of CO, CH4, CO2 and N2O. *Atmospheric Chemistry and Physics*, 6:1953–1976. ISSN: 1680-7316

Dozier, J. (1981). A method for satellite identification of surface temperature fields of subpixel resolution. *Remote Sensing of Environment*, 11: 221-229. ISSN: 0034-4257

Giglio, L & Kendall, J.D., (2001). Application of the Dozier retrieval to wildfire characterization. A sensitivity analysis. *Remote Sensing of Environment*, 77, 34-49. ISSN: 0034-4257

Giglio, L. & Justice, C.O. (2003). Effect of wavelength selection on characterisation of fire size and temperature. *Int. Journal of Remote Sensing*, 24,3515-3520. ISSN:0143-1161

Giglio, L., Descloitres, J., Justice, C.O. & Kaufman, Y.J. (2003). An enhanced contextual fire detection algorithm for MODIS. *Remote Sensing of Environment*, 87:273-282. ISSN: 0034-4257

Hunt, E.R. & Rock, C.R. (1989). Detection of changes in leaf water content using near and medium infrared reflectances. *Remote Sensing of Env.*, 30:43-54. ISSN:0034-4257

Ichoku, C., Kaufman, Y.J., Giglio, L., Li, Z., Fraser, R.H., Jin, J-Z & Park, W.M. (2003). Comparative analysis of daytime fire detection algorithms using AVHRR data for the 1995 fire season in Canada: perspective for MODIS. *International Journal of Remote Sensing*, 24, 1669-1690. ISSN: 0143-1161

Illera, P. Fernández, A., Calle, A. and Casanova, J.L. (1996). Evaluation of fire danger in Spain by means of NOAA-AVHRR images. *EARSeL Journal Advance in Remote Sensing*, 4-4:33-43. ISSN: 1017-4613

Justice, C.O & Malingreau, J.P.(editors). (1993). The IGBP satellite fire detection algorithm workshop technical report, IGBP-DIS Working paper 9, NASA/GSFC, Greenbelt, Maryland, USA, February, 1993.

Khalil, M. A. K, Pinto, J. P. and Shearer, M. J. (1999). Atmospheric carbon monoxide. Chemosphere: *Global Change Science*, 1, IX –XI. ISSN: 1465-9972

Kaufman, Y. & Justice, C. (1998). MODIS Fire Products. MODIS Science Team. EOS ID#2741

Kaufman, Y.J., Justice, C., Flyn, L. Kendall, J. Prins, E., Ward, D.E., Menzel, P. & Setzer, A. (1998). Potencial global fire monitoring from EOS-MODIS. *Journal of Geophysical Research*, 103, 32215-32238. ISSN: 0148-0227

Langaas, S. (1995). A critical review of sub-resolution fire detection techniques and principles using thermal satellite data. *PhD thesis*, Department of Geography, University of Oslo, Norway.

Lasaponara, R. Cuomo, V. and Tramutoli, V. (1998). Satellite forest fire detection in the Italian ecosystems using AVHRR data. *XII Int. Conference on Forest Fire Research* Luso 16/20 nov. 1998, vol II, 2013-2028

Li, Z., Nadon, S., Chilar, J. & Stocks, B. (2000). Satellite mapping of Canadian boreal forest fires: Evaluation and comparison of algorithms. *International Journal of Remote Sensing*, 21, 3071-3082. ISSN: 0143-1161

Li, Z., Kaufman, Y.J., Ichoku, C, Fraser, R., Trishchenkp, A., Giglio, L. Jin, J and Yu, X. (2001). A review of AVHRR-based active fire detection algorithms: Principles, limitations and recommendations in Global and Regional vegetation fire monitoring from space: Planning a coordinated international effort, SPB Academic Publishing, The Hague, Netherlands, pp. 199-225.

Liu, J., Drummond, J.R., Jones, D.B.A., Cao, Z., Bremer, H. Kar, J. Zou, J., Nichitiu, F. and Gille, J.C. (2006). Large horizontal gradients in atmospheric CO at the synoptic scale as seen by spaceborne Measurements of Pollution in the Troposphere. Journal *of Geophysical Research*, 111, D02306, doi:10.1029/2005JD006076.

Matson, M. and Dozier, J. (1981). Identification of sub-resolution high temperatures sources using a thermal IR sensor. *Photogrametric Engineering and Remote Sensing*, 47(9), 1311-1318. ISSN: 0099-1112

Nemani, R.R. and Running, S.W., (1989). Estimation of regional surface resistance to evapotranspiration from NDVI and thermal IR AVHRR data. *Journal of Applied Meteorology*, 28 (4): 276-274. ISSN: 0894-8763

Price, J. C. 1984, Land surface temperature measurements from the split window channels of the NOAA 7 AVHRR, *Journal of Geophysical Research*. D5:7231-7237. ISSN: 0148-0227

Prins, E.M. and Menzel, W.P. 1994. Trends in South American burning detected with the GOES VAS from 1983-1991. *Journal of Geophysical Research*, 99 (D8), 16719-16735. ISSN: 0148-0227

Prins, E., Govaerts, Y. and Justice, C.O. (2004), Report on the Joint GOFC/GOLD Fire and CEOS LPV Working Group Workshop on Global Geostationary Fire Monitoring Applications, GOFC/GOLD Report No. 19. 23-25 March 2004. EUMETSAT, Darmstadt, Germany.

Robinson, J.M., (1991). Fire from space: Global fire evaluation using infrared remote sensing. *International Journal of Remote Sensing*, 12: 3-24. ISSN: 0143-1161

Roldán-Zamarrón, A., S. Merino-de-Miguel, F. González-Alonso, S. García-Gigorro, and J. M. Cuevas (2006), Minas de Riotinto (south Spain) forest fire: Burned area assessment and fire severity mapping using Landsat 5-TM, Envisat-MERIS, and Terra-MODIS postfire images, *Journal of Geophysical Research*, 111, G04S11, doi:10.1029/2005JG000136.

Seiler, W. and Crutzen, P. J. (1980). Estimates of gross and net fluxes of carbon between the biosphere and the atmosphere from biomass burning. *Climate Change*, 2:207- 247. ISSN:0165-0009

Wooster, M.J., Zhukov, B & Oertel, D. (2003). Fire radiative energy for quantitative study of biomass burning: derivation from the BIRD experimental satellite and comparison to MODIS fire products. *Remote Sensing of Environment*, 86, 83-107. ISSN: 0034-4257

Wooster, M.J., Roberts, G., Perry, G.L.W and Kaufman, Y.J (2005). Retrieval of biomass combustion rates and totals from fire radiative power observations: FRP derivation and calibration relationships between biomass consumption and fire radiative energy release. *Journal of Geophysical Research*, 110, D24311, doi: 10.1029/2005JD006318

10

Current Advances in Uncertainty Estimation of Earth Observation Products of Water Quality

Mhd. Suhyb Salama
Department of Water Resources, ITC, University of Twente, Hengelosestraat 99, 7500 AA
Enschede
The Netherlands

1. Introduction

Remote sensing data over a water body are related to the physical and biological properties of water constituents through inherent optical properties (IOPs). These IOPs characterize the absorption and scattering of the water column and are used as proxies to water quality variables. The scientific procedure to derive IOPs from ship/space borne remote sensing data can be divided into three steps: *i- forward modeling*, relates the radiometric data to the IOPs of the water column; *ii- parametrization*, defines the minimal set of IOPs whose values completely characterize the observed radiance; *iii- inversion*, derives the values of IOPs, and hence water quality variables, from radiometric data.

Reliable methods for uncertainty quantification of earth observation (EO) products of IOPs are important for sensor and algorithm validation, assessment, and operational monitoring. High accuracy in both observations and algorithms may reduce considerable ranges of errors. EO derived IOPs, however, have an inherent stochastic component. This is due to the dynamic nature of aquatic biogeophysical quantities, intrinsic fluctuations, model approximations, correction schemes, and inversion methods. Due to stochasticity of the measurements, as well as model approximations and inversion ambiguity, the retrieved IOPs are not the only possible set that caused the observed spectrum (Sydor et al., 2004). Instead, many other IOPs sets may be derived. Each of these sets has an unknown probability of being the derived product. The probability distribution of the estimated IOPs provides, therefore, all the necessary information about the variability and uncertainties of derived IOPs.

Generally, uncertainty assessment of EO-data falls under one of two methods, namely analytical deterministic or stochastic methods. Deterministic methods are based on gradient techniques and have been used to asses the uncertainty of IOPs as derived from EO-data. Duarte et al. (2003) analyzed the sensitivity of the observed remote sensing reflectance due to variable concentrations of water constituents. Maritorena & Siegel (2005) employed a deterministic technique for consistent merging of different products using their uncertainties. Wang et al. (2005) performed a detailed study on the uncertainties of model inversion related to fluctuations in each of the IOPs and their spectral shapes. Salama et al. (2009) studied the uncertainty of model-inversion using the gradient-based method. They found that the derived IOPs are linearly related to their errors. Lee et al. (2010) used analytical derivative of the quasi-analytical algorithm (Lee et al., 2002, QAA) to estimate the uncertainty of IOPs as derived from QQA. On the other hand, Salama et al. (2011) developed a gradient based method to estimate the accuracy of a specific model-parameterizations setup. The advantage of their method is that it does not require radiometric information, however on the cost of deriving detailed information. The main drawback of gradient-based methods is that they depend on the used EO-model to derive the IOPs and *a priori* knowledge on the radiometric uncertainty. On the other hand, stochastic methods are less dependent on the used EO-model and can deal with non-convex functions. The basic idea of stochastic methods is to systematically partition the region of feasible solutions into smaller subregions and move between them using random search techniques. Stochastic uncertainty techniques have been recently adopted to estimate the uncertainty of EO-derived IOPs. Salama & Stein (2009) proposed a stochastic technique to quantify and separate the source of errors of IOPs derived from EO data. The main objective of this chapter is to review the two families of error-estimation methods and inter-compare their results.

The reminder of this chapter is organized as follow: in Section (2) we describe the ocean color paradigm, i.e. used ocean color model, its parametrization and inversion. Deterministic methods for error derivation are described in Section (3), whereas the principles of stochastic methods are detailed in Section (4). The results of both families (deterministic and stochastic) are inter-compared in Section (5) whereas, in Section (6) we present an exercise to decompose the different sources of uncertainty. Error propagation exercise is detailed in Section (7) followed by a discussion on the advantages and limitations of error estimation methods in Section (8). We finalize the chapter by a summary and future developments in Section (9).

2. Ocean color model inversion

Remote sensing reflectance, the ratio of radiance to irradiance, above the water surface Rs_w can be related to the inherent optical properties (IOPs) using the ocean color model of Gordon et al. (1988):

$$Rs_w(\lambda) = \frac{t}{n_w^2} \sum_{i=1}^{2} g_i \left(\frac{b_b(\lambda)}{b_b(\lambda) + a(\lambda)} \right)^i. \tag{1}$$

Where $Rs_w(\lambda)$ is the remote sensing reflectance leaving the water surface at wavelength λ; g_i are constants taken from Gordon et al. (1988); t and n_w are the sea−air transmission factor and water index of refraction, respectively. Their values are taken from literatures (Gordon et al., 1988; Lee, 2006; Maritorena et al., 2002). The parameters $b_b(\lambda)$ and $a(\lambda)$ are the bulk backscattering and absorption coefficients of the water column, respectively. The light field in the water column is assumed to be governed by four optically significant constituents, namely: water molecules, phytoplankton green pigment chlorophyll-a (Chl-a), colored dissolved organic matter (CDOM) and detritus/suspended particulate matter (SPM). The absorption and backscattering coefficients are modeled as the sum of absorption and backscattering from water constituents:

$$a(\lambda) = a_w(\lambda) + a_{ph}(\lambda) + a_{dg}(\lambda) \tag{2}$$

$$b_b(\lambda) = 0.5 b_w(\lambda) + \eta b_{spm}(\lambda). \tag{3}$$

Where the subscripts on the right hand side of equations (2) and (3) denote water constituents: water w; phytoplankton green pigment ph; lumped absorption effects of CDOM and detritus dg and suspended particulate matter spm. η is the backscattering fraction, its value is estimated from Petzold's "San Diego harbor" scattering phase function as $\eta \sim 0.018$ (Petzold, 1977).

The absorption and scattering coefficients of water molecules, $a_w(\lambda)$ and $b_w(\lambda)$, are assumed to be constant. Their values are obtained from Pope & Fry (1997) and Mobley (1994), respectively. The total absorption of phytoplankton pigments $a_{ph}(\lambda)$ is approximated as in Lee et al. (1998),

$$a_{ph}(\lambda) \simeq a_0(\lambda)a_{ph}(440) + a_1(\lambda)a_{ph}(440) \ln a_{ph}(440), \tag{4}$$

where $a_0(\lambda)$ and $a_1(\lambda)$ are statistically derived coefficients of Chl-a, their values are taken from Lee et al. (1998).

The absorption effects of detritus and colored dissolved organic matter (CDOM) are combined due to the similar spectral signature (Maritorena et al., 2002) and approximated using the model of Bricaud et al. (1981),

$$a_{dg}(\lambda) = a_{dg}(440) \exp\left[-s(\lambda - 440)\right], \tag{5}$$

where s is the spectral exponent of combined effects of detritus and CDOM. The scattering coefficient of SPM $b_{spm}(\lambda)$ is parameterized as a single type of particles with a spectral dependency exponent y (Kopelevich, 1983):

$$b_{spm}(\lambda) = b_{spm}(550) \left(\frac{550}{\lambda}\right)^y. \tag{6}$$

Equation (1) is inverted to derive five parameters from the IOCCG data set and three parameters from the NOMAD data set. The derived parameters are called the set of IOPs and expressed in a vector notation as **iop**. The exponents s and y are assumed to be unknown (Salama et al., 2009) and are derived from the IOCCG data set as:

$$\mathbf{iop} = \left[a_{ph}(440), a_{dg}(440), b_{spm}(550), s, y\right]. \tag{7}$$

The numerical inversion is carried out using the constrained Levenberg-Marquardt Algorithm (LMA) (Press et al., 2002), where the constraints are set such that they guarantee positive and physically meaningful values: between 0 and 100 m^{-1} for $a_{ph}(440)$, $a_{dg}(440)$ and $b_{spm}(550)$, between 0 and 2.5 for y and between 0 and 0.03 for s. Optimization is started using the initial values of Lee et al. (1999) and $s = 0.021$ nm^{-1} and $y = 1.7$. Maximum number of iteration is set equal to 100.

3. Error estimation via deterministic method

3.1 Description

The uncertainty in the derived IOPs is attributed to the infinitesimal change of radiance in equation (1) as,

$$\Delta Rs_w(\lambda) = w_{ph}(\lambda)\Delta a_{ph}(440) + w_{dg}(\lambda)\Delta a_{dg}(440) + w_{spm}(\lambda)\Delta b_{spm}(550), \tag{8}$$

where $\Delta Rs_w(\lambda)$ represents the radiometric uncertainty at the wavelength λ; w_{ph}, w_{dg}, w_{spm} are the partial derivatives of Rs_w with respect to the derived IOPs. Equation (8) represents an over determined linear set of equations that can only be solved if the radiometric uncertainty is known in at least n wavelengths, with n being the number of derived IOPs.

Analytical expressions of partial derivatives in (8) are listed hereafter. To simplify the notations let us define the ratio w as,

$$w = \frac{b_b(\lambda)}{c_b^2}, \tag{9}$$

where $c_b = b_b(\lambda) + a(\lambda)$. The partial derivative w_{ph} is,

$$w_{ph} = \frac{\partial Rs_w(\lambda)}{\partial a_{ph}(440)} = \frac{t}{n_w^2} \zeta_{ph} \sum_{i=1}^{2} j_i w^i, \tag{10}$$

where ζ_{ph} is the spectral dependency of Chl$a$,

$$\zeta_{ph} = a_0 + a_1 \left[1 + \log a_{ph}(440) \right]. \tag{11}$$

The parameters j_i are $j_1 = -g_1$ and $j_2 = -2g_2c_b$. The term w_{dg} is expressed as,

$$w_{dg} = \frac{\partial Rs_w(\lambda)}{\partial a_{dg}(440)} = \frac{t}{n_w^2} \zeta_{dg} \sum_{i=1}^{2} j_i w^i. \tag{12}$$

The partial derivative w_{spm} is expressed as,

$$w_{spm} = \frac{\partial Rs_w(\lambda)}{\partial b_{bspm}(550)} = \frac{t}{n_w^2} \sum_{i=0}^{2} v_i w^i, \tag{13}$$

where $v_0 = g1/c_b$, $v_1 = 2g_2 - g_1$ and $v_2 = j_2$.

Based on the above theoretical formulation in equation (8), Lee et al. (2010) obtained the uncertainty of IOPs using the quasi analytical algorithm (Lee et al., 2002) and a prior information on the radiometric errors. Salama et al. (2011), on the other hand, proposed a method that produces a single (or ensemble) uncertainty measure for the collective errors in the derived IOPs relative to the radiometric uncertainty without the need for model inversion or prior information on the radiometric errors. In addition, the method provides the optimum accuracy which can be achieved by a model-parametrization setup. The method of Salama et al. (2011) is self-contained and is directly applicable to existing satellite based IOP products, we therefore, brief this method hereafter.

3.2 Ensemble uncertainty of IOPs

Applying Taylor series approximation of the second moment on equation (8) gives:

$$\sigma_r^2(\lambda) = w_{ph}^2(\lambda)\sigma_{ph}^2(440) + w_{dg}^2(\lambda)\sigma_{dg}^2(440) + w_{spm}^2(\lambda)\sigma_{spm}^2(550) \tag{14}$$

Where $\sigma_r^2(\lambda)$ is the radiometric variance and $\sigma_{ph}^2(440)$, $\sigma_{dg}^2(440)$, and $\sigma_{spm}^2(550)$ are the variances of the derived IOPs. The covariance terms in equation(14) is assumed to be zero, i.e. the IOPs are mutually independent. Knowledge on the radiometric uncertainty is now avoided by dividing both sides of equation (14) by the radiometric variance,

$$\sum_{i=1}^{i=n} w_i^2(\lambda)\psi_i^2(\lambda) = 1, \tag{15}$$

with $\psi_i^2(\lambda) = \sigma_i^2(\lambda_0)/\sigma_r^2(\lambda)$. The ensemble uncertainty of IOPs per radiometric error, $\Psi(\lambda)$, is derived from equation (15) by normalizing both sides by the squared sum of partial derivatives and taking its square-root:

$$\Psi(\lambda) = \left(\sum_{i=1}^{i=n} w_i^2(\lambda)\psi_i^2(\lambda) / \sum_{i=1}^{i=n} w_i^2(\lambda)\right)^{0.5} = \left(\sum_{i=1}^{i=n} w_i^2(\lambda)\right)^{-0.5}. \tag{16}$$

$\Psi(\lambda)$ represent the ensemble uncertainty of IOPs per unit error of remote sensing reflectance and have the unit of sr m^{-1}. The advantages of this methods is that it can be applied on the readily available earth observation products of IOPs (water quality proxies). Fig.(1) shows the climatology of the ensemble uncertainty relative to the sum of derived IOPs. These figures are generated by applying equation (16), to the monthly mean values of GSM-derived IOPs and then averaged for each year from 1997-2007 (the year 1997 is not shown). It is clear that there are persistent patterns of high values throughout the last decade in the subtropical gyres, whereas lower values are observed in most coastal areas. These results are in accordance to the global uncertainty maps of Chlorophyll-a presented by Mélin (2010) for the subtropical gyres, whereas the coastal waters show contrary patterns, i.e. very small error. The spatial distribution of the relative-ensemble uncertainty largely resembles the observed values of remote sensing reflectance at 443 nm.

3.3 Detailed uncertainty of IOPs

Based on equation(8), Bates & Watts (1988) devised an elegant method to quantify the uncertainties for each derived IOPs as,

$$IOP_{i\pm} = IOP_i \pm \sigma \left\|W \cdot R^{-1}\right\| t(N - m, \alpha/2) \tag{17}$$

Where $IOP_{i\pm}$ is the upper "+" and lower "-" bounds of the derived IOP; W is the matrix of partial derivatives; σ is the standard deviation of residuals between measured and model best-fit radiances; $t(N - m, \alpha/2)$ is the upper quantile for a Student's t distribution with $N - m$ degrees of freedom. N is the number of bands and m is the number of unknowns. R is the upper triangle matrix of QR decomposition of the jacobian matrix. equation (17) has widely been used to estimate the error of derived IOP (Salama et al., 2009; Van Der Woerd & Pasterkamp, 2008). The derivative term in equation (17), can be approximated as being the gradient of equation (1) with respect to the derived IOPs and is computed for model-best-fit to the observation. This approximation is derived as follows.

Observed remote sensing reflectance can be approximated as being the sum of the model best-fit $Rs_m(\lambda)$ and its deviations from the observed one $\epsilon(\lambda)$:

$$Rs(\lambda) = Rs_m(\lambda) + \epsilon(\lambda) \tag{18}$$

The term $Rs_m(\lambda)$ is obtained from fitting the model in equation (1) to the radiometric observation of ocean color or/and field sensors. The error $\epsilon(\lambda)$ is a lumped term that includes

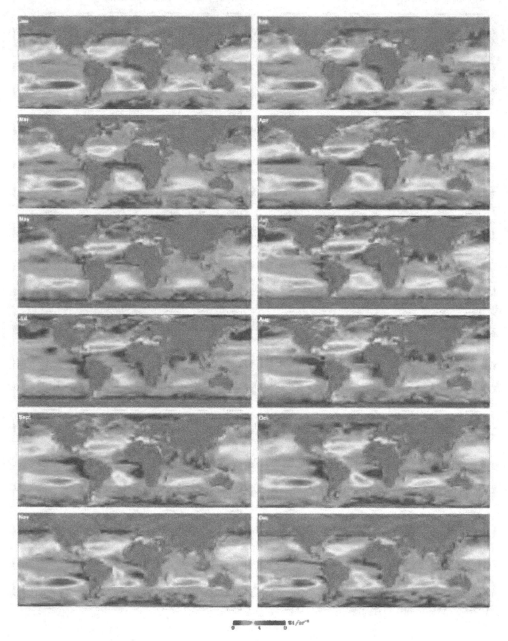

Fig. 1. Time series of ensemble-uncertainty of IOPs at 440 nm relative to the sum of derived IOPs.

model goodness-of-fit, measurements and atmospheric noises. For simplicity this term is assumed to be nearly independent the derived IOPs. The derivative of (18), with respect to the derived values, can then be written as:

$$\frac{\Delta Rs(\lambda)}{\Delta iop} = \frac{\Delta Rs_m(\lambda)}{\Delta iop} + \frac{\Delta \epsilon(\lambda)}{\Delta iop} \qquad (19)$$

By definition of the least square minimization that was used to derive model-best-fit $Rs_m(\lambda)$, we have:

$$\frac{\Delta \epsilon(\lambda)}{\Delta iop} \approx 0 \qquad (20)$$

Equation (19) can then be reduced to:

$$\frac{\Delta Rs(\lambda)}{\Delta iop} \approx \frac{\Delta Rs_m(\lambda)}{\Delta iop} \qquad (21)$$

The simplification in equation (21) implies that the gradient of measured remote sensing reflectance can be approximated by the gradient of the model in (1) which can easily be computed as in equation (21).

4. Error estimation via stochastic method

4.1 Description

In this section we summarize the method of Salama & Stein (2009) as it is the only stochastic method published so far in the field of ocean color.

Salama and Stein used prior information to obtain plausible ranges of the IOPs. These ranges are used in a log-normal distribution to generate a first-estimate of the probability distribution (PD) of the IOPs. This first-estimate PD is called the prior PD of the IOPs. The method, explained hereafter, uses the prior PD to converge to a "posterior" probability distribution that better describes the IOPs.

Prior information is obtained from known radiometric errors in Rs_w and model-inversion intrinsic errors. Radiometric errors are: (i) noise equivalent radiance of the sensor and (ii) error in aerosol optical thickness. Sensor equivalent radiance is known from sensor specifications and post-launch calibrations. Model approximation and inversion-accuracy can be quantified by evaluating the performance of the employed ocean color model against measurements and radiative transfer simulations. Atmospheric error, due to variation in aerosol optical thickness, can be evaluated from available measurements or by using standard atmospheric correction models. The error estimate algorithm will follow sequential steps as detailed hereafter.

An initial estimate of the confidence interval around water remote sensing reflectance can be computed using the method of (Bates & Watts, 1988, pp.59, cf. 1.36) or available knowledge on plausible fluctuations for model, noise and atmospheric residual respectively. The upper and lower bounds of this interval are then inverted to derive the corresponding two sets of IOPs iop_u, iop_l. These sets with the derived iop_{obs} from the water remote sensing reflectance, hereafter will be called the IOP-triplet: $(iop_l, iop_{obs}, iop_u)$ and denoted as ω. The value $\log iop_{obs}$ is assumed to approximate the mean of a first-estimate, i.e. prior, probability distribution (PD) of IOPs in the logarithmic space. The prior PD is first elicited using the IOP-triplet and prior knowledge on the log-normal shape of the IOPs as explained in section (4.2). The posterior probability distribution, or our gain in information, is then inferred by maximizing the expected utility (Bernardo, 1979; Carlin & Polson, 1991) as explained in section(4.3).

4.2 Prior probability distribution

Estimating the IOP-triplet, iop_l, iop_{obs} and iop_u, is the first step towards deriving the prior probability distribution of the IOPs. The use of flat or improper priors, e.g. uniform distribution, may invalidate the derivation of the posterior probability (Goutis & Robert, 1998). According to the maximum entropy principle (Jaynes, 1957a;b) a proper prior probability distribution should have the maximum entropy provided by the IOP-triplet. However applying the maximum entropy principle on the information provided by the IOP-triplet will give the probability values of iop_l, iop_{obs} and iop_u but not the whole probability distribution P(iop); for more detail one may consult Jaynes (1968). To overcome this limitation, in data values, we introduce the following method to elicit the prior distribution of IOPs assuming that they are log-normally distributed. The log-normal assumption is based on Campbell's work (Campbell, 1995) who pointed out that, in general, marine bio-geophysical quantities follow a log-normal distribution i.e. their log transform has a Gaussian distribution.

The IOP-triplet is first transformed to the log space, allowing us to use a Gaussian distribution to simulate the PD of IOPs. Second we assume that $\log iop_{obs}$ approximates the mean of the prior PD of the IOPs. The Gaussian distribution of the IOPs can be standardized to a N(0,1) distribution, i.e. normal distribution with zero mean and unity standard deviation. The standard Gaussian variate for $\log iop_u$ is,

$$\alpha_u = \frac{\log iop_u - \log iop_{obs}}{\sigma},$$ (22)

where α_u is a sample drawn from the N(0,1) that corresponds to iop_u. The parameters $\log iop_{obs}$ and σ are the expectation and the standard deviation of the population. From equation (22) and the second set in the IOP-triplet iop_l we can establish the ratio,

$$r_{u,l} = \frac{\alpha_u}{\alpha_l} = \frac{\log iop_u - \log iop_{obs}}{\log iop_l - \log iop_{obs}},$$ (23)

and for convenience we set $\log iop_u > \log iop_l$. The standardization of the IOPs distribution allows us to use the N(0,1) random number generator to simulate values of α as in equation (22). The ratios of these random values are also computed and compared to the ratio of the IOP-triplet in equation (23). The best fit allocates the two values α_u and α_l, hence the standard deviation of the prior distribution can be computed from equation (22). The prior probability distribution of the IOPs, is now known: N($\log iop_{obs}, \sigma$), i.e. a Gaussian distribution with $\log iop_{obs}$ mean and σ standard deviation.

Random values (1000) are generated from the N(0,1) distribution such that they satisfy an imposed acceptance-rejection condition. This condition requires that the ratio in equation (23) defines a unique ordered pair of α. This is to enable the use of a simple searching method with a fast convergence to the best-fit ratio. The uniqueness in this sense implies that the squared difference between the computed ratio, from the IOP-triplet, and the best-fit is a global minimum resolvable by the searching method and the used computer processor. Three look-up tables (LUTs) are then created from the generated values. These LUTs correspond to the following three scenarios:

$$\log iop_{obs} > \log iop_u \quad > \log iop_l$$
$$\log iop_{obs} < \log iop_l \quad < \log iop_u \tag{24}$$
$$\log iop_l \quad < \log iop_{obs} < \log iop_u$$

The generated N(0,1) values are, first, subdivided into two sets containing positive and negative values. The ratios of the first and second LUTs are, then, computed from the ordered descending sets as; x_i / x_{i+1}. The third LUT is generated from all possible combinations of the unordered positive and negative sets. This will results in ratio values between 0 and 1, > 1 and < 0 for the first, second and third LUT respectively. The ratio in equation (23) is first estimated from IOP-triplet. Based on the values of this triplet (equation 24) a lookup table is selected and searched to find the best-fit value to the computed ratio (equation 23). This best-fit is found either by direct search or interpolated. One of the corresponding pair is then used in equation (22) to compute the standard deviation of the prior PD $P(iop)$.

4.3 Posterior probability distribution

In section (4.2) we derived a proper prior distribution of the IOPs. This first-estimate, i.e. prior distribution, is converged to a posterior distribution that better describes the IOPs using the concept of Entropy. Entropy is a numerical measure of error associated with probability distribution of derived IOPs or any hydrological parameter (Singh, 1998). For a population with N sets of IOPs it is expressed as the Shannon entropy (Shannon, 1948):

$$H\{P(iop)\} = -\sum_{1}^{N} P(iop) \cdot \log P(iop) \tag{25}$$

where $P(iop)$ is the prior probability distribution (PD) of the derived set of IOPs iop.

If we design a function D that measures the information, e.g. equation (25), between the prior and the posterior PD, then we can derive the posterior PD such that it maximizes the expected information to be gained in D (Bernardo, 2005; Christakos, 1990). In other words, maximizing the function D will maximize the gained information from the posterior PD (Bernardo, 1979). The Kullback-Leibler divergence (Kullback & Leibler, 1951), or cross-entropy, belongs to this type of utility functions (Johnson & Geisser, 1985). It measures the divergence between the posterior $P(iop|\omega)$ and the prior $P(iop)$ probability distribution as:

$$D_{KL}\{P(iop|!)|P(iop)\} = \sum_{1}^{N} P(iop|\omega) \cdot \log \frac{P(iop|\omega)}{P(iop)} \tag{26}$$

where $P(iop|\omega)$ is the posterior probability of iop given the IOP-triplet ω. Equation (26) can be rewritten in view of equation (25) as:

$$D_{KL}\{P(iop|\omega)|P(iop)\} = H\{P(iop|\omega), P(iop)\} - H\{P(iop|\omega)\} \tag{27}$$

where $H\{P(iop|\omega), P(iop)\}$ is expressed as:

$$H\{P(iop|\omega), P(iop)\} = -\sum_{1}^{N} P(iop|\omega) \cdot \log P(iop) \tag{28}$$

Maximizing the cross-entropy in equation (26) or the corresponding expression in (27) is equivalent to minimizing the entropy (uncertainty) of the posterior probabilities distribution, i.e. maximizing gained information. The errors can then be estimated from the reconstructed posterior probability distribution of IOPs $P(\mathbf{iop}|\omega)$.

The posterior probability distribution is inferred by maximizing the utility function, i.e. Kullback-Leibler divergence (equation 26). The maximum is found by iteration through a sequential updating of the posterior using the prior parameters mean μ and variance σ^2 (Rubinstein & Kroese, 2004). The corresponding log-normal mean m and variance v are computed as Kendall & Stuart (1987):

$$m = e^{\mu} e^{0.5\sigma^2} \tag{29}$$

$$v = e^{2\mu} e^{\sigma^2} \left(e^{\sigma^2} - 1\right) \tag{30}$$

The following steps describe the algorithm, as implemented, to derive the posterior PD $P(\mathbf{iop}|\omega)$:

1. From the water remote sensing spectrum estimate the initial radiometric confidence interval using the method of (Bates & Watts, 1988, pp.59, cf. 1.36) or prior information on atmospheric and noise-induced radiometric fluctuations.
2. Invert the ocean color model in equation (1) to derive the IOPs from the water remote sensing spectrum and the upper and lower bounds. This will results in three sets of IOPs: $\mathbf{iop}_l$, $\mathbf{iop}_{obs}$, $\mathbf{iop}_u$; IOP-triplet.
3. Based on the order of this IOP-triplet allocate the suitable LUT using equation (24).
4. Search for the best-fit ratio calculated from equation (23).
5. Use equation (22) to estimate the standard deviation of the prior PD.
6. Use the standard deviation and $\log \mathbf{iop}_{obs}$ to generate the prior PD.
7. Use initial values of the mean and standard deviation to generate n Monte Carlo samples of PD.
8. Select the population that have the maximum Kullback-Leibler divergence (equation 26), and update the initial values.
9. Repeat step 7 to 8 till convergence.
10. Update the prior PD with the resulting posterior PD (from the pervious step: 9), and iterate steps 7 to 10 till convergence.

The convergence is defined by a threshold as follow. Keep track of the best ten candidates which maximize equation (26). The system converges if the variance of these ten values is less than 10^{-4}.

5. Inter-comparison between deterministic and stochastic methods

The inter-comparison between the deterministic method, described in Section (3), and the stochastic method, described in Section (4), is carried out using two data sets. The first, is radiative transfer simulations of synthetic IOPs obtained from the International Ocean Color Coordination Group (IOCCG), report-5 (Lee, 2006, IOCCG data set). The second consists of concurrent observations from the Sea viewing Wide Field-of-view Sensor (SeaWiFS) and measured inherent and apparent optical properties, retrieved from the NASA bio-Optical Marine Algorithm Data set (NOMAD) Version 1.3 (Werdell & Bailey, 2005, SeaWiFS matchup data set).

5.1 IOCCG

IOCCG data set (Lee, 2006) of synthesized IOPs and their radiative transfer simulations at 30° sun zenith angle are used to inter compare the results of the deterministic and stochastic methods. IOCCG simulated spectra, between 400 nm and 750 nm at 10 nm interval, are inverted using the ocean color model in equation (1) to derive five variables. These variables are: Chlorophyll-a absorption at 440 nm $a_{ph}(440)$, detritus and CDOM absorption at 440 nm $a_{dg}(440)$ and their spectral dependency s, SPM scattering at 550 nm $b_{spm}(550)$ and SPM spectral dependency y, as shown in equation (7).

The standard deviation of the posterior PD represents the error/confidence of the derived value iop_{obs}. The deviation of the posterior PD from known IOPs is measured using root-mean-square of errors (RMSE). These two values, RMSE and standard deviation, are related through the bias, i.e the actual difference between derived and measured IOPs. Figure (2) shows estimated errors, expressed as standard deviation using equation (30), against the known root-mean-square of errors (RMSE). The actual RMSE is estimated from the posterior PD and the known IOPs. The reproduced errors for the IOPs other than $a_{ph}(440)$ have a high accuracy with r^2 values between 0.77 and 0.96. Estimated errors of $a_{ph}(440)$ have the lowest r^2 and n values. It is worth noting that the determinacy method of Bates & Watts (1988) generally underestimates model-errors of the IOPs with lower r^2 values than the presented stochastic method. This is apparent at an almost threefold difference for the error values of $a_{ph}(440)$. On the other hand, the stochastic method has a tendency to overestimate the errors of the IOPs with a better fit and improved capability, in the sense that it can be applied to populations of any bio-geophysical variable.

5.2 NOMAD

Due to the limited number of available visible bands in this data set we reduced the number of unknowns to three only. The first three IOPs in equation (7) are derived from SeaWiFS spectra using the ocean color model (equation 1) and the constrained LMA technique. The values of s and y are set to 0.021 nm^{-1} and 1.7 respectively. The actual RMSE values are computed from the posterior PD and measured IOPs. The total error on derived IOPs is estimated by applying the stochastic method using (Bates & Watts, 1988, pp.59, cf. 1.36) radiometric confidence interval. The estimated errors are expressed as standard deviation using equation (30) and plotted against RMSE values in figure (3). The reproduced total error values are strongly correlated to the known RMSE values with r^2 between 0.67 and 0.9 and >90% of valid retrievals. Estimated errors from the deterministic technique (Bates & Watts, 1988), however, did not correspond to the actual values of RMSE.

Errors are computed for the ocean color model and SeaWiFS visible bands centered at [412, 443, 490, 510, 555, 670] nm. The average values of the derived standard deviation are 1.7802, 1.1431 and 1.6177 m^{-1} for $a_{ph}(440)$, $a_{dg}(440)$ and $b_{spm}(550)$, respectively.

6. Uncertainty sources

6.1 Description

The total remote sensing reflectance received at the sensor altitude can be written as the sum of several components (Gordon, 1997):

$$Rs_t(\lambda) = Rs_{path}(\lambda) + T(\lambda)Rs_{sfc}(\lambda) + T(\lambda)Rs_w(\lambda) \tag{31}$$

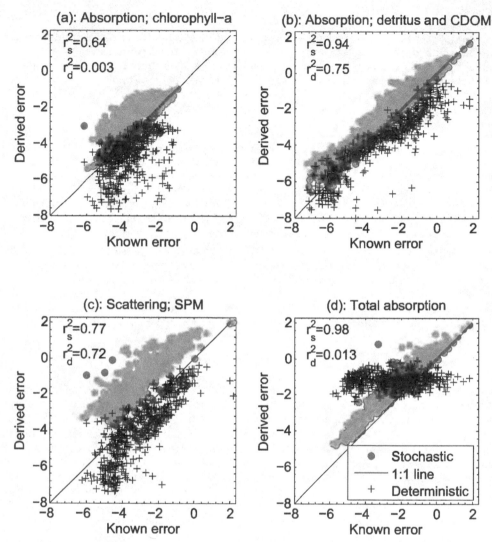

Fig. 2. Derived versus known errors of the IOPs estimated from the IOCCG data set for: (a) Chl-a absorption at 440 nm; (b) CDOM and detritus absorption at 440 nm; (c) SPM scattering at 550 nm; and (d) the total absorption at 440 nm. The data on the plots are log transformed. The coefficients of determination r_s^2 and r_d^2 are for stochastic and deterministic method respectively.

The subscript of the remote sensing reflectance Rs represents the contribution from: (i) the atmosphere (*path*), i.e. air molecules and aerosol multiple scattering; (ii) sea-surface (*sfc*); and (iii) water (*w*). $T(\lambda)$ is the diffuse transmittance.

The contribution of air molecules, i.e. the Rayleigh scattering, to the atmospheric path is well described in terms of geometry and atmospheric pressure (Gordon et al., 1988).

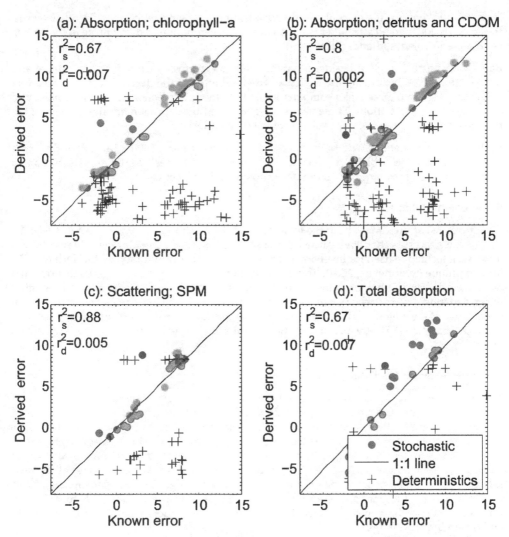

Fig. 3. Derived versus known errors of the IOPs estimated from the NOMAD data set for: (a) Chl-a absorption at 440 nm; (b) CDOM and detritus absorption at 440 nm; (c) SPM scattering at 550 nm; and (d) the total absorption at 440 nm. The data on the plots are log transformed. The coefficients of determination r_s^2 and r_d^2 are for stochastic and deterministic method respectively.

The contribution of sea-surface reflectance Rs_{sfc} can be estimated using the probabilistic formulations of Cox & Munk (1954) and ancillary data on wind field. Gaseous transmittance can be calculated from ancillary data on ozone and water vapor concentrations using the transmittance models of Goody (1964) and Malkmus (1967). For viewing angles < 60° the diffuse transmittance T is weakly dependent on aerosol and can be approximated following Gordon et al. (1983). Following the aforementioned approximations will basically leave two unknowns; the aerosol and the water remote sensing reflectance. In other words, the errors in Rs_w can be attributed to errors in aerosol estimation and any noise in the sensor, i.e. noise equivalent radiance (NER).

Radiometric errors in Rs_w, beside to model-inversion intrinsic errors, will accumulate and propagate to the IOPs during the retrieval. The total error of the derived IOPs can therefore be decomposed into three major components, namely model-inversion error, sensor noise and error in aerosol estimation. These errors are originated by various mechanisms during the processing chain of ocean color data as explained hereafter.

Each error component, x, will be expressed as the variance σ_x^2 of IOPs caused by this error x. The subscript x will be replaced by inv, ner and a to represent the contribution of model-inversion, noise equivalent radiance and aerosol, respectively.

6.2 Model-inversion error, σ_{inv}^2:

The employed approximations in the forward-model (equation 1) may not precisely describe the optical processes that have caused the observed signal (Zaneveld, 1994). Moreover, the numerical technique used for inversion provides an ambiguous solution, i.e. the derived IOPs are not unique (Sydor et al., 2004). These assumptions and ambiguity will generate error that is, at the one hand, inherent to the employed ocean color forward model and, on the other hand, dependent on the accuracy of the inversion scheme which could be related to the optical complexity of the water. Model-inversion error is quantified as a lumped sum of errors due to the approximation in (1), the parametrization of IOPs and inversion and abbreviated as model error.

6.3 Noise equivalent radiance, σ_{ner}^2:

Noise equivalent radiance (NER) depends on sensor specifications and performance over time, i.e. sensor degradation. This fluctuation could either increase or decrease the observed remote sensing reflectance and could also be wavelength dependent or random. The effects of NER is inversely proportional to the value of signal-to-noise ratio. Sensor degradation, i.e. sensitivity losses over time, will cause decrease in the signal-to-noise ratio of the sensor leading to low signal reading. Low signal can also be observed over clear water at the near infrared part of the spectrum or over turbid water, with high CDOM, detritus and Chl-a contents, at the blue part of the spectrum. The propagated error from NER to IOPs will therefore be dependent on sensor specification, sensor degradation over time, water turbidity and observing wavelength.

6.4 Variations of aerosol type and optical thickness, σ_a^2:

Atmospheric correction errors are, generally, caused by unknown aerosol type and optical thickness (AOT). The residual signals from atmospheric correction will have spectral and spatial dependency. The spectral dependency is due to the error about the aerosol type e.g. absorbing aerosol, while the spatial dependency is, on the one hand, related to the error about AOT spatial variations and, on the other hand, to water turbidity (Hu et al., 2004). It is assumed that aerosol optical thickness has a higher spatial variability than aerosol type, so that aerosol type can be assumed to be known and homogenous. Within the validity of this assumption, the residual signals from atmospheric correction will be caused by errors in estimating the aerosol optical thickness.

6.5 Decomposition

The total error of the derived IOPs, expressed as the variance $\sigma_t^2 \left(\sigma_{inv}^2, \sigma_{ner}^2, \sigma_a^2 \right)$, is thus described as a function of the three error components, σ_{inv}^2, σ_{ner}^2 and σ_a^2. Assuming that this function is continuous in its variables, we can approximate it by a first order Taylor series as:

$$\sigma_t^2 \approx \sigma_{t0}^2 + \frac{\partial \sigma_t^2}{\partial \sigma_{inv}^2} \sigma_{inv}^2 + \frac{\partial \sigma_t^2}{\partial \sigma_{ner}^2} \sigma_{ner}^2 + \frac{\partial \sigma_t^2}{\partial \sigma_a^2} \sigma_a^2 \qquad (32)$$

where, σ_{t0}^2 is the value of the function $\sigma_t^2(0,0,0)$. According to the assumption that the total error is caused by three components, the value of σ_{t0}^2 is negligible, i.e. $\sigma_{t0}^2 \simeq 0$. In other words, if we have perfect measurements, accurate atmospheric correction and exact model parameterizations and inversion then the total error on the derived IOPs will be negligible. The total error of the derived IOPs can thus be approximated as a weighted sum of the individual error components as:

$$\sigma_t^2 \approx w_{inv}^2 \sigma_{inv}^2 + w_{ner}^2 \sigma_{ner}^2 + w_a^2 \sigma_a^2 \qquad (33)$$

where the weights w_{inv}, w_{ner}, and w_a are the partial derivatives in equation (32). The functionality in σ_t^2, however, is commonly unknown and it is therefore difficult to find proper estimates of the weights w_{inv}, w_{ner} and w_a. An intuitive approach would be setting all the weights in equation (33) to unity and check its validity:

$$\sigma_t^2 \approx \sigma_{inv}^2 + \sigma_{ner}^2 + \sigma_a^2 \qquad (34)$$

Figure (4) depicts the relationship between the sum of the righthand side of equation (34) and the total error on the derived IOPs. On the X axis is the total error of the IOPs as calculated from all possible error sources σ_t^2. We then calculated each error component apart and summed their variances in the Y axis as: $\sigma_a^2 + \sigma_{ner}^2 + \sigma_a^2$. As anticipated from equation (33) there is a linear relationship between the actual variance and the linear sum of individual variances with R^2 values above 0.75 for the absorption coefficients of Chl-a and detritus-CDOM. The value of R^2 decreases to 0.69 for SPM scattering and 0.64 for the total absorption. The dispersion value as measured with RMSE is large for all IOPs. The results in figure (4) indicate that the linear sum in equation (34) is an acceptable approximation to the total variance. Due to the large values of RMSE in figure (4), the computed relative contribution should be treated with caution.

While model-induced error can directly be estimated from the techniques described in Brad (1974) and Bates & Watts (1988), noise and atmospheric-induced errors should be inferred from the available information. This information forms the prior knowledge that we will use in the following section to derive the error of the IOPs. Prior information is obtained from known sensor's noise, variation in aerosol optical thickness and ocean-color model's approximations and inversion accuracy.

6.5.1 IOCCG

The noise is estimated based on NER values of the Medium Resolution Imaging Spectrometer (MERIS) (Doerffer, 2008; Hoogenboom & Dekker, 1998). The variation in aerosol optical thickness (AOT) is set to be ±0.02. This value is estimated from the variation of recorded aerosol optical thickness by a newly calibrated sunphotometer (CIMEL) and cloud free

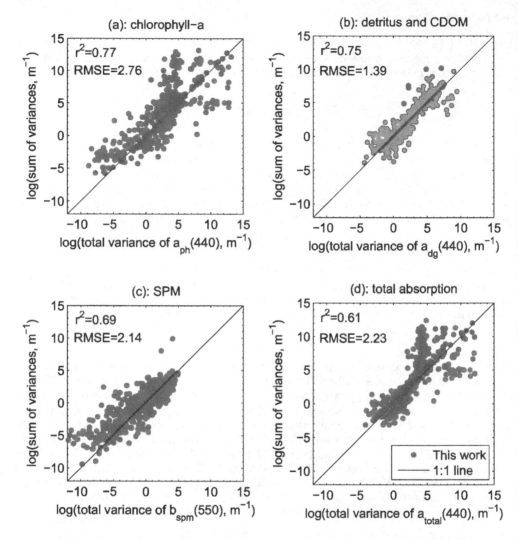

Fig. 4. Sum of variances versus the total variance of the IOCCG data set for: (a) Chl-a absorption at 440 nm; (b) absorption of detritus and CDOM at 440 nm; (c) SPM scattering at 550 nm; and (d) total absorption coefficient at 440 nm.

condition (Holben et al., 2000). The values of aerosol optical thicknesses are obtained from sunphotometer measurements situated at (51.225 N, 2.925 E) at the 8th of June 2006. The atmospheric paths are estimated with radiative transfer computation (Vermote et al., 1997) using maritime aerosol model with a nadir looking sensor at 30° sun-zenith and 203° sun-azimuth angles.

The relative contribution of model, noise and atmospheric errors are shown in table (1) and quantified for each of the derived IOP as follow. First we computed the total error, i.e. the total error in $Rs_w(\lambda)$ is due to aerosol estimation and sensor noise, inversion error will add up during the inversion. The same step is repeated for each error source in three steps: (i) model error is estimated from the error-free $Rs_w(\lambda)$; (ii) atmospheric-induced error σ_a^2 is computed from $Rs_w(\lambda)$ that contains errors due to aerosol estimation only; (iii) noise error is calculated from $Rs_w(\lambda)$ that contains sensor noise only. Note that model error will add up during the inversion in the last two steps. Now we can use equation (34) to estimate the relative contribution of each error component to the total error of the IOPs.

Errors due to atmospheric correction are the major source of errors in the derived IOPs. Imperfect atmospheric correction, due to the variability of aerosol optical thickness, is responsible for more than 50% of the total error and up to 82%. One fifth of the total errors on derived IOPs (except for the SPM scattering: one tenth) is attributed to noise-error. Model-error has the lowest contribution ($\approx 7\%$) to the total error on derived $b_{spm}(550)$ values, but it has a significant contribution ($\approx 16\%$) to y. This can be attributed to the assumed parametrization. On the one hand, the absorption of other constituents than water molecules is negligible at the near infrared (NIR) which will cause stability (one-to-one relation) in the derived SPM scattering coefficient, leading to a significant contribution from the atmosphere at the NIR region. On the other hand, the error in $b_{spm}(\lambda)$ will decrease towards the NIR region due to the assumed exponential spectral dependency. In general, model-induced errors are large for the spectral shape coefficients y and s. Note that the spectral shape of chlorophyll-a absorption is imbedded in coefficient $a_{ph}(440)$.

	error components		
IOPs	Model	Noise	Aerosol
$a_{ph}(440)$	17	22	61
$a_{dg}(440)$	9	19	72
$b_{spm}(550)$	7	11	82
y	16	24	60
s	28	21	51

Table 1. The average relative contribution (%) of error components on IOCCG data set.

6.5.2 NOMAD

The total error on estimated IOPs from the NOMAD data is derived from the values presented in figure (3). Model induced errors are subtracted from the total error using equation (33) to deduce atmospheric and noise-induced errors. The results are shown as percentages in table (2). Main uncertainty is due to atmospheric and noise-induced errors for $a_{ph}(440)$ and $b_{spm}(550)$, while model inversion is the main source of error to $a_{dg}(440)$ in this data set. These results are within the validity of the linear assumption expressed in equation (33) and the imposed values of s and y.

	error components	
IOPs	Model	Aerosol and Noise
$a_{ph}(440)$	10	90
$a_{dg}(440)$	57	43
$b_{spm}(550)$	19	81

Table 2. The average relative contribution (%) of error components on SeaWiFS observations in the NOMAD data set.

6.5.3 Model-sensor error table

The linear sum of individual variances in equation (34) can describe about 70%, the value of R^2, of the total variance of the IOPs. This linearization of the total variance is a simple yet effective approach. It allows us to estimate the relative contribution of the different error components to the total error budget on the IOPs. The relative contribution of model, noise and atmospheric errors to the total error budget using IOCCG data set are 20-40%, 10-25% and 40-80%, respectively. Model-induced errors, due to approximation and inversion, are inherent to the derived IOPs and inversely proportional to model-inversion degree-of-freedom, while atmospheric-induced errors are the major contributor to the total error budget on IOPs. These results are for assumed levels of noise and atmospheric fluctuations. This suggests that error table can be generated for specific model, sensor and range of IOPs. This model-sensor error table can serve as a benchmark to estimate the atmospheric-induced errors in the derived IOPs. The merit of this argument is based on the fact that the computations of model and noise-induced errors can be quantified using water radiative transfer simulations, for a specific range of IOPs, and known sensor's NER. The magnitude of these errors are in principle known for the ocean color model and the used sensor. An example of such a table for the MERIS sensor and the ocean color model is shown in table (3). This table is computed from table (1) for the MERIS visible bands centered at [412, 443, 490, 510, 560, 620, 665, 708, 778] nm, i.e. we simply reduced the spectral bands of IOCCG data set to fit those of MERIS. Table (3) shows that the reduced number of spectral bands for MERIS setup has increased model contribution to the total error approximately two fold. This will reduce noise and atmospheric contribution to the total error, since the relative contributions of all error components should sum to 1. Note that for weak radiometric signals, the lower bound might end up to negative values which will lead to further reduction in the number of bands (negative values are set to zero). This approach is demonstrated for ocean color observations obtained from NOMAD data set. Model and noise-induced errors are simulated from the IOCCG data set and subtracted from the total error of IOPs estimated from the NOMAD data set. The simulation is carried out simply by selecting IOCCG wavelengths that correspond to NOMAD spectral set-up. The simplicity of this approach can pose a limitation on the accuracy of equation (34). On the one hand, the method shows that model approximation and inversion are main contributors, $\approx 57\%$, to the total error of $a_{dg}(440)$. On the other hand, the presented stochastic method quantified these errors most efficiently. Atmospheric and noise-induced errors are significant for $a_{ph}(440)$ and $b_{spm}(550)$. This may suggests that model-induced errors are better quantified with the current method. However, errors of SPM scattering coefficient, which are mainly due to atmospheric residuals, are reproduced with high accuracy.

IOPs	error components		
	Model	Noise	Aerosol
$a_{ph}(440)$	40	13	47
$a_{dg}(440)$	41	13	46
$b_{spm}(550)$	45	5	50
y	42	19	39
s	42	16	42

Table 3. The average relative contribution (%) of error components on derived IOPs using the ocean color model (equation 1) and simulated MERIS bands from IOCCG data set.

7. Spectral propagation of errors and error correlation

The presented errors of IOPs were for two wavelengths: 440 nm for the absorption coefficients and 550 nm for the scattering coefficient, as defined by equation (7). We can use the parameterizations in equations (4), (5) and (6) to derive analytical description of error propagation to other wavelengths. Here below we provide an analytical derivation of error propagation and numerical examples for two wavelengths one at the blue, 400 nm, and the other at the red, 680 nm.

The errors of the IOPs will propagate to shorter and longer wavelengths following the parameterizations in equations (4), (5) and (6). For example, the error in $b_{spm}(550)$ has two components; one in $b_{spm}(550)$ itself and the other in the spectral shape y. Using the parametrization in equation (6) we will have:

$$\Delta b_{spm}(\lambda) = \frac{\partial b_{spm}(\lambda)}{\partial b_{spm}(550)} \Delta b_{spm}(550) + \frac{\partial b_{spm}(\lambda)}{\partial y} \Delta y + \frac{\partial b_{spm}(\lambda)}{\partial \lambda} \Delta \lambda \tag{35}$$

Carrying the derivation of the palatial derivatives, equation (35) can be written as:

$$\begin{aligned}
\Delta b_{spm}(\lambda) = & \left(\frac{550}{\lambda}\right)^y \Delta b_{spm}(550) \\
& + b_{spm}(550) \left(\frac{550}{\lambda}\right)^y \ln \frac{550}{\lambda} \Delta y \\
& - \frac{y}{\lambda} b_{spm}(550) \left(\frac{550}{\lambda}\right)^y \Delta \lambda
\end{aligned} \tag{36}$$

In this exercise we will neglect the error in the wavelength, i.e. $\Delta \lambda \approx 0$ and we will show that the derivative $\partial/\partial \lambda$ is negligible.

Let us take the two reference wavelengths: the blue 400 nm and the red 680 nm and assume $y = 1.7$, we will have:

$$\Delta b_{spm}(400) = 1.718 \Delta b_{spm}(550) + 0.547 b_{spm}(550) \Delta y \tag{37}$$
$$\Delta b_{spm}(680) = 0.697 \Delta b_{spm}(550) - 0.148 b_{spm}(550) \Delta y \tag{38}$$

The wavelength variation term $\partial/\partial \lambda$ in equations (37) and (38) is neglected. It takes the values, with λ expressed in nanometer, $7.3 \times 10^{-3} b_{spm}(550)\Delta\lambda$ and $1.74 \times 10^{-3} b_{spm}(550)\Delta\lambda$ for the blue and the red wavelengths, respectively.

Equations (37, 38) show that the error in SPM scattering coefficient at the blue wavelength is larger than that at the red wavelength if the relative error in the scattering coefficient satisfies the condition:

$$\frac{\Delta b_{spm}(550)}{b_{spm}(550)} > -0.681 \Delta y \tag{39}$$

In a similar approach we can quantify the propagated errors of $a_{dg}(440)$ to other wavelengths:

$$\begin{aligned}
\Delta a_{dg}(\lambda) = & \exp\left[-s\left(\lambda - 440\right)\right] \Delta a_{dg}(440) \\
& - a_{dg}(440)\left(\lambda - 440\right) \exp\left[-s\left(\lambda - 440\right)\right] \Delta s \\
& - s \times a_{dg}(440) \exp\left[-s\left(\lambda - 440\right)\right] \Delta \lambda
\end{aligned} \tag{40}$$

If we assume the value $s = 0.021$ nm^{-1} and take our reference bands to be the blue (400 nm) and red (680 nm) wavelengths we will have, with λ in meter:

$$\Delta a_{dg}(400) = 2.316\Delta a_{dg}(440) + 92.654 \times 10^{-9} a_{dg}(440)\Delta S \tag{41}$$

$$\Delta a_{dg}(680) = 6.47 \times 10^{-3}\Delta a_{dg}(440) - 1.554 \times 10^{-9} a_{dg}(440)\Delta S \tag{42}$$

The wavelength variation term $\partial/\partial\lambda$ is also negligible. It takes the values $-4.86 \times 10^{-11} a_{dg}(440)\Delta\lambda$ and $-1.36 \times 10^{-13} a_{dg}(440)\Delta\lambda$ for the blue and the red wavelengths, respectively. The error at the blue will be larger than that at the red if the relative error of $a_{dg}(440)$ satisfies the condition (from equations 41 and 42):

$$\frac{\Delta a_{dg}(440)}{a_{dg}(440)} > 4.08 \times 10^{-8}\Delta s \tag{43}$$

The parametrization of Chl-a absorption is based on the tabulated values a_0 and a_1, see equation(4). These tabulated values are taken to be constant per wavelength, i.e. $a_{ph}(\lambda)$ is function of $a_{ph}(440)$ only. The error in $a_{ph}(440)$ will propagate to other wavelengths following the derivative of equation (4):

$$\Delta a_{ph}(\lambda) = a_1(\lambda) + a_0(\lambda) + a_1(\lambda) \log a_{ph}(440)\Delta a_{ph}(440) \tag{44}$$

For the two reference bands, 400 nm and 680 nm, we will have:

$$\Delta a_{ph}(400) = 0.731 + 0.012 \log a_{ph}(440)\Delta a_{ph}(440) \tag{45}$$

$$\Delta a_{ph}(680) = 0.945 + 0.149 \log a_{ph}(440)\Delta a_{ph}(440) \tag{46}$$

From equations (45, 46) it can be shown that the error at the blue band is larger than that at the red if the following condition is satisfied:

$$\log a_{ph}(440)\Delta a_{ph}(440) < -1.562 \tag{47}$$

The analytical expressions in equations (36), (40) and (44) show that the errors are related to absolute values of the IOPs. Therefore, the three error components are expected to be correlated to water turbidity, and hence to each others. The results of the numerical examples also demonstrate that the errors of $b_{spm}(\lambda)$ and $a_{dg}(\lambda)$ will be larger at the blue than that at the red if the relative errors of $b_{spm}(550)$ and $a_{dg}(440)$ satisfy equations (36) and (40), respectively. Whereas the error in $a_{ph}(440)$ will propagate to other wavelengths following equation (44) and will be larger at the blue if the condition in (47) is satisfied.

8. Advantages and limitations of error estimation methods

Estimated errors from the deterministic method (Bates & Watts, 1988) did not correspond to the actual values of RMSE. This is due to the atmospheric and noise radiometric fluctuations. These fluctuations are imbedded in the observed signal and do not vary with IOPs values, i.e. different response function. Their large fluctuations may cause an ill-conditioned Jacobian matrix that produces erroneous estimates, see (Bates & Watts, 1988, pp.59, cf. 1.36). It should, nevertheless, be emphasized that the deterministic method is a well established technique to estimate retrieval errors. It can be used for the quantification of the combined accuracy of ocean color models and the parameterizations of IOPs, or model-parametrization setup. Its application produces a single (or ensemble) uncertainty measure for the collective errors in the derived IOPs relative to the radiometric uncertainty without the need for model inversion or prior information on the radiometric errors.

Error decomposition exercise shows that atmospheric and NER induced errors can be better quantified when prior knowledge is available. This is important for ocean color band ratio or single band algorithms, e.g. (Austin & Petzold, 1981; Salama et al., 2004). These algorithms are empirical in nature, i.e. Jacobian matrix is not available. In this case, deterministic methods to derive the error are not applicable. In contrary, the presented stochastic method is generic and can be applied to quantify the error of any derived bio-geophysical parameter regardless of the used derivation method. This is true if, beside to the derived quantity, two other values are known a priori so the IOP-triplet can be constructed.

The prior values were inferred from the quantiles of the populations. In practice this information is not available but it could be estimated from historical measurements or high temporal observations. The later, high temporal sampling, can be realized using sensors on board of geostationary satellites to quantify marine bio-geophysical parameters. For instance, the visible band of the Spinning Enhanced Visible and Infrared Imager (SEVIRI) on board of the METEOSAT second generation satellite (MSG) can be used to quantify the concentrations of SPM (Neukermans et al., 2008). With MSG 15 minutes of repeated sampling cycle, the stochastic method can be applied on three consecutive acquisitions, i.e. each 45 minutes, to produce SPM concentration and related-error maps. This error map provides vital input to the recently developed SPM assimilation model (Eleveld et al., 2008). Moreover it can be used, as weights, for ocean color products merging (Pottier et al., 2006). This generality aspect of the presented stochastic method expands its applicability to different fields other than ocean color. For example, Velde van der et al. (2008) developed a basis for Synthetic Aperture Radar (SAR)-based soil moisture downscaling methodologies.

One limitation of the presented stochastic method is the choice of the acceptance-rejection method. Although it facilitates the search for a unique pair of $N(0,1)$ values, the derived σ become sensitive to the ratio in (23), i.e. sensitive to the lower and upper pair (iop_u, iop_l) in the IOP-triplet. This may caused the 7~10% failure to reproduce the values of the standard deviation. This can be attributed to the small values of $\alpha \ll 1$ which produce large values of σ. These large values will further be magnified by equation (30).

Using equation (Bates & Watts, 1988, pp.59, cf. 1.36) to estimate the total error as a linear sum of all other error components is another limitation. Atmospheric or noise radiometric fluctuations can be interpreted, by model inversion, as high/low IOPs values with high goodness-of-fit. Using the same reasoning, bad fit to very complex signal (turbid water with high SPM, CDOM and Chl-a contents) can be attributed to atmospheric and sensor noise errors, although the observed signal might be error-free.

Model-sensor error tables were simulated from IOCCG data set without accounting for sensor's band width and response function. A more detailed simulations that includes band width, response function of the sensor and a specific range of the IOPs should be carried out to establish a more accurate model-sensor error tables.

Although we showed that equation (34) is an acceptable approximation to the total variance, the computed relative contribution of errors should be treated with caution.

9. Summary and future developments

In this chapter we reviewed the recent advances in uncertainty estimation of the earth observation products of water quality. Both deterministic and stochastic methods are presented and their results are inter- compared. The stochastic method is more appropriate to estimate actual errors of ocean color derived products than the deterministic methods, however, it is still limited to few studies and as the deterministic approach requires prior information. The uncertainties could be decomposed only if additional information is provided a priori. Using a simple exercise it was shown that atmospheric-induced errors are major contributors to the total error of IOPs whereas model-induced errors are inherent to the derived IOPs depending on the used derivation method and number of spectral bands.

The error in this chapter was estimated as the difference between ground truth measurement and satellite derived products. Direct matching between earth observation data and just above the water field measurements imbed, however, an inherent scale difference. This scale difference between *in-situ* observation and a pixel of ocean color satellite is at least three to four orders of magnitude for nadir match-up sites and much larger for off-nadir ones. This huge scale difference, means that point measurement is sampling a tiny fraction of the water body which is observed by a satellite pixel. Few studies were carried out to address the scale difference between point and aerospace measurements directly. Most of these studies have used re-sampling to smooth out the scale differences in the match-up sites, see (Bailey & Werdell, 2006; Bissett et al., 2004; Harding Jr. et al., 2005; Hu et al., 2000). For example, Hyde et al. (2007) applied a correction algorithm to SeaWiFS products of chlorophyll-a to overcome the mismatch which was partially due to sampling size differences. Although this assumption of spatial homogeneity have resulted in good matches for most open ocean matchup data (Carder et al., 2004; Garcia et al., 2005; Karl & Lukas, 1996; McClain et al., n.d.), it lowers the percentage of usable match-up points considerably (Hooker & McClain, 2000; Mélin et al., 2005) and should be avoided for productive waters (Chang & Gould, 2006; Darecki & Stramski, 2004; Harding Jr. et al., 2005). Salama & Su (2010; 2011), used the differences between the earth observation products and *in situ* data to quantify the sub satellite pixel spatial viabilities using both the deterministic and stochastic methods, respectively and neglecting the error. In principle the mismatch between earth observation derived products and *in situ* measured quantities is attributed to the scale difference and errors due to noise, correction and retrieval accuracy. Current uncertainty estimation methods do not consider the spatial dependency of errors and their relationships to the actual distribution of IOPs. Understanding the spatial characteristics of errors is necessary to resolve the smallest sub-scale variability of the IOPs. This aspect should be investigated in the future to define spatial-thresholds of measurable physical processes based on their errors. Moreover, the dependency of both deterministic and stochastic methods on the radiometric uncertainties limit their accuracy and application to cases where such data are available with an acceptable degree of confidence. A self-consistent and operational method is still required to estimate the uncertainties of IOPs without additional inputs or assumptions on the radiometric fluctuations.

10. Acknowledgment

The authors would like to thank NASA Ocean Biology Processing Group and individual data contributors for maintaining and updating the SeaBASS database.

11. References

Austin, R. & Petzold, T. (1981). The determination of the diffuse attenuation coefficient of seawater using the Coastal Zone Color Scanner, *in* J. Gower (ed.), *Oceanography from Space*, Plenum, New York, pp. 239–256.

Bailey, S. & Werdell, J. (2006). A multi-sensor approach for the on-orbit validation of ocean color satellite data products, *Remote Sensing of Environment* 102.

Bates, D. & Watts, D. (1988). *Nonlinear Regression Analysis and Its Applications*, John Wiley and Sons, NY.

Bernardo, J. (1979). Expected information as expected utility, *The Annals of Statistics* 7(3): 686–690.

Bernardo, J. (2005). Reference analysis, *in* D. Dey & C. Rao (eds), *Bayesian Thinking, Modeling and Computation*, Vol. 25 of *Handbook of Statistics*, Elsevier, pp. 17–90.

Bissett, W., Arnone, R., Davis, C., Dickey, T., Dye, D., Kohler, D. & Gould, R. (2004). From meters to kilometers: A look at ocean-color scales of variablity, spatial coherence, and the need for fine-scale remote sensing in coastal ocean optics, *Oceanography* 17(2): 32–43.

Brad, Y. (1974). *Nonlinear parameter estimation*, Academic Press.

Bricaud, A., Morel, A. & Prieur, L. (1981). Absorption by dissolved organic-matter of the sea (yellow substance) in the UV and visible domains, *Limnology And Oceanography* 26(1): 43–53.

Campbell, J. (1995). The log-normal distribution as a model for bio-optical variability in the sea, *Journal of Geophysical Research* (100): 13,237–13,254.

Carder, K., Chen, F., Cannizzaro, J., Campbell, J. & Mitchell, B. (2004). Performance of the MODIS semi-analytical ocean color algorithm for chlorophyll-a, *Advances in Space Research* 33: 1152–1159.

Carlin, B. & Polson, N. (1991). An expected utility approach to influence diagnostics, *Journal of the American Statistical Association* 86(416): 1013–1021.

Chang, G. & Gould, R. (2006). Comparisons of optical properties of the coastal ocean derived from satellite ocean color and in situ measurements, *Optics Express* 14(22): 10149–10163.

Christakos, G. (1990). A bayesian/maximum-entropy view to the spatial estimation problem, *Mathematical Geology* 22(7): 763–777.

Cox, C. & Munk, W. (1954). Measurements of the roughness of the sea surface from photographs of the sun glitter, *Journal of Optical Society of America* 44: 838–850.

Darecki, M. & Stramski, D. (2004). An evaluation of modis and seawifs bio-optical algorithms in the baltic sea, *Remote Sensing of Environment* 89: 326–350.

Doerffer, R. (2008). Analysis of the signal/noise and the water leaving radiance finnish lakes, *Technical report*, Brockmann Consult.

Duarte, J., Vélez-Reyes, M., Tarantola, S., Gilbes, F. & Armstrong, R. (2003). A probabilistic sensitivity analysis of water-leaving radiance to water constituents in coastal shallow waters, *Ocean Remote Sensing and Imaging*, Vol. 5155, SPIE.

Eleveld, M., van der Woerd, H., El Serafy, G., Blaas, M., van Kessel, T. & de Boer, G. (2008). Assimilation of remotely sensed observations in a sediment transport model, *Ocean Optics*, Vol. XIX; Extended abstract, Halifax: Lewis Conf. Services Int. Inc., Barga, Italy.

Garcia, C., Garcia, V. & McClain, C. (2005). Evaluation of SeaWiFS chlorophyll algorithms in the southwestern atlantic and southern oceans, *Remote Sensing of Environment* 95(1): 125–137.

Goody, R. (1964). *Atmospheric radiation 1, theoretical basis*, Oxford University Press.

Gordon, H. (1997). Atmospheric correction of ocean color imagery in the earth observing system era, *Journal Of Geophysical Research* 102(D14): 17081–17106.

Gordon, H., Brown, J. & Evans, R. (1988). Exact rayleigh scattering calculation for the use with the nimbus-7 coastal zone color scanner, *Applied Optics* 27(5): 862–871.

Gordon, H., Brown, O., Evans, R., Brown, J., Smith, R., Baker, K. & Clark, D. (1988). A semianalytical radiance model of ocean color, *Journal Of Geophysical Research* 93(D9): 10,909–10,924.

Gordon, H., Clark, D., Brown, J., Brown, O., Evans, R. & Broenkow, W. (1983). Phytoplankton pigment concentrations in the Middle Atlantic Bight: Comparison of ship determinations and CZCS estimates, *Applied Optics* 22(1): 20–36.

Goutis, C. & Robert, C. (1998). Model choice in generalised linear models: A bayesian approach via kullback-leibler projections, *Biometrika* 85(1): 22–37.

Harding Jr., L., Magnusona, A. & Malloneea, M. (2005). SeaWiFS retrievals of chlorophyll in Chesapeake Bay and the Mid Atlantic Bight, *Estuarine, Coastal and Shelf Science* 62(1-2): 75–94.

Holben, B., Eckdagger, T., Slutsker, I., Sospedra, E., Caselles, V., Coll, C., Valor, E. & Rubio, E. (2000). Validation of cloud detection algorithms, *Remote Sensing in the 21st Century: Economic and Environmental Applications* pp. 119–123.

Hoogenboom, H. & Dekker, A. (1998). The sensitivity of medium resolution imaging spectrometer (MERIS) for detecting chlorophyll and seston dry weight in coastal and inland waters, *IEEE Proceeding on Geoscience and Remote Sensing*, Vol. 1 of *IGARSS*, IEEE, pp. 183 – 185.

Hooker, S. & McClain, C. (2000). The calibration and validation of SeaWiFS data, *Progress In Oceanography* 45(3-4): 427–465.

Hu, C., Carder, K. & Muller-Karger, F. (2000). How precise are SeaWiFS ocean color estimates? implications of digitization-noise errors, *Remote Sensing of Environment* 76(2): 239–249.

Hu, C., Chen, Z., Clayton, T., Swarzenski, P., Brock, J. & Muller-Karger, F. (2004). Assessment of estuarine water-quality indicators using modis medium-resolution bands: Initial results from Tampa Bay, FL, *Remote Sensing of Environment* 93(3): 423–441.

Hyde, K., ÓReilly, J. & Oviatt, C. (2007). Validation of SeaWiFS chlorophyll-a in Massachusetts Bay, *Continental Shelf Research* 27(12): 1677–1691.

Jaynes, E. (1957a). Information theory and statistical mechanics, *Physical Review* 106: 620–630.

Jaynes, E. (1957b). Information theory and statistical mechanics, *Physical Review* 108: 171–190.

Jaynes, E. (1968). Prior probabilities, *IEEE Transactions On Systems Science and Cybernetics* 4(3): 227–241.

Johnson, W. & Geisser, S. (1985). Estimative influence measures of the multivariate general linear model, *Journal of Statistical Planning Inference* 11: 33–56.

Karl, D. & Lukas, R. (1996). The hawaii ocean time-series (hot) program: Background, rationale and field implementation, *Deep Sea Research Part II: Topical Studies in Oceanography* 43(1-2): 129–156.

Kendall, M. D. & Stuart, A. (1987). *The advanced theory of statistics: Distribution theory*, Vol. 1, 5th ed., Griffin, London.

Kopelevich, O. (1983). Small-parameter model of optical properties of sea waters, *in* A. Monin (ed.), *Ocean Optics*, Vol. 1 Physical Ocean Optics, Nauka, pp. 208–234.

Kullback, S. & Leibler, R. (1951). On information and sufficiency, *The Annals of Mathematical Statistics* 22(1): 79–86.

Lee, Z. (2006). Remote sensing of inherent optical properties: Fundamentals, tests of algorithms, and applications, *Technical Report 5*, International Ocean-Colour Coordinating Group.

Lee, Z., Arnone, R., Hu, C., Werdell, J. & Lubac, B. (2010). Uncertainties of optical parameters and their propagations in an analytical ocean color inversion algorithm, *Applied Optics* 49(3): 369–381.

Lee, Z., Carder, K. & Arnone, R. (2002). Deriving inherent optical properties from water color: a multiband quasi-analytical algorithm for optically deep waters, *Applied Optics* 41(27): 5755–5772.

Lee, Z., Carder, K., Mobley, C., Steward, R. & Patch, J. (1998). Hyperspectral remote sensing for shallow waters. 1. A semianalytical model, *Applied Optics* 37(27): 6329–6338.

Lee, Z., Carder, K., Mobley, C., Steward, R. & Patch, J. (1999). Hyperspectral remote sensing for shallow waters: 2. Deriving bottom depths and water properties by optimization, *Applied Optics* 38(18): 3831–3843.

Malkmus, W. (1967). Random Lorentz band model with exponential-tailed s^{-1} line intensity distribution function, *Journal of Optical Society of America* 57(3): 323–329.

Maritorena, S. & Siegel, D. (2005). Consistent merging of satellite ocean color data sets using a bio-optical model, *Remote Sensing of Environment* 94(4): 429–440.

Maritorena, S., Siegel, D. & Peterson, A. (2002). Optimization of a semianalytical ocean color model for global-scale applications, *Applied Optics* 41(15): 2705–2714.

McClain, C., Feldman, G., Hooker & S.B. (n.d.). An overview of the seawifs project and strategies for producing a climate research quality global ocean bio-optical time-series, *Deep Sea Research*.

Mélin, F. (2010). Global distribution of the random uncertainty associated with satellite-derived chla, *IEEE Geoscience And Remote Sensing Letters* 7(1).

Mélin, F., Berthon, J. & Zibordi, G. (2005). Assessment of apparent and inherent optical properties derived from SeaWiFS with field data, *Remote Sensing of Environment* 97(4): 540–553.

Mobley, C. (1994). *Light and water radiative transfer in natural waters*, Academic Press.

Neukermans, G., Nechad, B. & Ruddick, K. (2008). Optical remote sensing of coastal waters from geostationary platforms: a feasibility study - mapping total suspended matter with seviri, *Ocean Optics*, Vol. XIX; Extended abstract, Halifax: Lewis Conf. Services Int. Inc., Barga, Italy.

Petzold, T. (1977). Volume scattering functions for selected ocean waters, *in* J. Tyler (ed.), *Light in the Sea*, Vol. 12, Dowden, Hutchinson and Ross, Stroudsburg, Pa. USA, pp. 150–174.

Pope, R. & Fry, E. (1997). Absorption spectrum (380-700nm) of pure water: II, Integrating cavity measurements, *Applied Optics* 36(33): 8710–8723.

Pottier, C., Garçon, V., Larnicol, G., Sudre, J., Schaeffer, P. & Le Traon, P.-Y. (2006). Merging SeaWiFS and MODIS/AQUA ocean color data in North and Equatorial Atlantic using weighted averaging and objective analysis, *IEEE Transactions on Geoscience and Remote Sensing* 44(11): 3436–3451.

Press, W., Teukolsky, S., Vetterling, W. & Flannery, B. (2002). *Numerical recipes in C++, The art of scientific computing*, Cambridge University Press.

Rubinstein, R. & Kroese, D. (2004). *The Cross-Entropy Method: A unified approach to combinatorial optimization, Monte-Carlo simulation, and machine learning*, Information Science and Statistics, Springer, New York.

Salama, M. S., Dekker, A. G., Su, Z., Mannaerts, C. M. & Verhoef, W. (2009). Deriving inherent optical properties and associated inversion-uncertainties in the Dutch lakes, *Hydrology and Earth System Sciences* 13(7): 1113–1121.

Salama, M. S., Mélin, F. & Van der Velde, R. (2011). Ensemble uncertainty of inherent optical properties, *Optics Express* 19(18): 16772–16783.

Salama, M. & Stein, A. (2009). Error decomposition and estimation of inherent optical properties, *Applied Optics* 48(26): 4947–4962.

Salama, M. & Su, Z. (2010). Bayesian model for matching the radiometric measurements of aerospace and field ocean color sensors, *Sensors* 10(8): 7561–7575.

Salama, M. & Su, Z. (2011). Resolving the subscale spatial variability of apparent and inherent optical properties in ocean color matchup sites, *IEEE Transactions On Geoscience and Remote Sensing* 49(7):2612–2622.

Salama, S., Monbaliu, J. & Coppin, P. (2004). Atmospheric correction of advanced very high resolution radiometer imagery, *internation Journal of Remote Sensing* 25(7-8): 1349–1355.

Shannon, C. (1948). A mathematical theory of communication, *The Bell System Technical Journal* 27(Reprinted with corrections): 379–423, 623–656.

Singh, V. (1998). *Entropy-based parameter estimation in Hydrology*, Vol. 30 of *Water Science and Technology Library*, Kluwer Academic Publishers, Dordrecht.

Sydor, M., Gould, R., Arnone, R., Haltrin, V. & Goode, W. (2004). Uniqueness in remote sensing of the inherent optical properties of ocean water, *Applied Optics* 43(10): 2156–2162.

Van Der Woerd, H. & Pasterkamp, R. (2008). Hydropt: A fast and flexible method to retrieve chlorophyll-a from multispectral satellite observations of optically complex coastal waters, *Remote Sensing of Environment* 112(4): 1795–1807.

Velde van der, R., Su, Z. & Ma, Y. (2008). Impact of soil moisture dynamics on asar so signatures and its spatial variability observed over the tibetan plateau, *Sensors* 8: 5479–5491.

Vermote, E., Tanre, D., Deuze, J., Herman, M. & Morcrette, J. (1997). Second simulation of the satellite signal in the solar spectrum, 6S: An overview, *IEEE Transactions on Geoscience and Remote Sensing* 35(3): 675–686.

Wang, P., Boss, E. & Roesler, C. (2005). Uncertainties of inherent optical properties obtained from semianalytical inversions of ocean color, *Applied Optics* 44(9): 4074–4084.

Werdell, J. & Bailey, S. (2005). An improved in-situ bio-optical data set for ocean color algorithm development and satellite data product validation, *Remote Sensing of Environment* 98(1): 122–140.

Zaneveld, R. (1994). Optical closure: from theory to measurement, *in* R. Spinrad, K. Carder & M. Perry (eds), *Ocean Optics*, Vol. 25 of *Oxford Monographs on Geology and Geophysics*, Oxford University Press, New York, p. 283.

Permissions

The contributors of this book come from diverse backgrounds, making this book a truly international effort. This book will bring forth new frontiers with its revolutionizing research information and detailed analysis of the nascent developments around the world.

We would like to thank Dr. Rustam B. Rustamov and Dr. Saida E. Salahova, for lending their expertise to make the book truly unique. They have played a crucial role in the development of this book. Without their invaluable contribution this book wouldn't have been possible. They have made vital efforts to compile up to date information on the varied aspects of this subject to make this book a valuable addition to the collection of many professionals and students.

This book was conceptualized with the vision of imparting up-to-date information and advanced data in this field. To ensure the same, a matchless editorial board was set up. Every individual on the board went through rigorous rounds of assessment to prove their worth. After which they invested a large part of their time researching and compiling the most relevant data for our readers. Conferences and sessions were held from time to time between the editorial board and the contributing authors to present the data in the most comprehensible form. The editorial team has worked tirelessly to provide valuable and valid information to help people across the globe.

Every chapter published in this book has been scrutinized by our experts. Their significance has been extensively debated. The topics covered herein carry significant findings which will fuel the growth of the discipline. They may even be implemented as practical applications or may be referred to as a beginning point for another development. Chapters in this book were first published by InTech; hereby published with permission under the Creative Commons Attribution License or equivalent.

The editorial board has been involved in producing this book since its inception. They have spent rigorous hours researching and exploring the diverse topics which have resulted in the successful publishing of this book. They have passed on their knowledge of decades through this book. To expedite this challenging task, the publisher supported the team at every step. A small team of assistant editors was also appointed to further simplify the editing procedure and attain best results for the readers.

Our editorial team has been hand-picked from every corner of the world. Their multi-ethnicity adds dynamic inputs to the discussions which result in innovative outcomes. These outcomes are then further discussed with the researchers and contributors who give their valuable feedback and opinion regarding the same. The feedback is then collaborated with the researches and they are edited in a comprehensive manner to aid the understanding of the subject.

Apart from the editorial board, the designing team has also invested a significant amount of their time in understanding the subject and creating the most relevant covers. They scrutinized every image to scout for the most suitable representation of the subject and create an appropriate cover for the book.

The publishing team has been involved in this book since its early stages. They were actively engaged in every process, be it collecting the data, connecting with the contributors or procuring relevant information. The team has been an ardent support to the editorial, designing and production team. Their endless efforts to recruit the best for this project, has resulted in the accomplishment of this book. They are a veteran in the field of academics and their pool of knowledge is as vast as their experience in printing. Their expertise and guidance has proved useful at every step. Their uncompromising quality standards have made this book an exceptional effort. Their encouragement from time to time has been an inspiration for everyone.

The publisher and the editorial board hope that this book will prove to be a valuable piece of knowledge for researchers, students, practitioners and scholars across the globe.

List of Contributors

Rustam B. Rustamov, Saida E. Salahova, Maral H. Zeynalova and Sabina N. Hasanova
Institute of Physics, Azerbaijan National Academy of Sciences, ENCOTEC –Engineering &
Consulting Technologies, Institute of Botany, Azerbaijan National Academy of Sciences,
Architecture and Construction University/ENCOTEC LLC, Baku, Azerbaijan

Ussanai Nithirochananont and Anuphao Aobpaet
Geo-Informatics and Space Technology Development Agency, Thailand

Marius Trusculescu, Mugurel Balan and Alexandru Pandele
Institute for Space Sciences, Romania

Claudiu Dragasanu and Marius-Ioan Piso
Romanian Space Agency, Romania

Andrea Baraldi
Department of Geography, University of Maryland, College Park, Maryland, USA

Juan C. Valdiviezo-N and Gonzalo Urcid
Optics Department, INAOE, Mexico

Cristina Tarantino, Maria Adamo and Guido Pasquariello,
Francesco Lovergine, Palma Blonda and Valeria Tomaselli, National Council of Researches
(CNR), Italy

Meghan F. Cronin
NOAA Pacific Marine Environmental Laboratory, Seattle WA, USA

Robert A. Weller
Woods Hole Oceanographic Institution, Woods Hole, MA, USA

Richard S. Lampitt
National Oceanography Centre, Southampton, UK

Uwe Send
Scripps Institution of Oceanography, University of California, San Diego, La Jolla, CA, USA

Abel Calle and José Luis Casanova
University of Valladolid, Spain

Mhd. Suhyb Salama
Department of Water Resources, ITC, University of Twente, Hengelosestraat 99, 7500 AA
Enschede, The Netherlands

Printed in the USA
CPSIA information can be obtained
at www.ICGtesting.com
JSHW011440221024
72173JS00004B/883

9 781632 392312